MANUAL

OF

CHEMICAL ANALYSIS

AS APPLIED TO

THE EXAMINATION OF MEDICINAL CHEMICALS.

A GUIDE

FOR THE

DETERMINATION OF THEIR IDENTITY AND QUALITY,

AND FOR THE

DETECTION OF IMPURITIES AND ADULTERATIONS.

FOR THE USE OF

PHARMACEUTISTS, PHYSICIANS, DRUGGISTS, AND MANUFACTURING CHEMISTS, AND OF PHARMACEUTICAL AND MEDICAL STUDENTS.

BY

FREDERICK HOFFMANN, PH. D.,

PHARMACEUTIST IN NEW YORK.

NEW YORK:
D. APPLETON AND COMPANY,
549 & 551 BROADWAY.
1873.

PREFACE.

Although the preparation of most medicinal chemicals has passed away from the laboratory of the pharmaceutist, and is successfully conducted on a commercial scale in manufacturing establishments, yet the responsibility for the identity and quality of medicines, and of the substances used in their preparation, rests properly and legally with those who prepare, compound, and dispense them. It is therefore the duty of the pharmaceutist and the dispensing practitioner of medicine, as also, to a considerable extent, of the druggist and the manufacturing chemist, to examine the medicinal chemicals of commerce as to their identity, quality, and purity. In the exercise of this duty, they have frequent occasion to resort for information to references now widely scattered through chemical, pharmaceutical, and medical manuals and journals; since our literature, although of vast and increasing extent, and crystallizing more and more into distinct branches, is still wanting in a special guide for ready reference in the application of chemical analysis to such examinations.

In the present volume I have endeavored to supply this want, in a manner and to an extent which, it is hoped, will confine the work within the precise limits of requirement, without detracting from its general scope and its practical usefulness.

Since chemical tests and examinations bear upon and involve the methods of systematic chemical analysis, and as these cannot be described in each particular instance, I have deemed it expedient to preface the volume with a few notes on operations and reagents, and on a few important general tests, and to present a brief outline of a simple course of qualitative analysis for the systematic and progressive recognition of such substances as are met with in the medicinal chemicals. A brief guide has also been added, for the volumetric estimation of those compounds to which this mode of examination is especially applicable.

Upon these preliminaries is based the subsequent description of the physical and chemical properties and relations of the medicinal chemicals and their preparations, and of the methods employed for establishing their identity, and for ascertaining their quality and purity. It has been compiled with special reference to the fifth decennial edition of the United States Pharmacopœia, to the latest British Pharmacopœia, and to the Pharmacopœa Germanica of 1872. It has been brought within the briefest possible compass, with the view to furnish a concise and trustworthy guide, combining easy execution, simple apparatus, and economy of time, with the greatest attainable accuracy.

A glance at the table of contents will at once give an idea of the arrangement of the subject-matter, and of the scope as well as the limitations of the work. Chemical notation, although the most precise and definite, and at the same time the most comprehensive, method of expressing the constitution of compounds, has been omitted, for want of unanimity and absolute certainty in this particular point of chemical philosophy. The weights used are, with the exception of the volumetric tests and examinations, those of the troy scale, and the word "parts" means invariably parts by this weight, unless where volume-parts are expressly stated. Temperatures are denoted, in accordance with general adoption, by degrees of the centigrade scale. Comparative tables of the two scales, and of the troy and metric weights, are given at the end of the volume.

In preparing this compendium, I have consulted, and at times made free use of, a number of standard works, and periodicals of the kindred literature. I have, however, felt compelled, not without hesitation, to omit the introduction of references, which would have required much space, and would have greatly increased the size of the volume, without affording a corresponding advantage.

Though well aware of the shortcomings and imperfections of the work, I nevertheless venture to hope that it will meet with kind consideration, and will prove both serviceable and stimulating in a province not yet duly appreciated or deservedly cultivated. This hope is the stronger, as the work appears at a time when the rapid advance of both sciences and arts, the drift of public sentiment, and the consequently increasing obligations of the pharmaceutist and the physician, all tend toward higher qualifications, and necessitate also, among other attainments, a more extended exercise of knowledge and skill in chemical and microscopical investigation. I trust, moreover, to the candor of critics, who, while it is easy to detect faults in a volume of this nature and scope, in which perfection would scarcely be attainable even under more favorable circumstances, can at the same time appreciate the difficulties of a work that has been written in a pursuit and position which exclude, almost wholly, the quiet reflection and undisturbed study requisite for the prosecution of scientific or literary labor.

In conclusion, I desire to express my obligations and thanks to Prof. D. S. Martin, of Rutgers Female College, for his kind assistance in the revision of the proofs.

NEW YORK, *February*, 1873.

CONTENTS.

PART FIRST.

PART SECOND.

MEDICINAL CHEMICALS AND THEIR PREPARATIONS.

PART FIRST.

OPERATIONS AND REAGENTS,

INCLUDING A BRIEF OUTLINE OF

A SYSTEMATIC COURSE OF CHEMICAL ANALYSIS,

AND OF

VOLUMETRIC ESTIMATION.

OPERATIONS AND REAGENTS.

OPERATIONS.

The operations in simple tests and chemical examinations must be supposed to be familiar to the pharmaceutist, the druggist, the pharmaceutical manufacturer, and the physician; the description of them, therefore, can properly be left to the manuals of analytical and pharmaceutical chemistry, and may be confined in this volume to a few practical hints in relation to some of the principal operations.

Solutions.—The common solvent, water, has to be used distilled, and this fact is to be understood throughout this work; neither rain-water nor spring-water, however pure it may appear to be, can be used indiscriminately as a solvent or for edulcoration in chemical investigations.

Solutions for testing are best prepared in test-tubes, or in small flasks or beaker-glasses.

Increase of the surfaces of contact by comminution, agitation, and increase of temperature, as is well known, aid and accelerate the process of solution, as well as of chemical reaction; and one or both of these auxiliaries may be employed, unless the nature of the substance or the effect of heat upon it is such as to exclude their application.

Precipitation.—The formation of an insoluble body from a solution can be effected either by a change or modification of the solvent, or by the production of one or more new bodies, insoluble in the solvent. An instance of the first case is an aqueous solution of barium chloride, which will be precipitated by the addition of concentrated hydrochloric acid, or a solution of calcium sulphate, which will be precipitated by alcohol; in both these instances the solvent power of the menstruum is lessened and solution may be reëstablished by the addition of a sufficient quantity of water. Instances of the second case of precipitation are a solution of calcium hydrate, precipitated by sodium carbonate, and a solution of magnesium sulphate, precipitated by barium hydrate.

Precipitation is resorted to as the most important mode of detecting and discriminating bodies by their physical and chemical properties.

The term "turbidity" designates the formation of a precipitate so insignificant in quantity, or so finely divided, or so light in weight, that the suspended particles only impair the transparency of the fluid, and require a certain amount of time to subside in the form of a precipitate.

In the course of analytical investigation, each precipitation must be complete; to insure this, the reagent must be added gradually, allowing the precipitate to subside after each addition, until no further precipitate is produced. In almost all cases separation of precipitates is aided by the application of gentle heat.

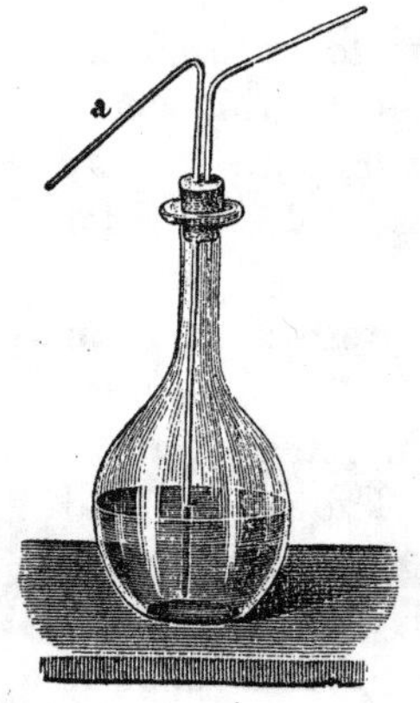

FIG. 1.

The separation of the supernatant menstruum from a precipitate is effected either by filtration and subsequent washing of the precipitate upon the filter by means of a washing-bottle (Fig. 1), or, where the precipitate speedily and completely subsides, by decantation. As a rule, funnels and filters must be small, and proportionate to the amount of the precipitate and the menstruum; filters should be cut so as not to project over the rim of the funnel, and it is also advisable to moisten the filter upon the funnel with distilled water, by means of the washing-bottle, previous to passing the fluid through it. Decantation is effected either by pouring off the supernatant clear part of the fluid by simply inclining the vessel and allowing the fluid to flow down a glass rod (Fig. 2), or by drawing it off by means of a small glass siphon or a pipette (Fig. 3).

Fig. 2.

In either mode of separation the precipitate, in most instances, must be thoroughly freed from the menstruum by washing with water, either on the filter or by decantation.

Fig. 3.

In all cases and operations of chemical examination, a reasonable economy with the substance and the solutions is advisable, so as to leave enough of the former for unseen contingencies and for confirmatory tests, as well as to repeat or verify any and all results of the examination. All tests and reactions are, therefore, performed on as small a scale as is reasonable and appropriate in the particular case; and all operations should proceed accordingly, and with constant observance of the principles and processes whereon they depend.

REAGENTS.

THE methods of chemical analysis and investigation consist in bringing the substances under examination into contact with other bodies of known properties, and observing the resulting phenomena. These phenomena consist in alterations, either in state of aggregation, form, or color, resulting from some chemical change. All bodies which are employed for this purpose are called *reagents*, and the ensuing phenomena *reactions*.

It is obvious, therefore, that a sufficient knowledge of theoretical chemistry in its details, and especially a familiarity with the deportment, properties, and relations of the common compounds and reagents, are indispensable to the pursuit of chemical tests and examinations. Upon such knowledge depend the conception and comprehension of the conditions necessary for the formation of new compounds, and for the manifestation of the various reactions, as well as the correct inference from the observations and results of all investigations; and without it they will remain unavailing and uncertain.

No special and definite rules can be assigned for the application of reagents in each instance, with respect to their proportion and quantity. These must depend upon the quantity and nature of the substance under examination and its solution, as well as upon the nature of the reagent, the strength of its solution, and the processes taking place in each particular reaction. Knowledge and reflection, as well as a ready comprehension of the object and aim of each test, of its issues, and of the possible incidents, and a correct inference from all phenomena, must decide at large, as well as in detail, not only what reagents should be employed, but also the amount and the conditions in each particular instance.

A common error, and an obstacle to the less skilled, is the use of an excess of reagents. There are reagents which in many cases admit a free application without disadvantage to the correctness of the result—as, for instance, hydrosulphuric acid, solution of calcium hydrate, etc.; but the majority of reagents need to be applied in common tests only by a few drops of their solutions—as, for instance, baric, ferric, cupric, and argentic solutions, etc. On the other hand, there are not unfrequently errors arising from an insufficient amount in the application of reagents, especially with dilute solutions, or in those cases in which the complete elimination of a substance by precipitation is required for the subsequent examination for other bodies: for instance, hydrosulphuric acid, applied in a limited quantity, produces a white precipitate with solutions of mercuric salts; applied in excess, it gives a black precipitate. There are other instances where an excess of the solution under consideration, as well as of the reagent, may redissolve, and consequently destroy, the precipitate whereon the reaction is based.

In operations of chemical analysis it must always be borne in mind and well understood that, in the processes and phenomena taking place between the reagents and the substances acted upon, as in all chemical changes and reactions, certain laws and definite limits exist between cause and effect, and that the ability of correctly applying knowledge, judgment, and skill, and of drawing the right inference from necessary as well as from casual reactions and phenomena, must rule and guide the methods and operations of the investigator, and carry them beyond mere conjecture and empiricism.

It is beyond the scope of this work to describe the mode of preparing the reagents, their use and application, and their deportment with the common compounds. For such information, reference must be had to the text-books of applied and analytical chemistry. From a practical point of view, only the usual strength of the solution of the reagents, as best suited for the common tests and examinations, and the mode of preparing a few of the rarer or special reagents, or of such as are not included among the medicinal chemicals considered in this volume, have been stated.

It hardly needs to be mentioned that all reagents must consist purely of their essential constituents, and must contain no admixture of any other substance; it must, therefore, be an invariable rule to test the purity of the reagents before they are employed.

The reagents and their solutions must be preserved according to their nature; of those whose solutions are liable to alteration or decomposition, only small quantities must be kept, and always in tightly-closed glass-stoppered bottles.

REAGENTS.

Concentrated Sulphuric Acid.—Spec. grav., 1.843; containing about 78.36 per cent. anhydrous sulphuric acid.

Strong Sulphuric Acid.—Spec. grav., 1.68; containing about 62.04 per cent. anhydrous sulphuric acid. Obtained by carefully mixing, in a flask or beaker, 1 part of water with 3 parts of concentrated sulphuric acid (Fig. 4).

FIG. 4.

In diluting concentrated sulphuric acid with water, the acid should invariably and gradually be added to the water, and in vessels which are either placed in cold water, or which are not liable to crack from the heat evolved.

Diluted Sulphuric Acid.—Spec. grav., 1.113–1.117; containing about 13.50 per cent. of anhydrous sulphuric acid. Obtained by mixing 1 part of concentrated sulphuric acid with 5 parts of water.

Sulphurous Acid.—An aqueous solution of sulphurous-acid gas, saturated at 15 C., containing about 37 times its volume of gas.

Concentrated Hydrochloric Acid.—Spec. grav., 1.16; containing about 32 per cent. of anhydrous acid.

Diluted Hydrochloric Acid.—Spec. grav., 1.038; containing about 8 per cent. of anhydrous acid. Obtained by mixing 4 parts of water with 1 part of concentrated hydrochloric acid.

Concentrated Nitric Acid.—Spec. grav., 1.42; containing 60 per cent. of anhydrous acid.

When concentrated acids are applied in small tests only by the drop, as, for instance, in testing alkaloids, etc., they are taken from the bottle by dipping a glass rod into the acid and allowing the drop or drops to fall upon the substance to be acted upon, or better by means of a kind of pipette consisting of a thin, strong glass tube, adjusted at one end to a small caoutchouc bulb, and contracted at the other extremity to a capillary end. The fluid is drawn into the tube, and delivered again by gentle pressure of the bulb (Fig. 5).

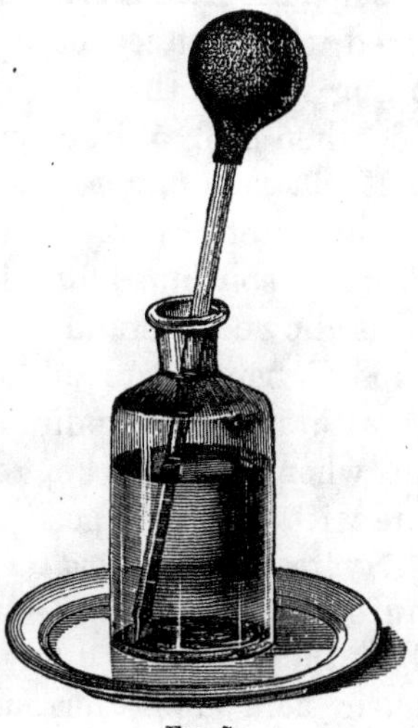

FIG. 5.

Diluted Nitric Acid.—Spec. grav., 1.068; containing 10 per cent. of anhydrous acid. Obtained by mixing 4 parts of water with 1 part of concentrated nitric acid.

Oxalic Acid.—Solution of 1 part of crystallized oxalic acid in 12 parts of water.

Acetic Acid.—Spec. grav., 1.047.

Diluted Acetic Acid.—Obtained by mixing 4 parts of water with 1 part of acetic acid.

Tartaric Acid.—Solution of 1 part of crystallized tartaric acid in 5 parts of water.

Tannic Acid.—Solution of 1 part of tannic acid in a mixture consisting of 8 parts of water and 2 parts of alcohol.

Indigo-Sulphuric Acid.—Solution of indigo in sulphuric acid. Five grains of finely-powdered indigo are digested in 2 drachms of concentrated sulphuric acid, in a corked test-tube, for 2 hours; subsequently, 4 ounces of strong sulphuric acid are added, and, after subsiding, the clear liquid is decanted, and carefully added to 2 ounces of water.

Indigo Sulphate.—Solution of 1 part commercial indigo carmine * in 15 parts of water.

Aniline Sulphate.—Solution of 5 drops of aniline in 1 ounce of diluted sulphuric acid.

Sulphuric-Acid Mucilage of Starch.—Half a drachm of starch, mixed with 1 ounce of cold water, is stirred into 3 ounces of boiling water; then half a drachm of diluted sulphuric acid, and, when cool, 5 drops of benzol are added.

Mucilage of Starch.—20 grains of starch, mixed with 1 ounce of cold water, are stirred into 3 ounces of boiling water, and the boiling is continued for a few minutes; when cool, the clear liquid is allowed to stand for 12 hours, is then decanted, and a few drops of benzol are added; or it is saturated with sodium chloride, when it may be kept a long time without decomposition.

Hydrosulphuric-Acid Gas (Sulphuretted Hydrogen).—Obtained by the action of diluted sulphuric acid upon ferrous sulphide. Among the several convenient forms of apparatus for the continuous preparation of the gas and to keep it ready for use, the one represented in Fig. 6 is preferable. It consists of three glass bulbs, the two lower ones being a single piece, and the upper one, prolonged by a tube reaching to the bottom of the lower, being ground airtight into the neck of the second. Through the tubulure of the middle bulb the ferrous sulphide is introduced, and the tubulure then closed by a cork containing a wide glass tube

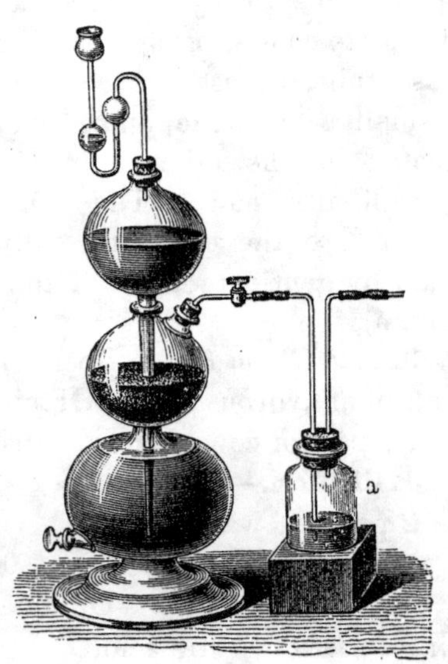

Fig. 6.

* Obtained by precipitation of a diluted solution of indigo in Nordhausen oil of vitriol by a concentrated solution of pure potassium carbonate in excess; the deep-blue precipitate is collected upon a filter, washed with some water until the filtrate ceases to effervesce with concentrated acids, and is then dried. The obtained sulphindigotic potassium is soluble, with a deep-blue color, in 140 parts of water.

provided with a stop-cock, or with a rubber tube, closed by a Mohr's wire clamp (*see* p. 56). The acid is poured in through the safety-tube, runs into the bottom globe, and rises to overflow the ferrous sulphide in the middle one. When the air has been allowed to escape through the delivery-tube, and this is closed, the pressure of the accumulating hydrogen sulphide forces the liquid from the second bulb down into the lower, and thence into the upper bulb, thus stopping the action, and preserving a volume of the gas ready for use.

FIG. 7.

When no such apparatus is at hand, hydrosulphuric-acid gas may be generated, in small tests, from a little flask or a test-tube (Fig. 7), taking care that none of the contents of the flask pass through the delivery-tube into the liquid under examination.

Hydrosulphuric Acid.—A solution of hydrosulphuric-acid gas (sulphuretted hydrogen) in water, saturated at 15° C., or

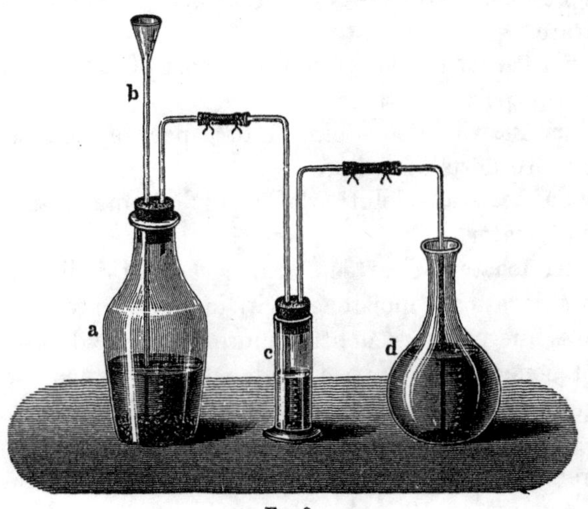

FIG. 8.

at a lower temperature, containing about four times its volume of the gas. The gas is likewise obtained by the action of diluted sulphuric acid upon ferrous sulphide, and is washed in water, contained in a small flask or cylinder (Fig. 8), before passing it into water for absorption.

In order to preserve the hydrosulphuric acid, it is advisable to fill the freshly-prepared saturated solution immediately into 2- or 4-ounce vials, and to place them, tightly corked, in an inverted position, in a jar filled with water.

When, in the course of a test, a solution has to be acted upon for some time by hydrosulphuric acid, a test-tube or flask may be employed, of such size as nearly to be filled by the liquid. It may then be tightly stoppered, allowing sufficient escape of air before corking, if it has to be warmed.

Chlorine Water.—A saturated aqueous solution of chlorine, containing about 0.4 per cent. by weight of gas. For analytical use, this solution is best preserved in 1-ounce vials, tightly corked and sealed, in a cool place.

Iodine Water.—Add 1 drop of tincture of iodine to 3 fluidounces of water.

Potassium Hydrate.—Solution of 1 part of fused potassium hydrate in 3 parts of water, containing 25 per cent. of potassium hydrate.

Liquor Potassæ, U. S. P.—Containing 5.8 per cent. of potassium hydrate.

Potassium Carbonate.—Solution of 1 part of pure potassium carbonate in 3 parts of water.

Potassium Bicarbonate.—Solution of 1 part of potassium bicarbonate in 10 parts of water.

Potassium Acetate.—Solution of 1 part of potassium acetate in 5 parts of water.

Potassium Iodate.—Fourteen grains each of iodine and of potassium chlorate are moistened with a little water, and triturated to a fine powder; a little more water, acidulated with 5 drops of concentrated nitric acid, is then added, and the whole digested in a flask at a gentle heat, until discolored; the heat is then increased to boiling, and the contents are subsequently transferred to a porcelain capsule, and evaporated to complete dryness at 100° C. The residue is dissolved in 5 ounces of water, and filtered.

Potassium Permanganate.—Solution of 1 grain of potassium permanganate in 5 ounces of water.

Potassium Bichromate.—Solution of 1 part of potassium bichromate in 10 parts of water.

Potassium Antimoniate.—A cold-saturated, aqueous solution of potassium antimoniate.

Potassium Cyanide.

Potassium Ferrocyanide.—Solution of 1 part of potassium ferrocyanide in 12 parts of water.

Potassium Ferri Cyanide.—Solution of 1 part of potassium ferri cyanide in 20 parts of water.

Potassium Sulphocyanide.—Obtained by boiling, in a flask, for 15 minutes, 1 drachm of fused potassium cyanide and 1 drachm of sublimed and washed sulphur in 6 drachms of water; then the liquid is filtered, and so much water added as to make the quantity 6 fluid drachms.

Potassium Iodide.—Solution of 1 part of potassium iodide in 10 parts of water.

Commercial potassium iodide generally contains traces of potassium iodate, and this should be eliminated by dissolving the salt in boiling alcohol, to saturation, filtering the hot solution, and, when cool, collecting and drying the separated salt.

Iodinized Potassium Iodide.—Eight grains of potassium iodide and 2 grains of iodine, dissolved in 2 ounces of water.

Potassium Mercuric Iodide.—A solution of 11 grains of mercuric iodide and 8 grains of potassium iodide in 10 drachms of water.

Potassium Mercuric Iodide with Potassium Hydrate (Nessler's test).—A solution obtained by gradually adding mercuric iodide, mixed in a test-tube with some water, to a warm solution of 20 grains of potassium iodide in 1 drachm of water, until the mercuric iodide ceases to be dissolved (about 20 grains will be required); when cooled, 1 drachm of water is added, and, after allowing the solution to stand for 24 hours, it is filtered, the filter washed with a few drops of water, and 5 drachms of Liquor Potassii, U. S. P., added.

Sodium Hydrate (Liquor Sodii, U. S. P.).—Spec. grav., 1.071; containing 5.7 per cent. of sodium hydrate.

Sodium Carbonate.—Dehydrated by exsiccation.

Sodium Carbonate.—Solution of 1 part of crystallized sodium carbonate in 10 parts of water.

Sodium Bicarbonate.—Saturated aqueous solution of sodium bicarbonate.

Sodium Bitartrate.—Saturated aqueous solution of sodium bitartrate.

Sodium Acetate.—Solution of 1 part of crystallized sodium acetate in 5 parts of water.

Sodium Phosphate.—Solution of 1 part of crystallized sodium phosphate in 5 parts of water.

Sodium Hyposulphite.—Solution of 1 part of crystallized sodium hyposulphite in 10 parts of water.

Ammonium Hydrate (Aqua Ammoniæ fortior).—A nearly saturated aqueous solution of ammonium hydrate. Spec. grav., 0.900; containing about 26 per cent. of the gas by weight.

Aqua Ammoniæ, U. S. P.—Spec. grav., 0.960; containing 10 per cent. of ammonium hydrate.

Ammonium Sulphydrate.—A solution of ammonium sulphydrate in water containing some ammonium hydrate; it is obtained by saturating, at 15° C. or a lower temperature, 4 parts of Aqua Ammoniæ fortior, U. S. P., with hydrosulphuric acid gas, and by subsequent addition of 1 part of aqua ammoniæ. It is best preserved in 1-ounce vials, tightly corked, and in a cool place. This solution, being concentrated, has to be employed, in the common test, only in small quantities, mostly by drops.

When hydrosulphuric acid gas is at hand, ammonium sulphydrate may, in many of its applications, be produced by saturating the liquid under examination with the gas, and by subsequent addition of ammonium hydrate; or, in ammoniated solutions, if dilution does not interfere with the reaction, by addition of hydrosulphuric acid.

Ammonium Carbonate.—Solution of 1 part of crystallized ammonium sesqui carbonate in 8 parts of water and 2 parts of aqua ammoniæ.

Ammonium Chloride.—Solution of 1 part of crystallized ammonium chloride in 10 parts of water.

Ammoniated Magnesium Sulphate.—Solution of 2 parts of magnesium sulphate, 1 part of ammonium chloride, and 1 part of aqua ammoniæ, in 8 parts of water.

Ammonium Oxalate.—Solution of 1 part of crystallized ammonium oxalate in 20 parts of water.

Calcium Hydrate (Lime-water).—Saturated aqueous solution of calcium hydrate.

Calcium Chloride.—1 part of precipitated calcium carbonate is shaken with 8 parts of water, and so much concentrated hydrochloric acid added as nearly to dissolve the carbonate; the solution is then filtered, and 2 parts of water are added.

Calcium Sulphate.—Saturated aqueous solution of calcium sulphate, containing about 0.2 per cent. of the salt.

Barium Hydrate.—Saturated aqueous solution of barium hydrate, containing about 5 per cent. of the salt.

Barium Nitrate.—Solution of 1 part of crystallized barium nitrate in 16 parts of water.

Barium Chloride.—Solution of 1 part of crystallized barium chloride in 16 parts of water.

Magnesium Sulphate.—Solution of 1 part of crystallized magnesium sulphate in 10 parts of water.

Ferrous Sulphate.—Solution of 1 part of ferrous sulphate, obtained by precipitation with alcohol, in 4 parts of water.

Ferrous sulphate is best obtained by pouring an aqueous solution of freshly-prepared crystallized ferrous sulphate, saturated at the boiling-point, into strong alcohol, collecting the precipitate upon a filter, washing with a little alcohol, drying by pressing between filtering-paper, and by immediately filling the humid salt into small warm vials, which are corked and sealed while warm. The absence of ferric sulphate may be ascertained by testing the solution with potassium ferro-cyanide. No blue turbidity, or only a very slight one, should occur.

Ferric Sulphate, obtained by diluting 1 part of Liquor Ferri Persulphatis, U. S. P., with 3 parts of water.

Ferric Chloride, obtained by diluting 1 part of Liquor Ferri Perchloridi, U. S. P., with 3 parts of water.

Ferri dinitrosulphide, obtained by adding, drop by drop, a solution of ferric chloride or sulphate, with constant stirring, to a mixture consisting of equal parts of strong solutions of potassium nitrate and ammonium sulphydrate, heating the liquid to boiling for a few minutes, and filtering while hot from the sulphur. The deep-colored liquid deposits, on cooling, black, needle-shaped rhombic prisms of ferri dinitrosulphide; these are dissolved 1 part in 10 parts of water, to give the required solution.

Cobaltous Nitrate.—Solution of 1 part of cobaltous nitrate in 10 parts of water.

Stannous Chloride.—Saturated solution of real and pure tinfoil in concentrated hydrochloric acid, with subsequent addition of a little concentrated hydrochloric acid.

Plumbic Acetate.—Solution of 1 part of crystallized plumbic acetate in 10 parts of water.

Plumbic Nitrate.—Solution of 1 part of crystallized plumbic nitrate in 10 parts of water.

Cupric Sulphate.—Solution of 1 part of crystallized cupric sulphate in 10 parts of water.

Fehling's Solution.—4½ drachms (17.33 grammes) of pure crystallized cupric sulphate are dissolved in 2 ounces of water; 2½ ounces and 2 scruples (86.25 grammes) of pure crystallized potassium and sodium tartrate are dissolved in 5 ounces of water, and 2¼ ounces (70 grammes) of Liquor Sodæ of 1.3 spec. grav. are added. Then the cupric solution is gradually added to the latter one, and the liquid passed through a filter, and so much water added as to make the whole measure 16 fluidounces (500 cubic centimetres).

Mercuric Chloride.—Solution of 1 part of crystallized mercuric chloride in 20 parts of water.

Argentic Nitrate.—Solution of 1 part of crystallized argentic nitrate in 20 parts of water.

Platinic Chloride.—Solution of 1 part of platinic chloride in 20 parts of water.

Nascent Hydrogen is a very delicate means of detecting arsenic; this test depends upon the production of hydrogen arsenide, whenever arsenic is present in any soluble form, in a solution in which hydrogen is being evolved by the action of zinc upon hydrochloric or sulphuric acid.

There are two methods of the application of this test; the one long known as Marsh's test, and preëminently adapted for forensic and quantitative detection and estimation of arsenic, the other of recent device, and preferable for the ready and prompt qualitative detection of arsenic.

A complete but simple form of apparatus for *Marsh's* test is represented in Fig. 9; it consists of a generating-flask *a*, of about 4 to 6 ounces' capacity, provided by means of a twice-perforated cork, or rubber stopper, with a funnel-tube and a delivery-tube, which last is connected by rubber corks with a wider glass tube, *c*, loosely filled with fragments of calcium-chloride; to the opposite end of this drying-tube is attached a long piece of narrow tubing of hard German glass, *d*, drawn out into a capillary end. and bent up so as to form a vertical jet.

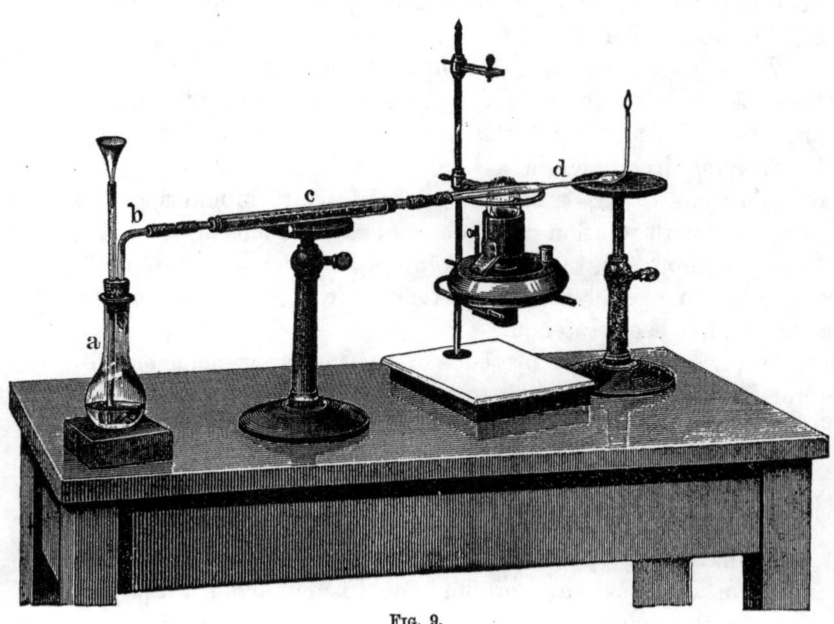

FIG. 9.

The test consists in introducing into the flask *a* pure granular zinc or magnesium, and a little water; when the apparatus is tightly fitted, hydrochloric or sulphuric acid is added through the funnel-tube.

It is necessary in each case first to ascertain the purity of these substances. After the evolution of gas has continued long enough to expel the atmospheric air, the reduction-tube, *d*, is heated to redness for about ten minutes, the escaping gas is lighted, and a piece of white porcelain held in the flame. If no dark deposit takes place, either in the tube or on the porcelain, the liquid to be tested may then be added through the funnel-tube, and the operation continued in the manner described. If arsenic-spots are obtained in the tube, a number of them may be produced by heating the tube in at least two places, at distances of about three inches, or if an approximately quantitative estimation of the arsenic is desired, all the arsenic may be obtained by the employment of a longer reduction-tube and several flames.

The obtained arsenic mirrors may be examined for identification or quantitatively determined by subsequently cutting off the tube with a file.

Hager's test, requiring a simpler apparatus and less time, depends upon the action of hydrogen arsenide upon soluble argentic salts, forming black argentic arsenide, and consists in evolving hydrogen in a long test-tube, and allowing the escaping gas to pass through a cover of white bibulous paper moistened with solution of argentic nitrate, or closing the tube loosely with a cork, into the lower end of which a strip of stiff white blotting-paper paper, moistened with the argentic nitric solution, has been fastened.

Since, however, the presence of the lower oxy-acids of sulphur and selenium give rise to the formation of hydrosulphuric acid, which also blackens argentic salts, this test has invariably to be preceded by a test for the absence of such acids, for which purpose plumbic acetate is employed, which is not decomposed by hydrogen arsenide.

A mixture of 1 part of concentrated hydrochloric acid with 2 parts of water, or, when sulphuric acid is employed, a mixture of 1 part of concentrated sulphuric acid with 4 to 5 parts of water, is introduced into a long test-tube (Fig. 10), taking care that it fills only about one-tenth of the capacity of the tube, and that the upper parts of the interior of the tube remain dry; a few pieces of pure zinc are then placed in the acid, and a small bunch of cotton, moistened with a solution of plumbic acetate, is inserted into the orifice of the tube, which is then allowed to stand for half an hour; if, after that time, the cotton remains colorless, the reagents are free from lower oxy-acids of sulphur and selenium, which are liable to form hydrosulphuric acid, and to impair the correctness of the test.

The operation may then be repeated with the addition of the liquid to be tested, and with the precaution to ascertain, notwithstanding the purity of the reagents, the absence of hydrosulphuric acid, by the simultaneous exposure of argentic nitrate and plumbic acetate to the escaping gas; this is done by employing the cotton moistened with solution of plumbic acetate and pressing it a little into the tube; the opening is then loosely tied with white blotting-paper, and this moistened with one drop of solution of argentic nitrate; or a few fragments of cork are placed upon the cotton, and a little gun-

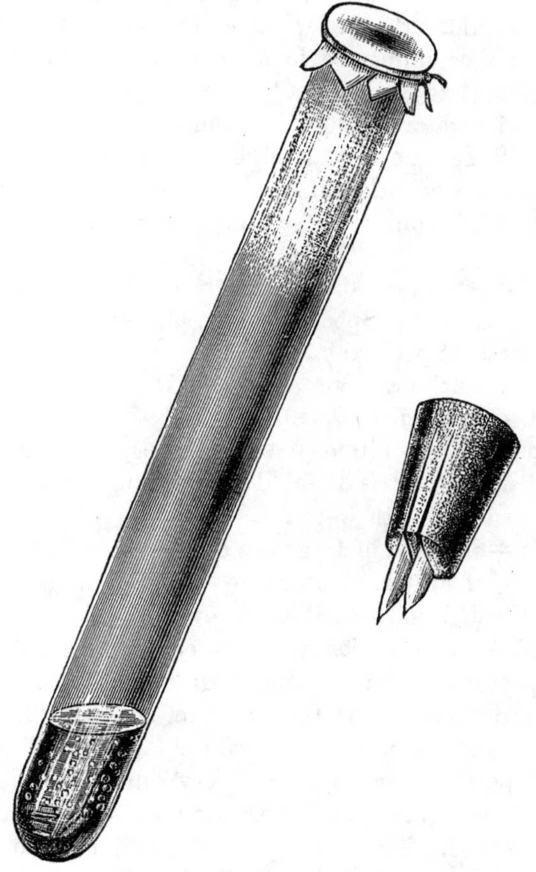

Fig. 10.

cotton moistened with solution of argentic nitrate is placed above. Instead of cotton and paper cover, a cork may be employed, provided with two small strips of stiff, white blotting-paper, which are moistened immediately before the operation is instituted, the one with solution of argentic nitrate, the other with solution of plumbic acetate; the cork is then loosely fitted into the test-tube, and this allowed to stand for half an hour.

If arsenic be present, the paper cover, or strip, or the gun-cotton, containing the argentic solution, will become black; the strip moistened with plumbic acetate must remain white. As an additional evidence that the brown or black reaction is argentic arsenide, the paper or cotton containing the dark reaction may be immersed in an aqueous solution of potassium cyanide (1 to 8) in a test-tube, and this gently warmed, by dipping it in water of 50° C.; the dark spots of argentic arsenide will become a little lighter, but will not disappear, while spots of argentic sulphide disappear at once, and spots of argentic phosphide or antimonide are dissolved gradually.

Ether.—Spec. grav., 0.728.

Absolute Alcohol.—Spec. grav., 0.793.

Chloroform.—Spec. grav., 1.480.

Carbon Bisulphide.—Spec. grav., 1.272.

Benzol.—Spec. grav., 0.680 to 0.700.

Albumen.—The white of one chicken-egg is agitated with 3 ounces of water, and is then filtered through clean tow, previously moistened with water.

Gelatin.—Solution of 10 grains of ichthyocolla, in shreds, in each ounce of water, obtained by digestion and subsequent filtration through moistened clean tow.

Neutral Litmus-Solution is prepared by digesting, for 24 hours, 1 part of powdered commercial litmus in about 10 parts of water; the solution is then filtered, and the filter washed with 4 parts of water. The filtrate obtained is divided into two equal parts; to one of them very dilute phosphoric acid is added, drop by drop, with constant stirring, until the liquid turns faintly red. The red liquid is then added to the blue portion, 2 parts of alcohol are added, and the whole kept in small, well-corked bottles.

Blue Litmus-Paper is prepared by drawing unsized white paper (Swedish filtering-paper) through the above neutral liquid.

Red Litmus-Paper is prepared by drawing unsized white

paper (Swedish filtering-paper) through the acidulated reddened part of the litmus-solution, as obtained and described above, in the preparation of neutral litmus-solution.

The paper obtained is dried in warm air, and, for ready use, is cut into strips of about ⅛ of an inch wide and 4 inches long, and preserved enclosed in paraffin-paper, or in tightly-corked bottles.

In reactions of neutralization, where carbonic-acid gas is evolved, this substance acts on litmus, and may impair the correctness of the test; in such operations it is therefore better, if admissible, to operate on warm solutions, in order quickly to expel the carbonic-acid gas; if heat be incompatible, turmeric-paper may be used instead of litmus-paper.

Turmeric-Solution, obtained by digestion of 1 part of powdered rhizomas of turmeric in a mixture of 4 parts of alcohol and 3 parts of water. After one or two days, the liquid is filtered off and preserved.

Turmeric-Paper is prepared from this tincture by steeping in it white unsized paper (Swedish filtering-paper). The paper need not be preserved from the action of the atmosphere, since it remains unchanged by carbonic acid.

Alkanet-Paper is prepared like litmus-paper, by saturating unsized paper with a solution of the alkanet-red. This is obtained by extracting dry alkanet-root with ether; the filtered solution is ready for use.

The blue paper may be obtained from the red one by dipping it into an aqueous solution of sodium carbonate (1.500). A neutral paper, answering for the alkaline as well as the acid test, may be prepared by dividing the ethereal solution of alkanet-red into two equal parts; to one is added, drop by drop, an aqueous solution of sodium carbonate, until the red is just changed to a distinct blue tint; then both liquids are mixed and used for the preparation of the paper.

BRIEF COURSE

OF

QUALITATIVE CHEMICAL ANALYSIS.

CHEMICAL tests and examinations must be founded upon a thorough knowledge of the nature and relations of the reagents, and of their deportment with the common compounds, and also upon a certain fixed order and methodical system in their application. These attainments, and the necessary skill, experience, and judgment, are requisite for every one who enters upon testing and investigation with a chance or claim of accuracy or certainty.

The following brief outline of a simple progressive course of qualitative chemical analysis depends, first, upon the successive elimination of groups of elementary compounds which possess certain common chemical properties, and, finally, upon the recognition of each member of such groups; it may therefore serve as a guide whenever, in the course of investigation, recourse is to be had to such a systematic method of analysis.

For obvious reasons, only those elementary bodies and their compounds have been taken into consideration which are met with in the examination of medicinal chemicals and their preparations. For any more comprehensive or elaborate information, the standard works of analytical chemistry should be consulted.

When the object of the examination is only to establish the presence or absence of some particular substance, the characteristic reagent may be employed at once; but, if a qualitative analysis is required, the substance, if a solid body, may be subjected first to a preliminary examination in the dry way, by which means approximate information as to its composition may be obtained; after this, it is dissolved, and examined. The course of qualitative analysis, therefore, consists of three parts:

I. Preliminary examination in the dry way.
II. Solution, or conversion into the liquid form.
III. Analysis of the solution.

I. PRELIMINARY EXAMINATION.

This consists in an accurate observation of the physical properties of the substance, its form, color, hardness, gravity, and odor, and of its deportment at a high temperature, either alone, or in contact with some chemical compound which produces decomposition.

1. THE SUBSTANCE IS HEATED IN A DRY NARROW TUBE (Fig. 11).

(*a*) Organic compounds *carbonize* and *blacken*, evolving empyreumatic, inflammable gases.

(*b*) The *substance remains unaltered;* indicating absence of organic matter, of salts containing water of crystallization, and of volatile compounds.

(*c*) The *substance fuses*, *expelling aqueous vapors*, which condense in the cooler parts of the tube; indicating salts with water of crystallization (these will generally re-solidify after the expulsion of the water), or decomposable hydrates, which often give off their water without fusing.

(*d*) *Gases or fumes are evolved:* smell of iodine from iodine compounds; smell of sulphurous acid from decomposition of sulphates; smell of nitric oxides from the nitrates; smell of ammonia from ammonium salts, from cyanides, or from nitrogenous organic compounds, in which latter case carbonization takes place, and either cyanogen or empyreumatic fumes escape with the ammonia.

(*e*) *Sublimates are formed* by volatile substances, as sulphur, and compounds of ammonium, mercury, arsenic, and antimony. In this case the sublimate is removed to the bottom of the test-tube, and, together with the substance, is covered with a

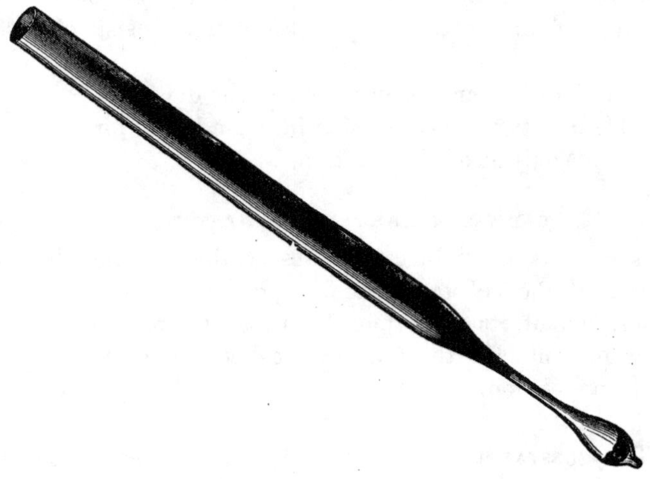

FIG. 11.

few small pieces of charcoal, and again heated; mercury and arsenic form metallic sublimates, the latter with the character-

FIG. 12.

istic garlic odor, the former without. In another tube part of the substance is heated, and the sublimate is moistened with solution of potassium hydrate; mercurous chloride turns black; mercuric chloride, red; and ammonium salts evolve the odor of ammonia.

2. THE SUBSTANCE IS MIXED WITH DRIED SODIUM CARBONATE, AND HEATED ON CHARCOAL IN THE REDUCING FLAME OF THE BLOW-PIPE (Fig. 12).

(*a*) *Fusion and absorption* into the coal indicate alkalies.

(*b*) An *infusible white residue*, either at once or after previous fusion in the water of crystallization, indicates compounds of calcium, barium, strontium, magnesium, aluminium, zinc, or tin.

(*c*) A *reduction to the metallic state* takes place, *without* formation of a *peripheric incrustation* upon the charcoal. Compounds of tin, silver, and copper, give malleable shining scales. Compounds of iron, manganese, cobalt, and nickel, are reduced to a gray infusible powder; all visible upon cutting the fuse from the coal, and triturating and levigating it in an agate mortar.

(*d*) *Reduction with incrustation:* Antimony compounds give a brittle metallic globule and a white incrustation; bismuth, a brittle globule and a brown-yellow incrustation; lead, a malleable globule and a yellow incrustation; zinc and cadmium are not reduced, but give, the former a white incrustation, not volatile in the oxidizing flame, the latter a brown-red incrustation.

(*e*) Arsenic compounds give the smell of garlic.

(*f*) Borates and aluminates swell up.

(*g*) Sulphur compounds give an alkaline sulphide, which, when moistened, leaves a black stain upon a clean silver plate.

II. SOLUTION OF SOLID BODIES.

After having ascertained, by the preliminary examination, to what class of bodies the substance under consideration belongs, it has then to be brought into the liquid form—in other words, to be dissolved. The usual solvents which are employed are water, hydrochloric and nitric acids, and aqua regia. The finely-powdered substance is first boiled with from 12 to 20 times its weight of distilled water, in order to ascertain its complete or partial solubility, or its insolubility therein. If it be not completely dissolved, the portion insoluble in water is collected upon a filter, and is then treated successively with dilute and concentrated hydrochloric acid; by this process carbonates evolve carbonic-acid gas, with effervescence; peroxides, chromates, and chlorates, evolve chlorine; cyanides give hydrocyanic acid; many sulphides, hydrosulphuric acid; sulphites and hyposulphites, sulphurous acid.

If hydrochloric acid does not completely dissolve the substance, it generally effects the separation of one or more of its constituents; for this reason the solution should be separated from the residue and examined apart. The residue may consist of compounds undecomposable by hydrochloric acid, which existed in the original substance; or of insoluble compounds formed by the decomposition of the original substance by hydrochloric acid. Thus sulphur is separated from polysulphides, and pulverulent or gelatinous silica from silicates; or, if lead, silver, or subsalts of mercury be present, insoluble chlorides of these metals will be formed. In this latter case, argentic chloride may be distinguished by its solubility in aqua ammoniæ, and mercurous chloride by its reduction to black mercurous oxide.

If the substance is not completely soluble in hydrochloric acid, the insoluble residue is treated successively with nitric acid and aqua regia, which either act as mere solvents or exert an oxidizing action.

When a finely-powdered substance is not dissolved by successive treatment with either of these solvents, it must be rendered soluble by other means, in order that its constituents may be determined. This is generally accomplished by fusion with 3 to 4 parts by weight of alkaline carbonates, in the case of the sulphates of barium, strontium, calcium, and lead, and also of silica and silicates, or by fusion with potassium bisulphate in the case of alumina or aluminates.

III. QUALITATIVE ANALYSIS OF SOLUTIONS.

1. Examination for bases.

The most important groups of bodies are the simple compounds of the so-called metallic elements, which contain generally but one base and one acid, or one metallic and one non-metallic element. The elimination of the former depends upon the different solubility of their sulphides, and, upon this fact, they are divided into three groups, which afterward are successively discriminated; by this proceeding, the detection of each individual member of such group is considerably facilitated.

Group I.—*Metals whose sulphides are insoluble in water or diluted acids.* They are all precipitated from their acid solutions by hydrosulphuric acid. They are divided into two subdivisions:

(A) Metals whose sulphides are sulpho-acids, forming with sulpho-bases (ammonium, potassium, and sodium sulphides) soluble sulpho-salts. These are: arsenic, antimony, tin, and gold.

(B) Metals whose sulphides do not possess acid properties, not combining with, and therefore insoluble in, alkaline sulphides. These are: lead, silver, mercury, bismuth, copper, and cadmium. (Sulphide of mercury is soluble in potassium or sodium sulphide, and sulphide of copper is somewhat soluble in ammonium sulphydrate.)

Group II.—*Metals whose sulphides form soluble sulpho-salts*, which consequently are not precipitated by hydrosulphuric acid from neutral or acid solutions, but partly from alkaline solutions; which, however, are completely precipitated by ammonium sulphydrate from neutral as well as from alkaline solutions. These are, again, subdivided into two groups:

(A) Metals which are precipitated as *sulphides:* iron, manganese, zinc, nickel, and cobalt.

(B) Metals which are precipitated as *oxy-hydrates:* aluminium and chromium.

Group III.—*Metals whose sulphides and hydrates are soluble in water;* and which, therefore, are not precipitated by hydrosulphuric acid nor by ammonium sulphydrate from any solution. This group includes the so-called alkaline and alkaline-earthy metals. They are subdivided according to their deportment with ammonium carbonate in the presence of ammonium chloride.

(A) Metals which are precipitated by ammonium carbonate, their normal carbonates being insoluble in water or in ammonium chloride; these are: calcium, barium, and strontium.

(B) Metals which are not precipitated by ammonium carbonate; these are: magnesium, potassium, sodium, and ammonium. Magnesium carbonate is insoluble in water, but soluble in ammonium chloride; the carbonates of the alkaline metals are soluble in water.

If a solution is supposed to contain several or all of these metals, it is obvious that, by the successive application of these general reagents, we shall separate, *first*, by hydrochloric acid, those metals whose chlorides are insoluble in water or in dilute acids; *secondly*, by hydrosulphuric acid, those metals whose sulphides are insoluble in diluted acids; *thirdly*, by ammonium sulphydrate, those remaining metals whose sulphides or hydrates are insoluble in neutral or alkaline liquids; and, finally, by ammonium carbonate, those metals whose carbonates are insoluble: so that at last only the alkaline metals are left in solution.

Hereupon the following course and method of investigation may be employed, with the observation of the following rules:

The precipitation by each general reagent must be complete. To insure this, the reagent must be added gradually, allowing the precipitate to subside between each addition, until no further precipitate is produced. In the case of hydrosulphuric acid, the precipitation is complete when the solution, after agitation, still smells strongly of the gas.

Each group, when precipitated, must be thoroughly freed, by washing with water, from all members of the subsequent groups, which may be contained in the solution. After the precipitation of each group, it is advisable to ascertain the presence or absence of any members of the succeeding groups, by evaporating on platinum-foil a few drops of the filtrate; if, after ignition, there is no distinctly visible residue, non-volatile substances need not be looked for further. It is obvious that, if complete precipitation and thorough washing be neglected, metals belonging to one group are liable to be found among those of another group; and, consequently, as the analysis proceeds, reactions will be obtained which will be a source of perplexity and errors.

To part of the solution of the substance under examination, acidulated with hydrochloric or nitric acid, add hydrosulphuric acid:

Precipitate No. 1.

Yellow precipitate: arsenic, tin, and cadmium.
Orange precipitate: antimony.
Brown precipitate: gold and subsalts of tin.
Black precipitate: mercury, lead, bismuth, silver, and copper.

Filter, preserve the **filtrate**, and mark it **No. 1.**

Wash precipitate No. 1 on a filter, and treat with ammonium sulphydrate; filter, and preserve the **residue,** marked **No. 1,** on the filter. The filtered solution of sulphides is now over-saturated with hydrochloric acid: the ensuing precipitate may contain the sulphides of arsenic, antimony, tin, and gold; it is washed and treated with solution of ammonium carbonate, which dissolves arsenic; from this filtered solution the arsenic may be re-precipitated by hydrochloric acid, and may be identified as **arsenic** by the metallic mirror, and by the garlic-like odor when heated with charcoal in a glass tube. Those sulphides which are not soluble in ammonium carbonate are distinguished by their deportment when heated on charcoal in the blowpipe-flame; antimony oxidizes in the oxidizing part of the flame, giving white fumes and an incrustation of **antimonic** oxide; sulphide of tin, not being volatile, remains on the charcoal, and gives, when fused with potassium cyanide, malleable metallic globules of **tin.** An examination for gold is only required when the solution of the original substance had a yellow color; in this case the solution is tested with ammonium chloride, and then heated with ferrous sulphate; **gold** is reduced and precipitated as a fine brown powder.

Residue No. 1 is treated on its filter with warm nitric acid; sulphide of **mercury** remains undissolved; it is dissolved in aqua regia, and recognized by reduction with stannous chloride. Part of the nitric-acid solution of the sulphides is tested with sulphuric acid for **lead;** if this be present, it must be completely precipitated from the solution before testing the same with hydrochloric acid for **silver,** and this metal, if present, must also be completely precipitated; the filtered solution is next over-saturated with ammonium hydrate; an ensuing blue color indicates **copper,** and a white precipitate, **bismuth;** in the latter case the precipitate is collected upon a filter and dissolved in a few drops of strong nitric acid, which solution, when dropped into an excess of water, will give rise to the formation of white turbid clouds. Finally, the solution filtered from the bismuth is treated with ammonium carbonate; **cadmium,** if present, is precipitated, and may be confirmed by the formation of brown circular incrustations on the charcoal, when the precipitate is heated with exsiccated sodium carbonate in the reducing flame.

Filtrate No. 1, when an examination upon fixed constituents, by evaporation and heating upon platinum-foil, has an affirmative result, *is over-saturated with ammonium hydrate, and an excess of ammonium sulphydrate is added.*

Precipitate No. 2.

Black precipitate: iron, cobalt, and nickel.
Flesh-colored precipitate: manganese.
Bluish-green precipitate: chromium.
White precipitate: aluminium and zinc.
Filter, preserve the **filtrate,** and mark it **No. 2.**

Wash the precipitate on the filter, and dissolve in hydrochloric acid. Only the sulphides of nickel and cobalt remain undissolved; they are dissolved upon the filter with warm nitric acid; the ensuing solution is slightly over-saturated with ammonium hydrate, and subsequently treated with potassium hydrate: **nickel** is precipitated as green oxyhydrate; the solution filtered from the nickel is tested with hydrosulphuric acid; a black precipitate of **cobalt** sulphide is identified as such by fusing it before the blowpipe with a borax-bead, to which it imparts a blue color.

The hydrochloric-acid solution of precipitate No. 2, containing the metals of the second group, is boiled, after the addition of concentrated nitric acid, until the evolution and odor of hydrosulphuric acid cease, when it is filtered and over-saturated with ammonium hydrate: the ensuing precipitate may contain the hydrates of iron, aluminium, and chromium. They are collected upon a filter, and the **filtrate** is marked **No. 3.** In order to discriminate them, the precipitate is treated with a solution of potassium hydrate: aluminium and chromium oxides dissolve; the liquid is diluted with a little water, filtered, and heated to boiling; **chromium** hydrate is re-precipitated, the solution is then filtered, and boiled with ammonium chloride: **aluminium** hydrate is precipitated. The residue from the precipitate remaining undissolved in potassium hydrate is dissolved in a little hydrochloric acid, and recognized as **iron** by precipitation with potassium ferrocyanide.

Filtrate No. 3 is boiled with potassium hydrate: an ensuing precipitate, first white, and gradually turning brown, indicates **manganese,** which may also be recognized by the green fuse, when a part of this precipitate is fused with a mixture of sodium carbonate and potassium nitrate on platinum-foil. The solution, filtered from the precipitate, gives, on the addition of hydrosulphuric acid, a white precipitate of **zinc** sulphide, if that metal be present.

Filtrate No. 2, when evaporation upon platinum-foil indicates fixed constituents, *is over-saturated with hydrochloric acid,* and is heated until the odor of hydrosulphuric acid nearly or entirely ceases; when cold, it is filtered, *slightly over-saturated with ammonium hydrate, and completely precipitated by ammonium carbonate.* The carbonates of calcium, barium, and strontium, are thrown down. The precipitate is dissolved in hydrochloric acid, and the resulting solution is tested by agitation with a saturated solution of calcium sulphate: if a white precipitate be formed immediately, the presence of **barium** is shown; if it be formed after a while, **strontium** is indicated. If both these alkaline earth-metals be present, the solution is completely precipitated by sulphuric acid; the filtered solution is then neutralized with ammonium hydrate, and tested with ammonium oxalate; an ensuing white precipitate indicates **calcium.**

If the solution filtered from the earthy carbonates shows fixed constituents upon evaporation on platinum-foil, part of it is examined with sodium phosphate: an ensuing white precipitate of magnesium and ammonium phosphate indicates

magnesium. If this metal be present, the solution is exactly neutralized with hydrochloric acid, and is then agitated with barium hydrate, so that the magnesium may be completely precipitated as hydrate; the excess of barium is completely removed from the filtrate by sulphuric acid; the filtrate is then evaporated to dryness, the residue moistened with hydrochloric acid, and heated to redness; it is then examined upon platinum-wire in an alcohol-flame; **sodium** imparts a yellow, **potassium** a red, color to the flame; for verification, the rest of the residue may be dissolved in a few drops of water, and tested with platinum chloride; a yellow crystalline precipitate of potassium platinic chloride will be formed, if potassium be present.

Ammonium is always sought for in a separate portion of the original substance or solution, by heating with a concentrated solution of potassium or sodium hydrate: any ammonium compound evolves the characteristic odor of ammonia.

II. EXAMINATION FOR ACIDS.

The examination for the bases is followed by that for the acids and for chlorine, iodine, and bromine. The preliminary examination, as well as the nature of the substance and the bases found therein, will give information, in most cases, as to what acids cannot be contained in the substance, and what acids may be present therein or should especially be looked for. Thus the acids of arsenic, and chromic, carbonic, and hydrosulphuric acids, have already been indicated. With soluble substances containing earthy and metallic bases, the presence of carbonic, phosphoric, boric, and oxalic acids, is excluded; soluble substances, containing silver, lead, and mercurous compounds, exclude chlorine; soluble substances, containing lead, barium, strontium, and mercurous salts, exclude sulphuric acid.

In the examination for acids, a neutral solution is frequently required, and generally ammonium hydrate is used for neutralization; but, as most of the heavy metals, as well as some alkaline earthy salts, are precipitated when their solutions are neutralized by ammonium hydrate, it is often necessary to remove from the solution all metals, except those of the alkalies, before proceeding to search for acids. When this is not necessary, it is frequently requisite, according to the nature of the substance and its chemical relations, to substitute, instead of hydrochloric acid and its salts, nitric acid and the corresponding nitrates.

The general reagents employed in the examination for acids are barium chloride or nitrate, calcium chloride, argentic nitrate, ammonium and magnesium sulphate, ferric chloride, and indigo-solution. By these reagents the common acids are divided into certain groups:

1. *Acids which are precipitated by barium nitrate:*—a, *from acidulated solutions:* sulphuric acid;—b, *from neutral solutions* (the precipitate being soluble in acids): sulphurous, phosphorous, carbonic, silicic, oxalic, chromic, boric, tartaric, citric, and arsenic acids.

2. *Acids which are precipitated by calcium chloride:*—a, *from neutral solutions only:* phosphoric, arsenic, boric, carbonic, sulphurous, tartaric, citric acids, and ferrocyanides;—b, *from neutral or acetic-acid solutions:* sulphuric and oxalic acids.

3. *Acids which are precipitated by magnesium sulphate,* in presence of ammonium hydrate and ammonium chloride: phosphoric, arsenic, and tartaric acids.

4. *Acids which are recognized by ferric chloride:*—a, *are precipitated:* ferrocyanides, phosphoric, arsenic, tannic acids (from a neutral or acetic-acid solution), boric, benzoic, and succinic acids (from neutral solutions only);—b, *those which give only a colored reaction (in acid solutions):* ferricyanides (brown);—*in neutral solutions only:* acetic acid, sulphurous acid (red), and gallic acid (black).

5. *Acids which are precipitated by argentic nitrate:*—a, *from neutral solutions* (precipitate being soluble in dilute nitric acid): phosphoric, pyro- and metaphosphoric, arsenious, arsenic, chromic, oxalic, boric, tartaric, citric, sulphurous, and formic acids;—b, *from acid solutions also* (the precipitate being insoluble in dilute nitric acid): hydrochloric, hydrobromic, hydroiodic, hydrocyanic, iodic, hydrosulphuric acids, and ferro- and ferricyanides.

6. *Acids which decolorize indigo solution:* free chlorine and bromine, and their oxygen acids, when free; free nitric acid, if not too dilute, and alkaline sulphides. On the addition of sulphuric or hydrochloric acids, and heating, indigo-solution is also discolored by all chlorates, bromates, iodates, and nitrates, and, on addition of hydrochloric acid and heating, besides all the foregoing acids, also by chromates, permanganates, and all peroxides.

Hereupon the following mode of investigation for the most common acids is based:

Add to a part of the solution barium chloride, or, if silver, lead, or mercurous salts are present, barium nitrate; an ensuing precipitate, insoluble on the addition of hydrochloric or nitric acid, indicates **sulphuric acid**; if the precipitate be partly or entirely dissolved, the acids of arsenic, or phosphoric, boric, chromic, oxalic, or tartaric acid, may be present: in this case part of the original solution is over-saturated with ammonium hydrate, and is tested with ammoniated magnesium sulphate, and warmed; the occurrence of a white crystalline precipitate indicates phosphoric or **arsenic acid**; the latter one has already been indicated in the preliminary examination, and its presence may be confirmed, by fusing part of the original substance, or the residue of the evaporated solution, on charcoal with potassium cyanide.

If arsenic acid is found to be present in the solution, it has to be removed by reduction with sulphurous acid and by subsequent precipitation with hydrosulphuric acid. Then phosphoric acid may be recognized in the filtrate by the addition of ammoniated magnesium sulphate, and by redissolving the resulting precipitate in nitric acid and re-precipitation by argentic nitrate, which, when **phosphoric acid** is present, will give a yellow precipitate.

Boric acid is detected in the substance by its property of imparting a green color to the alcohol-flame, and by its reaction with turmeric-paper. When the solution of the original substance was yellow or red, and the precipitate with barium was also yellow, chromic acid may be present; in this case, the **chromic acid** has to be reduced to chromium oxide by alcohol and hydrochloric acid, and may then be recognized, as stated in the examination for bases (page 43).

Tartaric and **citric acids** have been already indicated in the preliminary examination. **Oxalic acid** is detected in neutral solutions by calcium sulphate, the white calcium oxalate being insoluble in acetic acid.

Part of the solution, prepared or acidulated with nitric acid, is precipitated with argentic nitrate; an ensuing precipitate, insoluble in nitric acid, would indicate hydrochloric or hydrocyanic acids, iodine, bromine, and ferro- and ferricyanides.

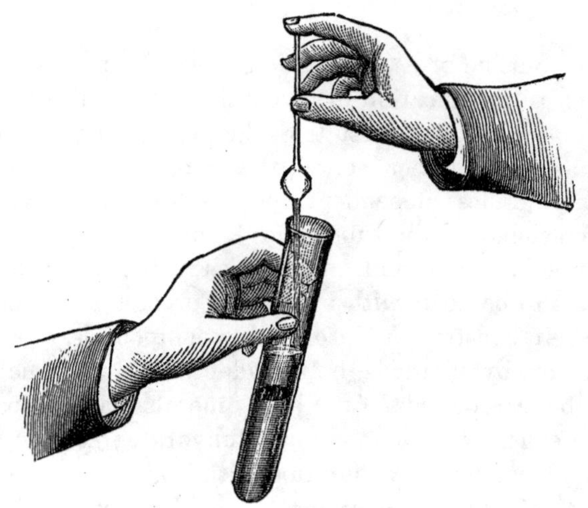

FIG. 13.

Hydrocyanic acid, besides being already indicated in the preliminary examination, may be detected by its reaction with ferrous and ferric salts. If the silver precipitate be yellowish, and insoluble in ammonium hydrate, it indicates **iodine:** in this case the original substance, on being heated in strong sulphuric acid, emits the vapor and odor of iodine. If the precipitate be curdy, and soluble in ammonium hydrate, it indicates, if no cyanogen compounds be present, **hydrochloric acid:** if this be the case, the cyanogen has to be destroyed by fusing the substance with a mixture of sodium carbonate and nitrate; the residue is extracted with water, and the filtered solution, acidulated with nitric acid, is then tested with argentic nitrate.

Nitric and chloric acids have already been detected in the course of the preliminary examination. The former may also be recognized by mixing the solution in a test-tube with an equal bulk of concentrated solution of ferrous sulphate, and pouring the mixture carefully upon concentrated sulphuric acid, so that the fluids do not mix (Fig. 13). If **nitric acid** be present, the line of contact between the two fluids will show a dark red-brown coloration. **Chloric acid** will be detected by the evolution of chlorine, when the original substance is warmed in strong sulphuric acid.

This brief outline of a systematic course of analysis is necessarily open to modification in its details; such will be designated in many cases by the results of the preliminary examination. As has already been stated, the search for the acids is especially more or less dependent upon the results of the previous examination, and the substances found.

Complex substances, and those which resist the common solvents, have to be dealt with by special processes in order to transform their constituents into soluble compounds, as stated on page 38, or by methods better adapted to their nature. Thus, insoluble compounds of the heavy metals may be decomposed by digestion with ammonium sulphydrate; sulphates of strontium and calcium, by digestion with sodium carbonate; in both cases, the filtrate contains the acid, together with an excess of the decomposing agent, while the metal is found in the residue. Insoluble salts of organic acids are decomposed by boiling with an alkaline carbonate: iron salts containing volatile organic acids, by digestion with ammonium hydrate; in both cases, the filtrate contains an alkaline salt of the acid. Sulphides, and all the lower oxygen acids of sulphur, yield sulphuric acid when digested with nitric acid, or any corresponding oxidizing agent.

TABLE

OF THE DEPORTMENT OF THE COMPOUNDS OF THE COMMON METALS WITH SOME OF THE GENERAL REAGENTS.

Oxides precipitated by HYDROSULPHURIC ACID from

Acid Solutions:	*Alkaline Solutions:*	
as Sulphides:	*as Sulphides:*	*as Oxyhydrates:*
As Sb Sn Au Pt Hg Bi Ag Cu Pb Cd.	Fe Mn Co Ni Zn.	Al Cr.

Sulphides soluble in

Nitric Acid:	*Aqua Regia:*	*Ammonium Sulphydrate:*
PbS BiS CuS AgS	HgS CoS NiS.	AsS SbS SnS
CdS FeS MnS ZnS.		AuS PtS.

Oxides precipitated by

Hydrochloric Acid:	*Sulphuric Acid:*	*Ammonium Hydrate:*
Pb Ag	Pb Hg_2O	Fe Co Ni Mn
Hg_2O.	Sb_2O_3 SnO_2	Zn Al Cr.
	Ba Ca.	

Ammonium Carbonate:	*Water:*
Ba St Ca.	Bi Cu_2O Hg_2O SnO Sb_2O_3.

Oxides soluble in

Potassium Hydrate:	*Ammonium Hydrate:*
Pb Sb_2O_3 Sn O_2 Zn Al Cr.	Ag Cd Cu Zn Co Ni.

Re-precipitated by boiling:

Zn Cr.

VOLUMETRIC ANALYSIS.

THE quantitative estimation of a number of medicinal chemicals and their preparations has been much simplified in practice by the volumetric method of chemical examination, which is based upon the fact that chemical substances combine in definite and equivalent proportions, and consists in noting the volume of a test-solution of known strength, required to produce by chemical reaction a certain visible effect when added to a known quantity of the substance under examination. On obtaining this effect, the quantity of the reagent being ascertained, and that of the substance being already known, an accurate estimate may readily be made by equation and simple calculation.

By the aid of this simple and rapid mode of examination, the proportion of the constituents of chemical compounds and their preparations may be at once quantitatively estimated. In the following part of this volume, in treating of all those chemicals and preparations in which quantitative determination of the principal constituents is required, and to which the volumetric mode of examination is best suited, either alone or as a confirmatory test, reference has been made to these pages, in which is stated the quantity of the volumetric test-solution requisite to produce, with a definite weight of the substance under examination, corresponding to its molecular weight, the exact reaction indicative of its officinal strength.

Volumetric determinations are principally based, either upon *Saturation*, in which the quantity of a base or an acid is measured by the quantity of acid or base which is necessary to convert it into a neutral salt; or, upon *Oxidation* and *Reduction*, in which the quantity of the substance to be determined is found by the quantity of chlorine, iodine, or oxygen, to which it is equivalent as an oxidant, or which it requires to pass from a lower to a higher stage of oxidation; or, upon *Precipitation*, in which case the quantity of the substance to be determined is derived from that of the reagent required to separate it in an insoluble state.

FIG. 15.

The quantities of the substances to be assayed volumetrically are submitted to examination by weight, generally coincident with their molecular weight, and are expressed either in grains, or, as is now generally preferred, in grammes;* those of the test-solutions by measure in cubic centimetres.†

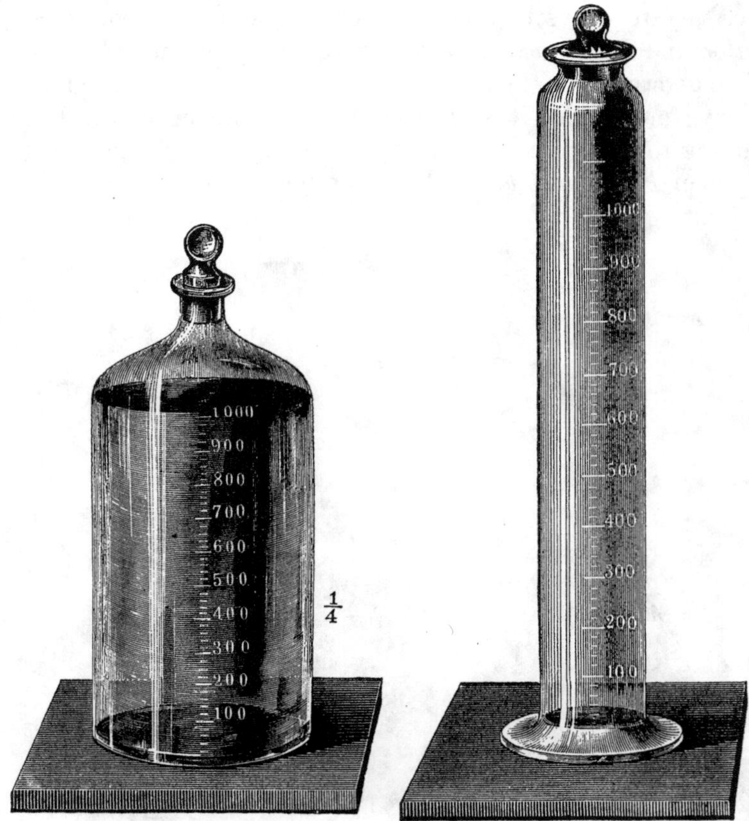

Fig. 16. Fig. 17.

* One gramme is equal to 15.4340 grains of Troy-weight.

Although in this work, as previously mentioned, the time-honored apothecary's weight has been employed, for all statements of quantities by weight, as still being in general use in pharmaceutical practice in the United States, Great Britain, and its colonies, yet the metric system of weights and measures has been preferred for volumetric analysis for the above reason, and because the instruments for volumetric analysis are generally constructed with metric divisions. It is obvious, however, that grain weights and corresponding proportions of measures can readily be employed instead of metric units.

† A cubic centimetre (Fig. 14) is the volume occupied by one gramme of distilled water at its point of greatest density, 4° C.; metric measurements, however, are uniformly taken at 15°.55 C. (60° F.).

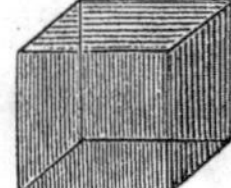

Fig. 14.

The apparatus required for volumetric analysis consists, besides the common utensils, as beakers, funnels, porcelain-capsules, crucibles, stirring-rods, balances, etc., of one or several

Fig. 19.

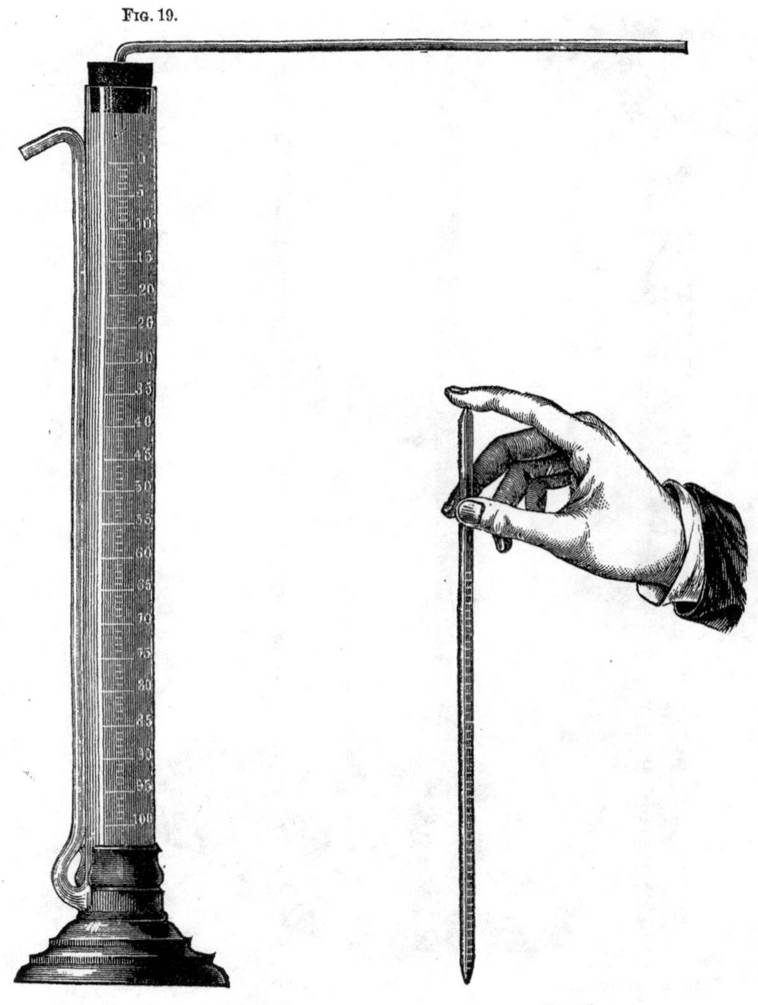

Gay Lussac's Burette.

Fig. 18.

litre-flasks for the preparation of the test-solutions (Fig. 15); these, when filled to a mark on the neck, have mostly a capacity of 1,000 cubic centimetres (1 litre) of distilled water;

some cylindrical graduated litre-jars (Figs. 16 and 17), divided into 100 or 1,000 centimetre-parts, and used for the preparation of test-solutions as well as for the admixture of parts of a litre; and one or several graduated tubes for the delivery and measure-

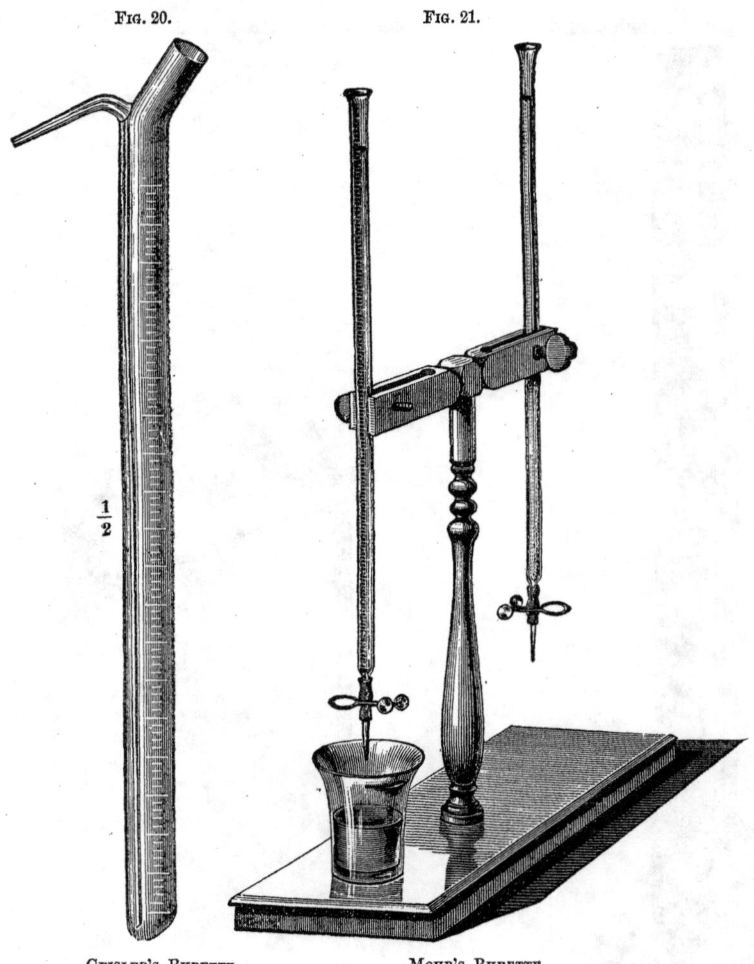

Fig. 20. Geisler's Burette.

Fig. 21. Mohr's Burette.

ment of the test-solutions; pipettes and burettes. The former are provided either with a single mark upon the narrow neck, or are graduated into a number of cubic centimetres or parts thereof. In using pipettes, they are filled by sucking and then closed above with the forefinger of the right hand (Fig. 18); the delivery is effected by a more or less gentle displacement of the finger. Burettes are preferable for delivery and measurement, and are now most generally employed; they hold to a certain mark 100 or less cubic centimetres, and are divided into a corresponding number of equal parts.

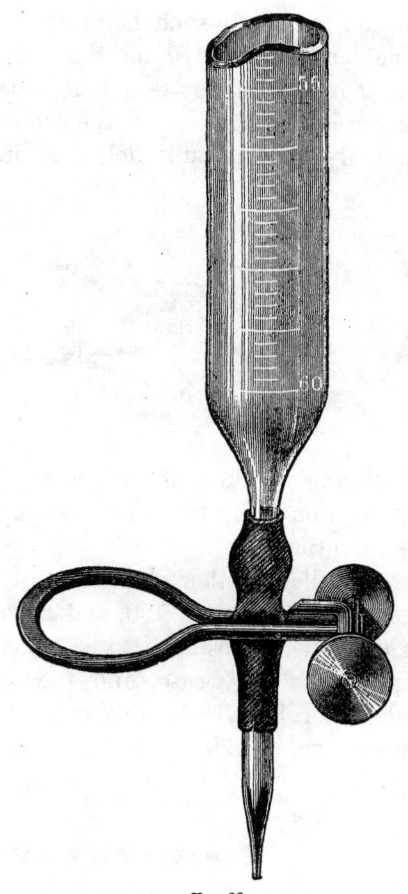

Fig. 22.

There are three kinds of burettes in use, which differ mainly in their construction for delivery: Gay Lussac's burette, Geisler's burette, and Mohr's burette.

Of these, Mohr's burette is now in general use; it consists of a graduated glass tube, of about half an inch inside width, and about 30 inches in length; to its contracted lower extremity is fitted (Fig. 22) a small piece of caoutchouc tube; into the other end of which a small piece of narrow glass tube, about 1 inch long, is tightly inserted. A strong wire clamp (Fig. 23) closes the rubber tube so that the fluid can only pass through, either in a stream or drop by drop, when the knobs of the clamp are pressed. Since the correctness of the test depends upon the accurate reading of the height of the test-solution in the burette, a small hollow glass float is employed for this purpose (Fig. 24); it is of such a width that it can move freely in the tube without undue friction, and of such a weight that it sinks to more than half its length into the test-liquid. A line is scratched around the centre of the float, serving to mark the height of the fluid in the burette.

Burettes are conveniently kept for ready use on a revolving stative (Fig. 25).

The tests are made by first filling the burette with the test-solution to exactly such height that the mark on the float is coincident with the 0 on the scale of the tube. The solution or mixture to be tested is placed in a beaker-glass under the burette (Fig. 21), and then so much of the test-solution is gradually and carefully delivered into the beaker, with gen-

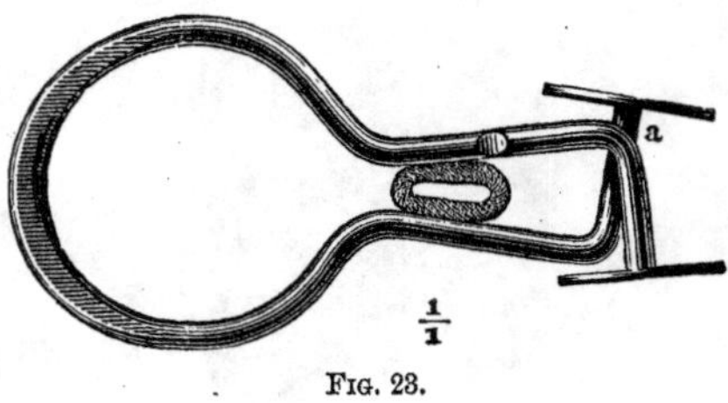

FIG. 23.

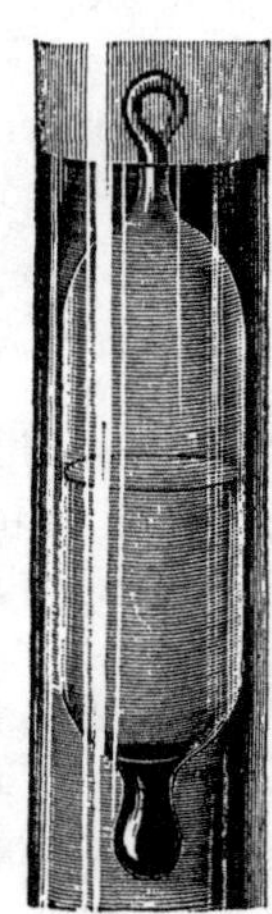
FIG. 24.

tle stirring with a glass rod, as to accomplish the reaction indicative of the completion of the operation.

A smaller number of cubic centimetres of the test-solution than 100, and a corresponding amount of substance to be tested, may be employed in any case, and the calculation made accordingly. In this case smaller and more convenient burettes, holding only 50 c. c., may be employed.

ANALYSIS BY SATURATION.

ESTIMATION OF ALKALIES.

Test-Solution of Oxalic Acid.

Sixty-three grammes of pure crystallized oxalic acid are dissolved in water; the solution is filtered into a litre-flask, and the filter washed with water until the exact volume of 1 litre, at about 16° C., is obtained.

One hundred cubic centimetres of this solution contain one-twentieth of the molecular weight, in grammes, of oxalic acid, and are, therefore, capable of neutralizing one-twentieth of the

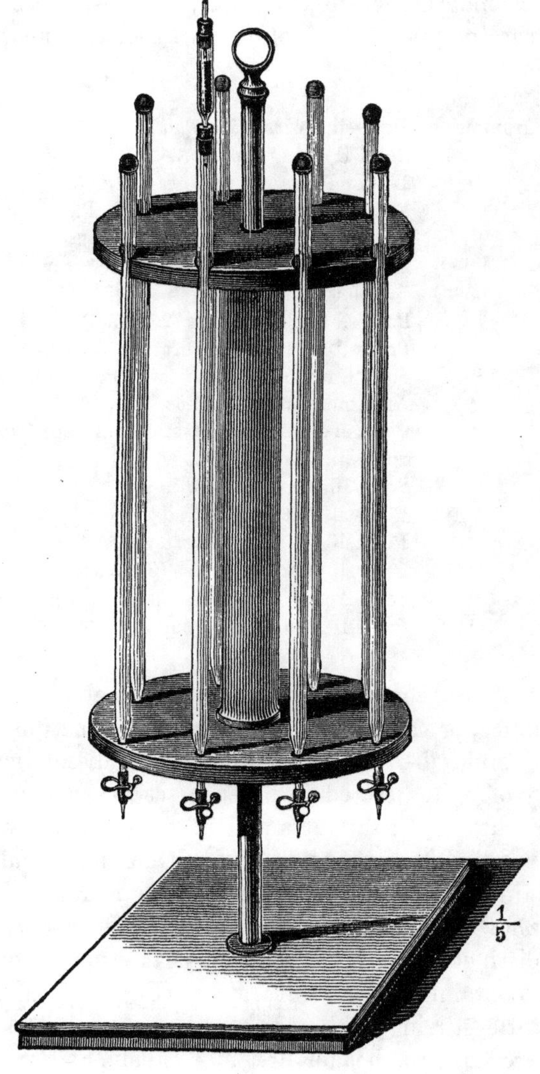

Fig. 25.

molecular weight in grammes of bivalent bases, or salts containing 2 atoms of univalent metals, or one-tenth of the molecular weight in grammes of salts containing one atom of univalent bases.

This test-solution is applied for the estimation of the alkaline hydrates, carbonates, acetates, tartrates, citrates, and borates: 100 cubic centimetres of the solution will exactly neutralize, if of officinal strength:

5.61	grammes of	Potassii Hydras (2 oz. aq.).
4.00	"	Sodii Hydras (2 oz. aq.).
96.00	"	Liquor Potassæ, spec. grav., 1.065.
6.54	"	Aqua Ammoniæ fortior, U. S. P., spec. grav., 0.900 (2 oz. aq.).
17.00	"	Aqua Ammoniæ, U. S. P., spec. grav., 0.960 (1 oz. aq.).
6.91	"	Potassii Carbonas pura (2 oz. aq.).
8.30	"	Potassii Carbonas depurata (2 oz. aq.).
14.30	"	Sodii Carbonas crystallisata (2 oz. aq.).
5.90	"	Ammonii Carbonas (2 oz. aq.).
10.01	"	Potassii Bicarbonas crystallisata (3 oz. aq.).
8.40	"	Sodii Bicarbonas (4 oz. aq.).
19.10	"	Sodii Biboras (3 oz. aq.).
11.30	"	Potassii Tartras (3 oz. aq.).
18.80	"	Potassii Bitartras (3 oz. aq.).
14.10	"	Potassii et Sodii Tartras (3 oz. aq.).
13.60	"	Sodii Acetas crystallisata (3 oz. aq.).
11.60	"	Potassii Acetas (2 oz. aq.).

The operation is conducted by weighing the above quantity of the substance, or the preparation to be estimated, placing it in a beaker, and, when required, diluting or dissolving it in the quantity of water stated in parentheses for each member of the above list. The tartrates and acetates have first to be completely reduced to carbonates, by ignition in a small covered porcelain crucible. When the solution is ready for the test, one or a few drops of litmus-solution are added, so as to impart to it a distinct bluish tint; then the beaker is placed under the burette containing the test-solution (Fig. 26), and, with constant gentle stirring with a small glass rod, the test-solution is delivered into the beaker, first in a stream, and, when approaching the point of saturation, drop by drop, until the blue assumes a violet hue; the solutions of carbonates have to be heated before approaching saturation, in order to expel the carbonic-acid gas from the solution as nearly as possible, and thereby to lessen its reaction upon litmus.

These operations require care and skill in every point, so as to avoid the slightest loss of either of the liquids, and a consequent error in the result.

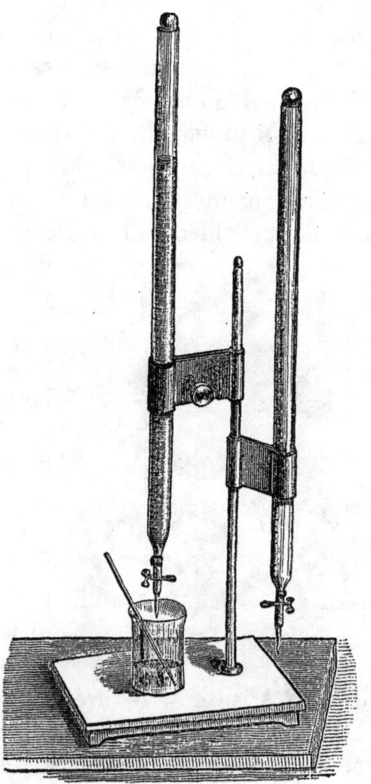

FIG. 26

When saturation is indicated by the violet-reddish color of the liquid, the process is complete, and the volume of test-acid employed is read off. The number of cubic centimetres employed, less than 100, indicates at once the percentage of impurities or of want of strength.

On the other hand, each cubic centimetre of the test-solution employed corresponds to one milligramme-molecule of hydrate or one-half milligramme-molecule of carbonate, i. e.:

1	cubic centimetre corresponds to	0.056	gramme	Potassii Hydras,
1	" "	0.069	"	Potassii Carbonas,
1	" "	0.040	"	Sodii Hydras,
1	" "	0.053	"	Sodii Carbonas (anhydrous),

and a simple equation gives the amount of alkaline hydrate or carbonate present. By operating on 100 times the half-milligramme-molecule, i. e., 6.9 grammes of potassium carbonate or 5.3 grammes of sodium carbonate, all calculation is dispensed with; for, as this amount, if present, would require 100 cubic centimetres of oxalic-acid test-solution for its saturation, the number of cubic centimetres actually required, at once indicates the percentage of alkaline carbonate.

When, in estimating alkaline carbonates, the amount of carbonic acid, namely, the percentage of real carbonate, has to be determined, the following mode is simple and accurate: Two small, light glass flasks with twice-perforated rubber-corks, are connected with a twice-bent tube c (Fig. 27); the flask A is provided with the tube a reaching to the bottom of the flask, and closed at its outer end with a globule of soft wax; that of B is provided with the short tube d. Fifty grains of the carbonate under examination are weighed, and introduced into the flask A, together with a little water; the flask B is half filled with concentrated sulphuric acid; the apparatus is then tightly fitted, and weighed. A small quantity of air is now sucked out of flask B by means of the tube d, whereby the air in A is likewise rarefied. On allowing the air to return, a quantity of the sulphuric acid ascends to the tube c, and flows over into flask A, causing a disengagement of carbonic-acid gas, which escapes through the tube d, after having been dried by passing through the acid in B. This operation is repeated until the whole of the carbonate is decomposed, and the process is terminated by opening the wax stopper, and drawing some air through the apparatus by sucking on tube d. The apparatus is then reweighed, and the difference of the two weighings expresses the quantity of carbonic acid in the 50 grains of carbonate under examination.

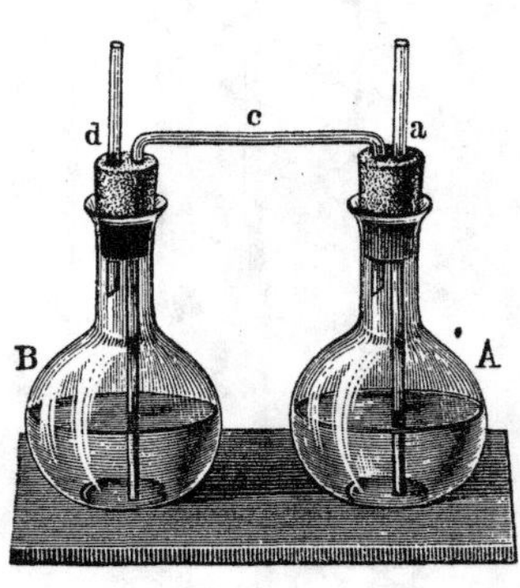

Fig. 27.

ESTIMATION OF ACIDS.

Test-Solution of Sodium Hydrate.

One hundred cubic centimetres of the test-solution of oxalic acid are placed in a beaker with a little neutral litmus-solution, the measure rinsed out, and the washings poured into the beaker. Then a strong solution of sodium hydrate is allowed to flow into the oxalic-acid solution until exact neutralization ensues; the quantity of the soda-solution is noted, and, to every similar quantity of the whole bulk of the sodium-hydrate solution, water is added until the whole measures 100 volume-parts. If, for instance, 93 cubic centimetres of soda-solution have neutralized the 100 cubic centimetres of acid, then, 7 cubic centimetres of water must be added to 93 cubic centimetres of the soda-solution, or 70 to 930 to make a litre. Simple proportion will show to what extent any other quantity is to be diluted.

One hundred cubic centimetres of this test-solution of sodium hydrate contain one-tenth of the molecular weight (= 4) of sodium hydrate taken in grammes, and will neutralize an equal quantity of an acid.

This test-solution is employed for the estimation of the following medicinal acids:

One hundred cubic centimetres of the solution will neutralize, if of officinal strength:

16.70	grammes of	Acidum aceticum, U. S. P., spec. grav., 1.047.
140.58	"	Acidum aceticum dilutum, U. S. P., 1.006.
7.00	"	Acidum citricum.
11.48	"	Acidum hydrochloricum, U. S. P., Spec. grav., 1.160.
47.77	"	Acidum hydrochloricum dilutum, U. S. P., Spec. grav., 1.038.
9.00	"	Acidum nitricum, U. S. P., Spec. grav., 1.420.
54.33	"	Acidum nitricum dilutum, U. S. P., Spec. grav., 1.068.
5.06	"	Acidum sulphuricum, U. S. P., Spec. grav., 1.843.
37.18	"	Acidum sulphuricum dilutum, U. S. P., spec. grav., 1.082.
7.50	"	Acidum tartaricum.

The concentrated acids are to be diluted with four or five times their volume of water, and the solid ones to be dissolved in about eight times their weight of water before being tested. To indicate saturation, and the consequent termination of the test, litmus-solution is used, as before; and the mode of operation is similar to that described on page 58. The number of cubic centimetres employed for saturation, less than 100, indicates at once the percentage of impurities or want of strength.

ANALYSIS BY PRECIPITATION.

ESTIMATION OF ACIDULOUS RADICALS PRECIPITATED BY ARGENTIC NITRATE.

Test-Solution of Argentic Nitrate.

Seventeen grammes of pure crystallized argentic nitrate are dissolved in water; the solution is filtered into a litre-flask, and the filter washed with so much water as to make up the exact volume of one litre.

One hundred cubic centimetres of this solution contain one one-hundredth of the molecular weight, in grammes, of argentic nitrate, and will decompose an equivalent quantity of a salt of any base which yields silver compounds insoluble in water.

This test-solution is employed for the estimation of the following officinal preparations, and 100 cubic centimetres of it will require, for complete decomposition of the argentic nitrate:

27.00	grammes of	Acidum hydrocyanicum dilutum, U. S. P.
1.19	"	Potassii Bromidum.
1.66	"	Potassii Iodidum.
1.50	"	Sodii Iodidum.

In testing hydrocyanic acid, it is, before delivering the test-solution into it, slightly over-saturated with a solution of potassium hydrate. The test-solution is then added, until, after constant stirring, a slight permanent turbidity remains. The quantity of argentic nitrate employed represents exactly half the amount of hydrocyanic acid present, and has therefore to be doubled in order to indicate the correct percentage.

The bromides and iodides are dissolved in one ounce of water, and the test-solution is added until, after agitation of the liquid and subsidence of the precipitate, a drop of the test-solution ceases to cause further precipitation. The quantity of argentic nitrate employed represents an equivalent amount of the bromide or iodide.

ANALYSIS BY OXIDATION.

Test-Solution of Iodine.

12.7 grammes of pure iodine, and 18 grammes of potassium iodide, are dissolved in about 12 ounces of water; the solution is filtered into a litre-flask, and the filter washed with so much water as to make up the volume of one litre.

One hundred cubic centimetres of this solution contain one one-hundredth of the atomic weight of free iodine in grammes, and, if water be present, will cause the oxidation of one two-hundredth of the molecular weight of arsenious acid in grammes, or one four-hundredth of the molecular weight of common white arsenic in grammes, arsenic acid being produced.

One hundred cubic centimetres of this test-solution require for oxidation of the following officinal preparations:

0.495	grammes of	Acidum Arseniosum.
68.30	"	Liquor Potassii Arsenitis, U. S. P.
2.48	"	Sodii Hyposulphis.

The arsenious acid is dissolved, together with about 2 grammes of sodium bicarbonate, in 2 ounces of boiling water; when cool, a little mucilage of starch is added, and the test-solution delivered into the arsenious solution until a permanent blue coloration ensues.

The Liquor Potassii arsenitis is boiled, after adding about 1½ grammes of sodium bicarbonate, and, when cool, is tested in like manner.

The Sodium hyposulphite is dissolved in 2 ounces of water, and a little mucilage of starch added.

ANALYSIS BY DEOXIDATION.

Test-Solution of Sodium Hyposulphite.

Thirty grammes of sodium hyposulphite are dissolved in nearly one litre of water; the solution is allowed to flow from a burette into a beaker containing exactly 100 cubic centimetres of the volumetric solution of iodine, until the brown color of the iodine just ceases—or, if a little mucilage of starch be added, until the blue iodide of starch is decolorized. The number of cubic centimetres of the solution required is noted, and so much water is added to the bulk of the solution, that 2.48 grammes of sodium hyposulphite are contained in every 100 cubic centimetres.

This test-solution contains one one-hundredth of the molecular weight of sodium hyposulphite in grammes, and will indicate the presence of one one-hundredth of the atomic weight of iodine in grammes, in any quantity of a liquid containing free iodine, or iodine liberated by an equivalent quantity of free chlorine.

100 cubic centimetres of this solution require for exact deoxidation:

1.27 grammes of Iodinum.
5.42 fluid-drachms of Tinctura Iodini, U. S. P.
54–58 grammes of Aqua Chlorini.
1.17 " Calx Chlorinata.

The iodine is dissolved in about 3 ounces of water, together with 1.5 grammes of potassium iodide; the test-solution is added until the brown color just disappears; or the solution is blued with a trace of mucilage of starch, and the test-solution added until the blue color disappears.

The chlorine-water is mixed with an equal bulk of water in which about 1½ grammes of potassium iodide have been dissolved. Mucilage of starch is also employed to indicate the termination of the test.

The chlorinated lime is mixed with about 3 ounces of water, in which 3½ grammes of potassium iodide have been dissolved; the mixture is subsequently acidulated with dilute hydrochloric acid, and a little mucilage of starch is added before delivering the test-solution into it, which is done until the blue color just ceases.

PART SECOND.

THE MEDICINAL CHEMICALS

AND THEIR PREPARATIONS,

THEIR PHYSICAL AND CHEMICAL CHARACTERISTICS,

WITH DIRECTIONS FOR THE

EXAMINATION OF THEIR QUALITY AND PURITY.

THE

MEDICINAL CHEMICALS

AND THEIR PREPARATIONS.

ACETUM.

Vinegar.

VINEGAR contains, on the average, from 4½ to 6 per cent. of mono-hydrated acetic acid; when of this strength, one fluid-ounce requires for saturation not less than 35 grains of crystallized potassium bicarbonate; and 50 cubic centimetres of the volumetric test-solution of sodium hydrate will saturate 70.29 grammes of vinegar, corresponding to 5 per cent. of anhydrous acetic acid.

Examination:

Mineral acids are indicated when about half an ounce of the vinegar is thoroughly mixed and boiled for five minutes with half a grain of starch; when cool, one drop of solution of iodinized potassium iodide is added; if any free mineral acid be present, it acts on the starch, transforming it into glucose, and consequently prevents the formation of blue iodine-starch.

Sulphuric and *hydrochloric acids*. A crude mode of detecting the presence of sulphuric acid in vinegar consists in evaporating, by a gentle heat, a small portion of it after the addition of a few grains of cane-sugar, in a porcelain capsule, to the consistence of a thick syrup; this will become almost black, if free sulphuric acid be present.

Since the water and the materials used for the preparation of vinegar generally contain traces of sulphates and chlorides, most vinegar yields a slight turbidity with barium and argentic nitrates. In order to ascertain the presence of either of these acids in a free state, about half an ounce of the vinegar is distilled from a small flask, the recipient being cooled in ice-water

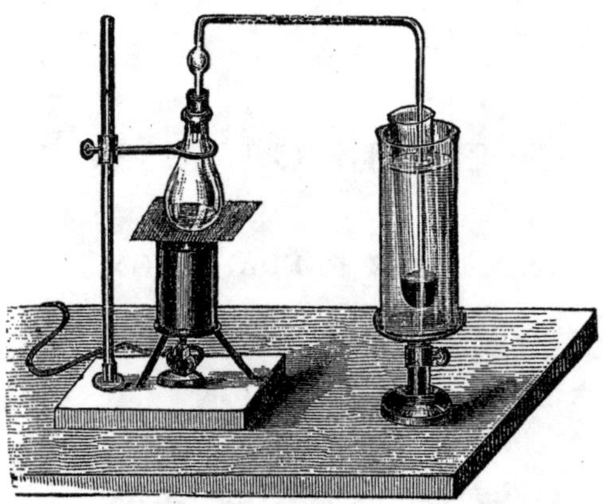

FIG. 28.

(Fig. 28); when about one drachm of distillate has been obtained, this is preserved for examination for aldehyde; another receiver is applied, and the distillation continued until almost the entire fluid has distilled over and the scanty residue commences to become dry and dark. The distillate is then acidulated with a few drops of nitric acid, and tested with barium nitrate for sulphuric acid, and with argentic nitrate for hydrochloric acid.

The presence of sulphuric acid may also be detected, or verified, by reducing about one ounce of the vinegar by evaporation, in a porcelain capsule, upon a water-bath, to the consistence of a thick syrup, which, when cool, is thoroughly mixed with about half a fluidounce of strong alcohol, and allowed to stand in a corked test-tube, with occasional agitation, for about an hour; the filtrate is then mixed with about half an ounce of distilled water, is acidulated with a few drops of diluted nitric acid, and tested with a few drops of solution of barium nitrate; an ensuing white precipitate, not disappearing on dilution with water, proves the presence of sulphuric acid.

Sulphurous acids and *sulphites* may be detected by a greenish coloration of the vinegar, when about two drachms of it are warmed with one drop of solution of potassium bichromate.

Nitric acid may be detected by adding one drop of neutral indigo solution to about half an ounce of the vinegar in a test-tube, and warming the mixture by dipping the tube into hot water; a discoloration of the bluish or bluish-greenish tint of the liquid will indicate free nitric acid; when the color remains, one or two drops of concentrated sulphuric acid are added while warm; if discoloration takes place now, nitrates are indicated.

Metallic impurities are detected by saturating the vinegar with hydrosulphuric-acid gas, and allowing the fluid to stand for a few hours; an ensuing white turbidity would indicate *tin* or sulphurous acid; a dark one lead and copper, which may be discriminated by collecting and washing the precipitate, obtained by a repetition of the test on a larger scale, upon a filter, and dissolving it in nitric acid; the solution is examined in two portions, the one by over-saturation with ammonium hydrate, the other by addition of one or two drops of diluted sulphuric acid; a blue coloration in the first instance indicates *copper*, a white precipitate, in the second one, *lead*.

Aldehyde may be detected in the first portion of the distillate obtained and preserved in the second test; it must be void of alcoholic odor or taste, and must not become yellow or brown, when mixed and warmed with an equal bulk of strong solution of potassium hydrate; such coloration would indicate aldehyde.

Acrid vegetable substances are detected by saturating a portion of the vinegar with magnesium carbonate, and by subsequent evaporation of the filtrate, at a gentle heat, to about one-third of its volume. Acrid vegetable substances may then be recognized by their odor and taste.

ACIDUM ACETICUM.

Acetic Acid.

Acetic-acid hydrate, or monohydrated acetic acid, contains one molecule (15 per cent.) of water, and forms, below 15° C., large, colorless, transparent crystals (glacial acetic acid), which, above 16 to 18° C., fuse to a thin, colorless liquid of a pungent odor and strong acid reaction, having a spec. grav. of 1.063 ; it boils at 120° C., yielding very pungent and acid inflammable vapors. Acetic acid hydrate is miscible in all proportions with water, alcohol, and ether, and dissolves albumen, fibrin, camphor, and many resins, gum-resins, and essential oils ; diluted with water, it forms the commercial and medicinal acetic acids, which maintain the character of acetic acid as long as the admixture of water does not exceed 18 to 19 per cent., beyond which dilution the acid loses more or less the character of a strong acid, and its solvent properties for the abovementioned substances.

Acetic acid is recognized, either free or combined, by yielding a deep-red color with solutions of ferric salts, and, when not too much diluted, by the odor of acetic ether, when heated with a mixture of equal parts of alcohol and concentrated sulphuric acid.

Two strengths of acetic acid are officinal: Acidum Aceticum of the spec. grav. of 1.047 (1.044 British Pharmac., 1.064 Pharm. German.), and Acidum Aceticum dilutum of the spec. grav. of 1.006 (1.040 Pharmac. German.). The strong acid of 1.047 spec. grav. contains 28 per cent. of anhydrous, or 36 per cent. of monohydrated, acetic acid, and 100 parts of it require 60 parts of crystallized potassium carbonate for saturation ; the diluted acid of 1.006 spec. grav. contains 3.63 per cent. of anhydrous acid, and 100 parts of it saturate 7.6 parts of crystallized potassium bicarbonate.

The strength of aqueous acetic acid cannot be correctly inferred from its specific gravity, since its density is increased to a certain extent by contraction, the spec. grav. of 1.0735 being the maximum density, below which it again decreases; for instance: a mixture of one molecule of monohydrated acetic acid with three molecules of water has the same specific gravity as a mixture of one molecule of the acid with nine molecules of water, namely, 1.0686. The strength of acetic acid may be determined by observing the exact quantity of crystallized potassium bicarbonate required to saturate a known weight of the acid, and by subsequent equation ; or by treating a known weight of the acid with an excess of barium carbonate, and by calculating its strength from the amount of carbonate decomposed, which is ascertained by deducting the weight of the remaining undissolved carbonate from the total amount used, and by dividing the remainder by 1.933; the quotient indicates at once the quantity of monohydrated acetic acid contained in that quantity of the acid.

The readiest mode of ascertaining the strength of acetic acid, is the volumetric estimation with test-solution of sodium hydrate (page 60). Sixty parts by weight (the molecular weight) of anhydrous acetic acid require for exact saturation 40 parts (the molecular weight) of sodium hydrate; since 1,000 cubic centimetres of the test-solution contain 40 grammes of sodium hydrate, these saturate exactly 60 grammes of anhydrous acetic acid. Upon this equation the calculation for the strength of acetic acid may be based: for instance, an acid, of which 182.0 grammes saturate 1,000 cubic centimetres of test-solution, contains 33 per cent. of anhydrous acid; an acid, however, of which the same quantity is saturated by 930 cubic centimetres of the test-solution, contains, therefore, only 30.7 per cent. of anhydrous acid.

In order to secure accurate results in the volumetric estimation of acetic acid, alkanet-tincture is preferably used as the indicator of saturation, since the alkaline acetates, though neutral in combination, have an alkaline reaction on litmus.

Examination:

Empyreuma betrays itself by its odor; small traces, however, may be recognized by neutralizing a little of the acid with Liquor Potassæ, and by tingeing it subsequently faintly with potassium permanganate, and by gently heating it.

Sulphuric and *hydrochloric acid* may be detected by testing separate portions of the diluted acid with barium nitrate for the former, and with argentic nitrate for the latter.

Sulphurous acid is indicated by a greenish coloration, when a little of the acid is heated with a few drops of solution of potassium bichromate; it may also be recognized when a mixture of about one drachm of the acid with two drachms of diluted hydrochloric acid and two drachms of water, is poured upon a little pure granular zinc in a large test-tube, and the mouth of the tube is at once loosely closed with a bunch of cotton moistened with solution of plumbic acetate (Fig. 29); if sulphurous acid be present, the plumbic solution will become dark, or black.

Nitric acid may be detected by discoloration when a little of the acid is tinged slightly blue with one drop, or part of a drop, of neutral indigo-solution and heated.

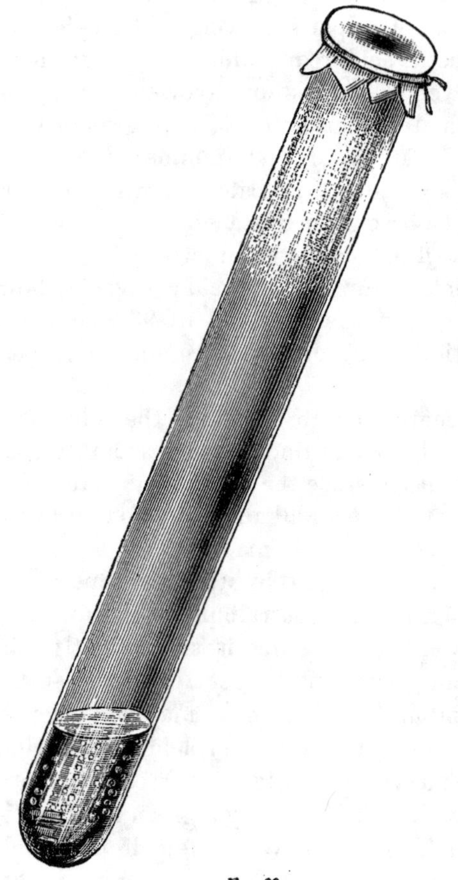

Fig. 29.

Metallic impurities are detected by mixing the acid with about two or three times its bulk of hydrosulphuric acid, or, when diluted acid is under examination, by saturating it with hydrosulphuric-acid gas, and allowing the liquid to stand in a corked test-tube in a warm place for several hours; an ensuing white precipitate would indicate *tin* or *zinc* (if no sulphurous acid be present, which also causes a white turbidity with the reagent), a black one *lead* or *copper;* in the latter case, the precipitate may be collected and washed upon a filter, and then dissolved in a few drops of warm, strong nitric acid, and then diluted with a few drops of water; this solution is tested in two separate portions, the one by over-saturation with ammonium hydrate, the other with one drop of dilute solution of sodium sulphate; a blue coloration in the first test would indicate *copper*, and a white precipitate, in the second one, *lead*.

ACIDUM ARSENIOSUM.

ACIDUM ARSENICOSUM. ARSENICUM ALBUM.

Arsenious Acid. White Arsenic.

Arsenious acid occurs in two polymeric crystalline forms, and in the amorphous state; the latter one being the commercial white arsenic. When freshly prepared, it forms heavy, transparent, glassy cakes, with smooth conchoidal fracture, and has a spec. grav. of 3.738; this becomes gradually opaque and porcelain-like by passing into the crystalline state, which change proceeds from the surface toward the inside; at the same time its density is slightly diminished (3.689), and its solubility in water increased. In consequence of the simultaneous occurrence of the amorphous and the crystalline modifications, and the difference in their solubility, the statements of the solubility of arsenious acid in water are slightly at variance. The amorphous glassy, and the commercial powdered, acid is soluble in 10 to 12 parts of boiling, and 30 parts of cold, water; it is almost insoluble in alcohol and insoluble in ether, but freely soluble in the alkaline hydrates and in warm diluted acids, especially in hydrochloric and tartaric acids, from which latter solutions it deposits, on cooling, in small transparent octahedral crystals.

Arsenious acid volatilizes at about 218° C. without fusion, forming a colorless, inodorous vapor which solidifies, on cooling, in small transparent and brilliant octahedrons; but when heated, in contact with reducing-agents, such as potassium cyanide, organic substances, and carbon, the acid is reduced to metallic arsenic, which sublimes and deposits in a brilliant metallic crust when the reduction is carried on in a glass tube (Fig. 30), emitting, at the same time, a peculiar and characteristic odor, somewhat similar to garlic.

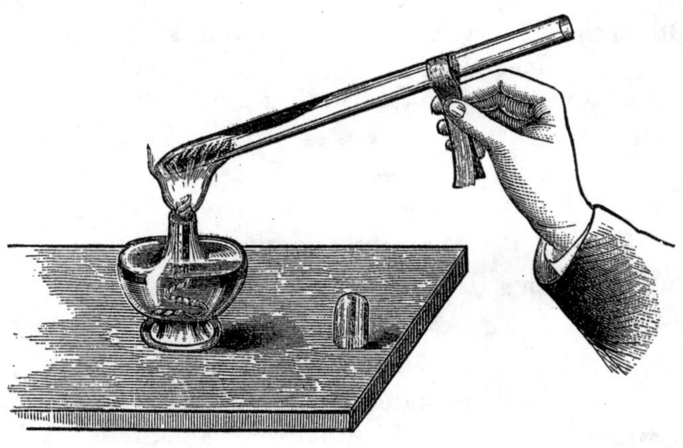

Fig. 30.

The aqueous solution of arsenious acid has a feeble acid reaction on litmus; it yields a white precipitate with lime-water, a yellow one with hydrosulphuric acid, soluble in aqua ammoniæ; argentic nitrate and cupric sulphate produce only a turbidity in solutions of arsenious acid; upon the addition of an alkali, however, a yellow precipitate is formed with the former reagent, and a brilliant green one with the latter, both precipitates being soluble in excess of ammonia, and the argentic one also in nitric acid.

Solution of arsenious acid, when mixed with either hydrochloric or sulphuric acid, and the mixture allowed to act on metallic zinc or magnesium, gives rise to the formation of hydrogen arsenide simultaneously with the formation of hydrogen sulphide, which has an alliaceous odor, and burns with a bluish-white flame, depositing a brilliant black spot on cold surfaces held in the jet (*Marsh's* test). When solution of arsenious acid is mixed with an excess of concentrated hydrochloric acid, and a piece of bright copper-foil immersed and boiled, it deposits the reduced arsenic upon the copper, coating it with a dark-gray metallic film (*Reinsch's* test); when the same test is made, taking tin-foil (real), or stannous chloride, instead of copper, the tin becomes equally coated with arsenic, and, at the same time, a voluminous brown deposit is formed (*Bettendorf's* test).

Examination of Powdered White Arsenic:

Mineral admixtures are recognized and left behind when a little of the white arsenic is either volatilized at a red heat, or dissolved in an excess of boiling liquor potassæ or hydrochloric acid. When an insoluble residue is left from either of these solutions, it is washed, and, when dry, is mixed and fused in a porcelain crucible with four times its weight of a mixture of equal parts of exsiccated carbonates of sodium and potassium; the obtained fused mass is triturated and boiled with a sufficient quantity of water, and the filtered solution tested with barium nitrate for sulphates (calcium and barium sulphates). The residue on the filter is washed and treated with warm diluted hydrochloric acid, and the filtrate subsequently tested with sodium sulphate for *barium*, and, in another portion, nearly neutralized with aqua ammoniæ, with ammonium oxalate for *calcium*.

One hundred grains of arsenious acid, when dissolved in boiling diluted hydrochloric acid, yield, upon complete precipitation with hydrosulphuric acid, a precipitate of arsenious sulphide, which, when collected upon a tared filter, washed, and dried, should weigh 124 grains.

ACIDUM BENZOICUM.

ACIDUM BENZOICUM SUBLIMATUM. FLORES BENZOES.

Benzoic Acid.

Colorless, soft, feathery needles, of a silky lustre, inodorous when cold and pure, but of a faint smell when gently warmed. The agreeable aromatic odor of the officinal benzoic acid is due to traces of essential oil. When derived from solutions, benzoic acid forms colorless, pearly needles or laminæ of six-sided prisms. When warmed, the acid begins to volatilize below 100° C.; melts at 121° C., forming a colorless liquid, which, on cooling, solidifies; at 145° C. it evaporates freely, and at near 240° C. it boils, without decomposition, emitting an acrid, irritating vapor, which readily inflames, burning away without residue.

Benzoic acid is soluble in about 200 parts of cold, and 25 parts of boiling, water, in 10 parts of glycerin, and freely in alcohol, ether, oil of turpentine, and many essential and fatty oils, and also in solutions of the alkaline hydrates. Concentrated nitric and sulphuric acids dissolve benzoic acid more or less without decomposition; on addition of water it is precipitated unaltered. A concentrated solution of benzoic acid in aqua ammoniæ, in which the ammonium hydrate is nearly neutralized, gives, on addition of a few drops of solution of a ferric salt, a copious reddish-yellow precipitate of ferric benzoate, which, on addition of hydrochloric acid, disappears, while the benzoic acid is precipitated, redissolving on addition of ether or alcohol.

Examination:

Hippuric acid, as is well known, is resolved by the action of hydrochloric acid, and other agents, into benzoic acid and glycocine, and much benzoic acid is obtained from this source. Such acid, however, differs in some of its chemical properties from the benzoic acid derived from the benzoic resin, especially in the degree of solubility of its salts in water and alcohol, and in its deportment with potassium permanganate; this difference extends very likely also to their physiological and therapeutical action, and the acid derived from hippuric acid cannot yet be considered as admissible for medicinal use.

Hippuric acid may be distinguished from, or recognized in, benzoic acid, by heating in a test-tube a mixture of about 20 grains of the acid with 50 grains of dry potassium hydrate, and about 50 drops of water; if hippuric acid is present, it will be indicated by a faint odor of ammonia, and by white vapors with a glass rod moistened with acetic acid and held over the orifice of the test-tube, as well as by blackening a strip of unsized paper moistened with solution of mercurous nitrate, and held in the mouth of the test-tube.

Hippuric acid, and the benzoic acid derived therefrom, may be distinguished from, or recognized in, true benzoic acid by agitating about 5 grains of the acid with 4 fluid-drachms of water, in a test-tube, and by adding so much solution of potassium permanganate as to tinge the mixture deep violet; it is then heated to from 50° to 60° C., by dipping the tube into water of that temperature; the violet color will remain for at least five minutes with true benzoic acid, while it will become distinctly red within that time with hippuric acid, or benzoic acid derived therefrom.

An additional test may be made by tingeing a solution of 5 grains of the acid, in 3 fluid-drachms of liquor potassæ, deep violet with solution of potassium permanganate, and by boiling this solution for a few minutes; the solution of true benzoic acid becomes green; that of hippuric-benzoic acid, yellow, or even colorless, with the separation of brown manganic oxide.

Cinnamomic acid may be detected by the odor of oil of bitter almonds, when a mixture of the acid with potassium bichromate and strong sulphuric acid, or of a concentrated aqueous solution of the acid, is heated with potassium permanganate.

Sugar is indicated by the odor of caramel, and by intumescence, when a few grains of the acid are gradually heated in an iron spoon.

Mineral substances, not readily volatilizable or soluble in alcohol, may at once be detected by a residue left on volatilization, as well as upon solution of the acid in alcohol. If any fixed residue is left, it may be dissolved in warm water acidulated with nitric acid, and tested with barium nitrate for sulphates, and with argentic nitrate for chlorides and phosphates.

Boric acid may be recognized by the green coloration of the flame of burning alcohol, previously saturated with the acid.

ACIDUM BORACICUM.

ACIDUM BORICUM.

Boracic Acid. Boric Acid.

Colorless, translucent, pearly, six-sided scales, somewhat unctuous to the touch. They contain 3 molecules (42.16 per cent.) of water, two of which are water of crystallization; when heated upon platinum-foil, the latter is readily evaporated, while the other molecule does not part unless heated to redness, when the anhydrous acid melts, forming a glassy, transparent fuse, which is a solvent for most metallic oxides, and which solidifies to a fissured glass on cooling.

Boracic acid is soluble in 26 parts of cold, and in 3 parts of boiling, water, which solution has but little taste and feebly affects blue litmus-paper, while it renders turmeric-paper reddish brown, quite different from the color produced by alkalies, and becoming more distinct after drying. The acid is also soluble in alcohol, and the solution burns with a green-edged flame.

Examination:

About 10 grains of the crystals of boric acid are added to 1 drachm of water in a test-tube, and heated; a clear and complete solution must take place, and, when part of the hot solution is dropped into alcohol, no turbidity or precipitate must ensue; otherwise the presence of admixtures insoluble or less soluble in water or alcohol is indicated.

Chlorides and *sulphates* are detected in the aqueous solution acidulated with nitric acid, by argentic nitrate and barium nitrate respectively.

ACIDUM CARBOLICUM.

ACIDUM PHENYLICUM.

Carbolic Acid, or Phenylic Acid. Phenol. Phenic Alcohol.

Pure carbolic acid forms, at temperatures below 41° C., rhomboidal needles or crystalline masses consisting of such; when not quite pure, they become frequently more or less reddish; at temperatures above 41° C., pure carbolic acid forms an oily, colorless liquid, which boils at about 182° C.; its spec. grav. is 1.060 to 1.065; it absorbs moisture on exposure to the air, and gradually deliquesces; it dissolves about one-tenth of its weight of water, and then remains liquid far below its melting-point.

Pure carbolic acid dissolves to some extent in water,* and is freely miscible with alcohol, ether, chloroform, glycerin, acetic acid, and the volatile and fatty oils; its aqueous solution does not react upon test-paper, coagulates albumen and collodion, and assumes a violet-blue color upon the addition of a few drops of a dilute solution of neutral ferric chloride.

Carbolic acid is miscible with concentrated sulphuric acid, with a slight evolution of heat, without apparent change; it yields a more or less violent reaction with concentrated nitric acid, and with chlorine and bromine; it is also soluble in the alkaline hydrates, and combines with them, and with other salifiable bases, forming neutral salts, which, however, maintain the alkaline reaction, and are decomposed by all acids.

The pure commercial carbolic acid contains mostly a slight quantity of cresol, and perhaps of other homologous aromatic alcohols, which do not modify the principal properties of pure phenol, except that they lower its melting-point, and apparently decrease its solubility in water, which may also be influenced with pure phenol by a slight quantity of water.

Examination:

The quality of carbolic acid is sufficiently recognized by the above characteristics, by its odor, and by its appearance; in order to detect admixtures, the fused acid may be tested by mixing it with twice its volume of liquor potassæ, and warming the mixture by immersing the test-tube in boiling water; a complete solution should take place, which, when cold, should remain limpid, and not separate any oily liquid upon dilution with three times its bulk of water.

* The United States Pharmacopœia states that carbolic acid is soluble in from 20 to 33 parts of water, the purest being the most soluble. The Pharmacopœa Germanica states the proportion of solubility as 1 part of the acid to from 50 to 60 parts of water.

ACIDUM CHROMICUM.

Chromic Acid.

Brilliant crimson-red acicular crystals or crystalline masses, of an acid, metallic taste; very deliquescent, and therefore frequently of a humid appearance. At a heat between 180° and 190° C., they melt into a reddish-brown liquid, which, on cooling, forms a red, opaque, and brittle mass.

Chromic acid is freely soluble in water and alcohol, forming orange-yellow solutions. It is a powerful oxidizing agent, and decomposes organic matters.

When its aqueous solution is heated, after the addition of a few drops of hydrochloric acid and a few drops of alcohol, the orange-red color of the liquid becomes yellowish green, with the evolution of ethereal vapors. When dissolved in an aqueous solution of sulphurous acid, it forms immediately a green solution.

Examination:

Sulphuric acid may be detected by boiling a diluted solution of the acid, to which a few drops of hydrochloric acid and a little alcohol have been added, until the liquid appears green. Part of it is then tested with barium nitrate; an ensuing white precipitate would indicate sulphuric acid.

Potassium bichromate may be detected in the rest of the green solution, by evaporating it, in a porcelain capsule, nearly to dryness, and by subsequent solution of the residue in a few drops of water; the filtered solution is tested with sodium bitartrate; an ensuing white crystalline precipitate would indicate potassium.

ACIDUM CITRICUM.

Citric Acid.

Colorless rhomboidal prisms with dihedral summits, permanent in the air, but slightly efflorescent in a dry and warm atmosphere, and becoming moist in a damp one. Exposed to a strong heat, the acid fuses, and is wholly dissipated, with the evolution of empyreumatic vapors.

Citric acid dissolves in three-fourths of its weight of cold, and in half its weight of boiling, water, and is also soluble in alcohol, forming sour solutions of an acid reaction; the aqueous solution, when exposed to the air, is subject to gradual and spontaneous change; it forms no precipitate with potassium salts (except the tartrates), and, when sparingly added to lime-water, so that the alkaline reaction still remains prevalent, it does not render it turbid (distinction from oxalic, tartaric, and racemic acids); when, however, the liquid is warmed and agitated, it becomes turbid, but transparent again upon cooling.

Sixty grains of citric acid neutralize 70.41 grains of potassium carbonate, 85.71 grains of potassium bicarbonate, 120.24 grains of crystallized sodium carbonate, 72 grains of sodium bicarbonate, and 40.92 grains of magnesium carbonate.

Volumetric Estimation, see page 61.

Examination:

In order to obtain an average sample of the crystallized acid for examination, it is advisable to grind several ounces of the crystals, and to make from a small portion of the powder two solutions: an aqueous one, in the proportion of 1 part of the acid to 4 parts of water; and an alcoholic one, in the proportion of 1 part of acid to 6 parts of alcohol. Both the solutions should be complete and clear.

Tartaric acid is detected as a granular white precipitate, when 2 fluid-parts of the aqueous solution and 1 fluid-part of the alcoholic solution are mixed together and agitated with 1 fluid-part of a strong solution of potassium acetate.

When many samples of the crystallized acid have to be examined, the following method is also applicable:

A large glass pane is placed upon ultramarine-blue or dark-brown paper on an horizontal table or board; a solution of potassium hydrate in diluted alcohol (15 grains of dry potassium hydrate in 6 drachms of distilled water and 2 drachms of strong alcohol) is then spread over the pane as thick as will remain stationary upon it; a number of crystals and fragments thereof are now placed a little apart from each other, the crystals of each sample separate. Instead of a glass pane, small plates may be employed. Agitation being carefully avoided, the citric-acid crystals, after several minutes' action of the alkaline solution, appear clearer and more transparent; if crystals of tartaric acid be present, they will be recognized by their cloudy and white appearance; after two or three hours, the crystals of citric acid are nearly or quite dissolved, and in their stead is frequently left a small, delicate, dust-like spot (due to traces of calcium salts); if crystals of tartaric acid be present, they will appear whitish, covered with a coat of small transparent acicular crystals, and surrounded by a deposit of small overlapping groups of similar crystals, or a thin, though broad, crystalline film (all crystals of potassium bitartrate).

Metallic impurities may be detected in the aqueous solution, by a dark coloration or turbidity with hydrosulphuric acid; if it be so considerable as to form a deposit, this is collected and washed upon a filter, and then dissolved in a few drops of warm nitric acid; the obtained solution is divided into two portions, one of which is examined by over-saturation with aqua ammoniæ for *copper*, which will be indicated by a blue coloration, and the other with one drop of diluted sulphuric acid for *lead*.

Sulphates may be detected in the diluted aqueous solution, to which a few drops of diluted nitric acid have been added, by a white precipitate with barium nitrate.

ACIDUM GALLICUM.

Gallic Acid.

Small acicular prisms or silky needles, or a crystalline powder, nearly colorless, or of a pale fawn-color. They contain 2 molecules (9.5 per cent.), of water of crystallization, which pass off at about 100° C.; at about 215° C., the acid is resolved, without fusing, into carbonic acid and pyrogallic acid, which latter sublimes in small crystalline plates; when exposed to a strong heat, with free access of air, gallic acid burns away without residue.

Gallic acid is soluble in 100 parts of cold, and in 3 parts of boiling, water; it is also soluble in cold, and freely in boiling, alcohol, but less soluble in ether and glycerin. The aqueous solution has an acidulous and astringent taste and an acid reaction, and is liable to spontaneous decomposition on exposure to the air; it gives no precipitate with solutions of ferrous salts, if free from ferric salt, but it gives a bluish-black precipitate with solutions of ferric salts, the color of which disappears when the liquid is heated, from the deoxidation of the ferric to ferrous salt, at the expense of the gallic acid. A solution of gallic acid, when dropped into lime-water, produces a white turbidity, which soon becomes blue, and passes through a greenish or violet tint to a purple color. The solution forms no precipitate with argentic nitrate, but reduces it to metallic silver, gradually at common temperature, at once when heated.

Solutions of the alkaline hydrates, as well as concentrated sulphuric and nitric acid, when poured upon dry gallic acid, dissolve it, with a deep-red color.

Examination:

Tannic acid may be detected by a white precipitate, when the solution of the acid is added to a dilute solution of gelatin.

Sugar and *dextrin* remain behind upon solution of the acid in strong alcohol.

Resinous admixtures will remain behind and separate on cooling, floating on the surface when a few grains of the acid are dissolved in boiling water.

ACIDUM HYDROCHLORICUM.

ACIDUM MURIATICUM.

Muriatic Acid. Hydrochloric Acid.

Concentrated hydrochloric acid is a colorless fuming liquid, of a pungent and suffocating odor and corrosive taste; its spec. gravity depends upon the quantity of hydrogen chloride held in solution, and varies in the strong acid between 1.160 and 1.120, corresponding to 32 and 24½ per cent. of the anhydrous hydrochloric acid. The crude commercial acid has generally a spec. grav. of from 1.160 to 1.180, containing 32 to 35 per cent. of the gas and various impurities, which impart a yellowish color to the acid.

Hydrochloric acid may further be recognized by the evolution of chlorine gas when heated in a test-tube with black manganic dioxide, and, when diluted, by a white curdy precipitate with argentic salts, which is insoluble in acids, but soluble in ammonium hydrate and in potassium cyanide, and which becomes black on exposure to the solar light.

Volumetric Estimation, *see* page 61.

Examination:

Fixed impurities are recognized by a residue, upon evaporation of the acid in a platinum spoon or watch-glass.

Sulphuric acid may be detected in the acid, after dilution with at least five times its volume of water, by a white precipitate when tested with a few drops of barium chloride.

Sulphurous acid may be detected in the filtrate of the preceding test, after the sulphuric acid, if such be present, has been completely eliminated, by mixing with it a little chlorine-water; an ensuing white turbidity would indicate sulphurous acid.* This may also be recognized or confirmed when a little of the hydrochloric acid, diluted with 4 or 5 parts of water, is poured upon pure granular zinc in a small flask (Fig. 31), and the evolved gas allowed to pass into a diluted solution of plumbic acetate in a test-tube; an ensuing black precipitate in the solution would indicate sulphurous acid; or the test may be made in a large test-tube, and the solution of plumbic acetate exposed to the escaping gas by closing the mouth of the tube with a bunch of cotton, moistened with the solution (Fig. 29; page 72).

FIG. 31.

Chlorine may be detected by the occurrence of a blue coloration, when the acid, diluted with about five times its bulk of water, is mixed with a few drops of solution of potassium iodide (free of iodate) and a little mucilage of starch.

Iodine and Bromine.—About eight volumes of the acid are

* Chlorine and sulphurous acids, when in contact with water, form hydrochloric and sulphuric acids; therefore the presence of either one of these impurities in hydrochloric acid excludes the other one.

agitated in a test-tube with one volume of chloroform; after subsidence, the stratum of chloroform will appear red when iodine, and yellowish when bromine, is present. If the chloroform, however, remains colorless, a few drops of chlorine-water are added; when, after agitation and subsequent subsiding, the chloroform still remains colorless, the absence of hydrobromic and of hydroiodic acids is also proved.

Metals are detected in the acid, diluted with at least four times its bulk of water, when tested with hydrosulphuric acid: a white turbidity would indicate sulphurous acid or ferric chloride, a yellow one, arsenious chloride, and a dark one, metals (copper, lead, or tin); in order to distinguish the latter, the precipitate is collected upon a filter, washed, and then treated with a little warm ammonium sulphydrate; the filtrate is oversaturated with a few drops of hydrochloric acid; a brown precipitate indicates *tin.* The precipitate upon the filter, insoluble in ammonium sulphydrate, is washed, and dissolved in a few drops of warm nitric acid; part of this solution is oversaturated with ammonium hydrate; a blue coloration indicates *copper;* another portion is diluted and tested with one or two drops of diluted sulphuric acid; a white precipitate indicates *lead.*

Arsenic.—The presence of arsenious or arsenic chlorides may be shown by adding about 25 drops of concentrated solution of stannous chloride to about three fluid-drachms of the concentrated acid, and then adding, drop by drop, about one fluid-drachm of concentrated sulphuric acid; the mixture becomes hot, but remains clear and colorless; if, however, traces of arsenic are contained in the acid, it becomes more or less turbid, and deposits a brown precipitate.

Another test is to introduce a mixture of one fluid-drachm of the acid, with four drachms of water wherein about three grains of cupric sulphate have been dissolved, into a long test-tube (Fig. 32), taking care that the liquid fills only about one-tenth of the tube, and that the upper interior surfaces of the tube remain dry; a few pieces of pure zinc are then added, and a small bunch of gun-cotton moistened with solution of argentic nitrate, or a cork provided with a strip of strong white blotting-paper moistened with the argentic solution, is inserted into the orifice of the tube, which is then allowed to stand, protected from the solar light, for half an hour.

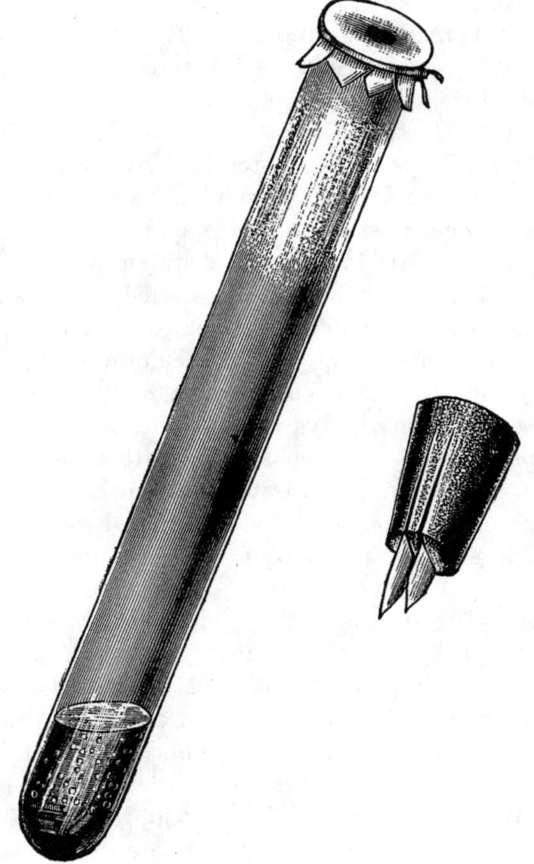

Fig. 32.

A dark coloration of the solution of argentic nitrate, not disappearing when immersed in a warm dilute solution of potassium cyanide, confirms the presence of arsenic.

TABLE

OF THE QUANTITY BY WEIGHT OF HYDROCHLORIC-ACID GAS CONTAINED IN 100 PARTS BY WEIGHT OF AQUEOUS HYDROCHLORIC ACID AT DIFFERENT DENSITIES.

TEMPERATURE 16° C.

Specific Gravity.	Percent. of Hydrochl. Acid.	Specific Gravity.	Percent. of Hydrochl. Acid.	Specific Gravity.	Percent. of Hydrochl. Acid.	Specific Gravity.	Percent. of Hydrochl. Acid.
1.2013	41	1.1551	31.25	1.1056	21.5	1.0573	11.75
1.2002	40.75	1.1539	31	1.1044	21.25	1.0561	11.5
1.1991	40.5	1.1526	30.75	1.1031	21	1.0549	11.25
1.1980	40.25	1.1513	30.5	1.1019	20.75	1.0537	11
1.1969	40	1.1501	30.25	1.1007	20.5	1.0524	10.75
1.1958	39.75	1.1488	30	1.0994	20.25	1.0512	10.5
1.1947	39.5	1.1475	29.75	1.0982	20	1.0500	10.25
1.1936	39.25	1.1462	29.5	1.0969	19.75	1.0488	10
1.1925	39	1.1450	29.25	1.0957	19.5	1.0475	9.75
1.1913	38.75	1.1437	29	1.0945	19.25	1.0463	9.5
1.1902	38.5	1.1424	28.75	1.0932	19	1.0451	9.25
1.1890	38.25	1.1412	28.5	1.0920	18.75	1.0439	9
1.1878	38	1.1399	28.25	1.0907	18.5	1.0427	8.75
1.1867	37.75	1.1386	28	1.0895	18.25	1.0414	8.5
1.1855	37.5	1.1373	27.75	1.0883	18	1.0402	8.25
1.1844	37.25	1.1361	27.5	1.0870	17.75	1.0390	8
1.1833	37	1.1348	27.25	1.0858	17.5	1.0378	7.75
1.1821	36.75	1.1335	27	1.0845	17.25	1.0366	7.5
1.1810	36.5	1.1323	26.75	1.0833	17	1.0353	7.25
1.1798	36.25	1.1310	26.5	1.0821	16.75	1.0341	7
1.1787	36	1.1297	26.25	1.0807	16.5	1.0329	6.75
1.1775	35.75	1.1284	26	1.0795	16.25	1.0317	6.5
1.1763	35.5	1.1272	25.75	1.0783	16	1.0305	6.25
1.1752	35.25	1.1259	25.5	1.0770	15.75	1.0292	6
1.1739	35	1.1246	25.25	1.0758	15.5	1.0280	5.75
1.1727	34.75	1.1234	25	1.0746	15.25	1.0268	5.5
1.1714	34.5	1.1221	24.75	1.0733	15	1.0256	5.25
1.1702	34.25	1.1208	24.5	1.0721	14.75	1.0244	5
1.1689	34	1.1196	24.25	1.0709	14.5	1.0231	4.75
1.1677	33.75	1.1183	24	1.0696	14.25	1.0219	4.5
1.1664	33.5	1.1170	23.75	1.0684	14	1.0207	4.25
1.1652	33.25	1.1157	23.5	1.0672	13.75	1.0195	4
1.1639	33	1.1145	23.25	1.0659	13.5	1.0170	3.5
1.1627	32.75	1.1132	23	1.0647	13.25	1.0146	3
1.1614	32.5	1.1119	22.75	1.0635	13	1.0122	2.5
1.1602	32.25	1.1107	22.5	1.0622	12.75	1.0097	2
1.1589	32	1.1094	22.25	1.0610	12.5	1.0073	1.5
1.1577	31.75	1.1081	22	1.0598	12.25	1.0048	1
1.1564	31.5	1.1069	21.75	1.0585	12	1.0024	0.5

The density of the aqueous acid being decreased by an increase of temperature, and increased by a decrease of temperature, the consequent change of the specific gravity amounts for each degree of the centigrade thermometer in either direction—

For acids of a specific gravity of 1.1739 to those of 1.1386 to about 0.0005
" " " 1.1335 " 1.0982 " 0.0004
" " " 1.0932 " 1.0635 " 0.0003

For instance: An acid of a specific gravity of 1.1234 at 16° C. containing 25 per cent. of hydrochloric-acid gas will have at 18.5° C. a specific gravity of (1.1234−0.004 × 2.5=) 1.1224, and at 13.5° C. a specific gravity of (1.1234+0.004 × 2.5=) 1.1244.

ACIDUM HYDROCYANICUM DILUTUM.

ACIDUM HYDROCYANATUM.

Diluted Hydrocyanic Acid.

Pure hydrocyanic acid is a thin, colorless, and exceedingly volatile and unstable liquid. Its odor is very powerful and characteristic, resembling that of peach-blossoms or oil of bitter almonds. It mixes with water and alcohol in all proportions. The officinal acid is a highly-dilute aqueous solution, containing only about two per cent. of anhydrous acid. It imparts a faint evanescent color to litmus, and forms a white precipitate with argentic nitrate, soluble in potassium cyanate, as well as in nitric acid, but less soluble in ammonium hydrate. Upon addition of ferric chloride to the dilute acid, and subsequent slight over-saturation with potassium hydrate, the liquid will acquire a blue color, and deposit a slight dark-blue precipitate after the addition of hydrochloric acid in excess. If a solution of mercurous nitrate is added, drop by drop, to the officinal hydrocyanic acid, a gray precipitate of metallic mercury is formed at once.

Examination:

Hydrochloric and *phosphoric acids* may be detected by shaking some of the acid with a little powdered sodium biborate; then the whole, without filtering, is evaporated in a porcelain capsule, at a gentle heat, to dryness; the residue is dissolved in dilute nitric acid, and part of the filtered solution is examined with argentic nitrate for *hydrochloric acid*, which is indicated by a white precipitate. To another part of the solution ammonium molybdate is added, and heated to boiling; an ensuing yellowish precipitate indicates *phosphoric acid.*

Formic acid, if present, will be detected by its property of reducing red oxide of mercury to gray metallic mercury, when a little of the acid is warmed and agitated with a few grains of the oxide.

Sulphuric acid is detected by the formation of a white precipitate upon addition of a few drops of barium nitrate to the acid.

Estimation of the strength of hydrocyanic acid.

There are two simple modes of ascertaining the quantity of anhydrous acid contained in hydrocyanic acid. The one depends upon the fact that one part of anhydrous hydrocyanic acid forms 5 parts of argentic cyanide; that, accordingly, 100 parts of the officinal acid should yield 10 parts of argentic cyanide.

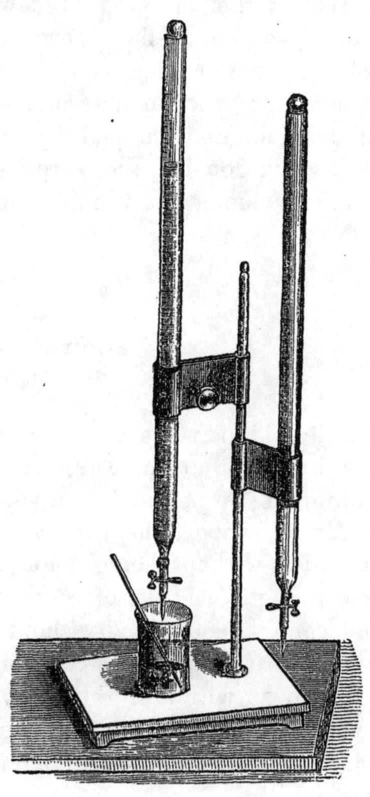
FIG. 33.

The second mode is the volumetric one, and depends upon the property of argentic cyanide to form a soluble compound with alkaline cyanides. When, therefore, the officinal hydrocyanic acid is converted into sodium or potassium cyanide by addition of sodium or potassium hydrate, no permanent precipitate will appear, upon the addition of nitrate of silver, until more than sufficient cyanide of silver is produced to form the soluble compound.

I. One hundred grains of the acid are completely precipitated by a solution of argentic nitrate. Then two filters of exactly the same size and paper are cut; through the one the liquid is filtered, the precipitate washed, and then both the empty filter and the one containing the argentic cyanide are dried, at a heat not exceeding 100° C. When the weight of the latter filter remains constant, both filters are weighed, the empty one serving as a counterpoise of the one containing the precipitate; the excess of weight of the latter is argentic cyanide, of which—

10	grains represent	2.015	per cent.	anhydrous	hydrocyanic	acid.
10.5	"	2.10	"	"	"	"
11	"	2.20	"	"	"	"

II. Twenty-seven grammes of the diluted hydrocyanic acid are converted into sodium cyanide, by adding solution of sodium hydrate until, after stirring, litmus-paper shows that the liquid has a slightly alkaline reaction. The test-solution of argentic nitrate is then allowed to flow in gradually (Fig. 33), until, after agitation, a slight permanent turbidity remains. When this occurs, the quantity of argentic nitrate added represents exactly half the amount of hydrocyanic acid present in the diluted acid, and the quantity of acid indicated denotes, therefore, when doubled, the correct percentage.

For volumetric estimation, *see* also page 62.

ACIDUM LACTICUM.

Lactic Acid.

A limpid, odorless, syrupy liquid, colorless, or of a pale-yellowish tint, of a sour taste, and of from 1.21 to 1.24 spec. grav. (containing 75 per cent. of hydrated lactic acid). It is miscible, in all proportions, with water, glycerin, alcohol, and ether, and also, without being colored, with concentrated sulphuric acid. Lactic acid dissolves zinc and iron, with effervescence, and cannot be distilled without undergoing partial decomposition. Heated upon platinum-foil, it emits inflammable vapors, which burn with a bright flame, and leaves a charred residue, which is completely dissipated at red heat. When heated with solution of potassium permanganate, lactic acid emits the odor of aldehyde.

One drachm of lactic acid requires for saturation not less than fifty grains of potassium bicarbonate.

Examination:

Gum, Mannite, and Glucose.—A few drops of the acid are diluted with water in a test-tube, and slightly over-saturated with sodium carbonate; to the clear liquid are added a few drops of Fehling's solution, and the whole is gently warmed; a blue coagulation, upon the addition of the cupric solution before warming, would indicate the presence of gum; a brick-colored precipitate, after heating, indicates glucose.

The presence of gum and mannite may also be recognized by the occurrence of a turbidity upon dropping the acid into a mixture of equal parts of alcohol and ether.

Glycerin may be detected by mixing, in a porcelain capsule, about 20 drops of the acid with 10 grains of oxide of zinc, previously triturated with a little water; the whole is then evaporated, upon a water-bath, to dryness; the residue is treated with strong alcohol, and the obtained alcoholic solution is evaporated upon a watch-glass; a neutral, syrupy, sweet residue would indicate glycerin.

Organic Acids.—Two drops of the lactic acid are added in a test-tube to so much lime-water that the alkaline reaction remains prevalent; if a turbidity takes place at once, *oxalic, tartaric*, or *phosphoric acid* is indicated; if the turbidity does not ensue before the liquid is heated to boiling, *citric acid* is indicated. *Acetic* and *butyric acids* are recognized by their respective odors when the acid is gently heated in a porcelain capsule.

Sulphuric, hydrochloric, and *phosphoric acids* may be detected in the diluted aqueous solution of the acid by testing it, in separate portions, with barium chloride for the former, and with argentic nitrate for the two latter.

Acid calcium phosphates would be indicated by a white turbidity of the dilute solution of the acid when tested with ammonium oxalate.

Metals are detected in the acid, when neutralized with aqua ammoniæ, and then tested with hydrosulphuric acid; a white turbidity would indicate *zinc*, a brown or black precipitate, *copper* or *lead.*

ACIDUM NITRICUM.

Nitric Acid.

Nitric acid, in its most concentrated form, is a colorless, fuming, corrosive liquid, having the spec. grav. of 1.517 at 15.5° C., and boiling at 84.5° C.; it consists of 54 parts (93 per cent.) of nitrogen pentoxide and 9 parts of water.

The crude commercial nitric acid is of two strengths: the so-called double acid has a spec. grav. of 1.36, containing about 50 per cent. of nitrogen pentoxide; and the single acid, of 1.22 spec. grav., containing about 30 per cent. of nitrogen pentoxide.

The officinal nitric acid has the spec. grav. of 1.420, and contains about 60 per cent. of nitrogen pentoxide, corresponding to about 70 per cent. of monohydrated nitric acid. The Acidum nitricum dilutum, of 1.068 spec. gravity, contains 11.66 per cent. of monohydrated acid.

Nitric acid is readily decomposed, and is a powerful oxidizing agent, acting violently upon most of the metals, and upon organic compounds, converting many non-nitrogenous vegetable substances into explosive bodies. From its tendency to decompose, nitric acid has frequently a yellowish color from nitrous and nitric oxides, held in solution, which, upon dilution of the acid with water, or upon heating, cause a further decomposition and consequent disengagement of nitric peroxide. Nitric acid may also be recognized by its property of dissolving copper-turnings to a blue solution, with the evolution of colorless nitric-oxide gas, which, however, at once unites with atmospheric oxygen, forming red fumes of nitric peroxide; by the ready decoloration of diluted solution of indigo; by its coloring pine-wood bright yellow; and by deep red or brown compounds with ferrous salts.

The salts of nitric acid are remarkable for being all soluble in water.

The characteristic reaction of nitric acid with ferrous salts extends also to the nitrates, when previously acted upon by strong sulphuric acid. The test is performed either by placing a crystal of ferrous sulphate in the liquid under examination, mixed with concentrated sulphuric acid, or by mixing the liquid with a concentrated solution of ferrous sulphate, and pouring this mixture carefully upon concentrated sulphuric acid in a test-tube, or by pouring it into a conical cylinder and placing concentrated sulphuric acid below it by means of a pipette (Fig. 34), so as to form in either case two supernatant layers. If a large quantity of nitric acid is present, the surfaces of the crystal, or the line of contact between the liquids, become black; if but a small quantity is present, they become reddish-brown or purple.

Volumetric Estimation, *see* page 61.

Examination:

Hydrochloric acid may be detected in the acid diluted with about five times its volume of water, by the formation of a white precipitate, when tested with argentic nitrate.

Fig. 34.

Sulphuric acid is detected in the diluted acid by a white precipitate, when the diluted acid is tested with barium nitrate.

Nitrous and *hyponitric acids* are detected in the diluted acid, by one or two drops of a very dilute (1:1000) solution of potassium permanganate: their presence is indicated by decoloration.

Iodine and Iodic Acid.—About one drachm of the acid (the most concentrated acid has to be diluted with an equal volume of water) is shaken in a test-tube with about one-half its bulk of chloroform, which, after subsiding, will appear yellowish or reddish, if free iodine be contained in the acid; when it remains colorless, a little hydrosulphuric acid is added, drop by drop, with gentle agitation; if a coloration of the chloroform takes place now, *iodic acid* is indicated.

A confirmatory test is, to mix the acid, after dilution, if strong acid is under examination, with a few drops of mucilage of starch; a bluish coloration will take place after a while, when iodine is present; if no reaction occurs, a few drops of solution of sulphurous acid may be added, drop by drop, when the blue color will appear, if iodic acid be present.

Metals may be detected by mixing the diluted acid with an equal bulk of hydrosulphuric acid, and by subsequent over-saturation with aqua ammoniæ.

TABLE

OF THE QUANTITY BY WEIGHT OF ANHYDROUS AND OF MONOHYDRATED NITRIC ACID CONTAINED IN 100 PARTS BY WEIGHT OF AQUEOUS NITRIC ACID AT DIFFERENT DENSITIES. (TEMPERATURE 17.5° C.)

Specific Gravity.	Percent. of NO^5.	Percent. of NO^5+H^2O	Specific Gravity.	Percent. of NO^5.	Percent. of NO^5+H^2O	Specific Gravity.	Percent. of NO^5.	Percent. of NO^5+H^2O
1.523	85	99.16	1.414	58.5	68.25	1.232	32	37.33
1.521	84.5	98.58	1.412	58	67.66	1.228	31.5	36.75
1.519	84	98.00	1.409	57.5	67.08	1.224	31	36.16
1.517	83.5	97.41	1.406	57	66.50	1.220	30.5	35.58
1.516	83	96.83	1.403	56.5	65.91	1.217	30	35.00
1.514	82.5	96.24	1.400	56	65.33	1.213	29.5	34.41
1.512	82	95.66	1.397	55.5	64.75	1.209	29	33.83
1.510	81.5	95.08	1.394	55	64.16	1.205	28.5	33.25
1.508	81	94.50	1.392	54.5	63.58	1.201	28	32.66
1.506	80.5	93.91	1.389	54	63.00	1.198	27.5	32.08
1.504	80	93.33	1.386	53.5	62.41	1.194	27	31.50
1.502	79.5	92.74	1.383	53	61.83	1.190	26.5	30.91
1.500	79	92.16	1.380	52.5	61.25	1.186	26	30.33
1.498	78.5	91.58	1.377	52	60.66	1.182	25.5	29.74
1.496	78	91.00	1.374	51.5	60.08	1.178	25	29.16
1.494	77.5	90.41	1.371	51	59.50	1.174	24.5	28.58
1.492	77	89.83	1.368	50.5	58.91	1.170	24	28.00
1.490	76.5	89.24	1.364	50	58.33	1.167	23.5	27.41
1.488	76	88.66	1.361	49.5	57.75	1.163	23	26.83
1.486	75.5	88.08	1.358	49	57.16	1.159	22.5	26.25
1.484	75	87.50	1.355	48.5	56.58	1.155	22	25.66
1.482	74.5	86.91	1.352	48	56.00	1.151	21.5	25.08
1.480	74	86.33	1.349	47.5	55.41	1.147	21	24.49
1.478	73.5	85.74	1.345	47	54.83	1.143	20.5	23.91
1.476	73	85.16	1.342	46.5	54.25	1.140	20	23.33
1.474	72.5	84.58	1.338	46	53.66	1.136	19.5	22.74
1.472	72	84.00	1.334	45.5	53.08	1.132	19	22.16
1.470	71.5	83.41	1.330	45	52.50	1.129	18.5	21.58
1.469	71	82.83	1.327	44.5	51.91	1.125	18	21.00
1.467	70.5	82.24	1.323	44	51.33	1.122	17.5	20.41
1.465	70	81.66	1.319	43.5	50.75	1.118	17	19.83
1.462	69.5	81.08	1.315	43	50.16	1.114	16.5	19.25
1.460	69	80.50	1.312	42.5	49.58	1.111	16	18.66
1.458	68.5	79.91	1.308	42	49.00	1.107	15.5	18.08
1.456	68	79.33	1.304	41.5	48.41	1.104	15	17.50
1.454	67.5	78.75	1.301	41	47.83	1.100	14.5	16.91
1.451	67	78.16	1.297	40.5	47.25	1.096	14	16.33
1.449	66.5	77.58	1.294	40	46.66	1.092	13.5	15.74
1.447	66	77.00	1.290	39.5	46:08	1.089	13	15.16
1.444	65.5	76.41	1.287	39	45.50	1.086	12.5	14.58
1.442	65	75.83	1.283	38.5	44.91	1.082	12	14.00
1.440	64.5	75.25	1.279	38	44.33	1.078	11.5	13.41
1.438	64	74.66	1.275	37.5	43.75	1.075	11	12.83
1.436	63.5	74.08	1.271	37	43.16	1.071	10.5	12.25
1.434	63	73.50	1.267	36.5	42.58	1.068	10	11.66
1.432	62.5	72.91	1.263	36	42.00	1.064	9.5	11.07
1.430	62	72.33	1.259	35.5	41.41	1.060	9	10.50
1.428	61.5	71.75	1.255	35	40.83	1.056	8.5	9.91
1.426	61	71.16	1.251	34.5	40.25	1.053	8	9.33
1.424	60.5	70.58	1.247	34	39.66	1.050	7.5	8.84
1.422	60	70.00	1.243	33.5	39.08	1.045	7	8.16
1.419	59.5	69.41	1.239	33	38.50	1.038	6	7.00
1.417	59	68.83	1.236	32.5	37.91	1.032	5	5.83

With the decrease and increase of temperature, the density of nitric acid suffers a corresponding increase or decrease, amounting for each degree of the centigrade thermometer in either direction—

For acids of a specific gravity of		to those of		to		in the average.
For acids of a specific gravity of	1.492	to those of	1.476	to	0.00213	in the average.
" " "	1.472	"	1.456	"	0.002	" "
" " "	1.454	"	1.434	"	0.00186	" "
" " "	1.430	"	1.412	"	0.00171	" "
" " "	1.406	"	1.383	"	0.00155	" "
" " "	1.377	"	1.352	"	0.00141	" "
" " "	1.345	"	1.315	"	0.00128	" "
" " "	1.308	"	1.279	"	0.00114	" "
" " "	1.271	"	1.239	"	0.001	" "
" " "	1.232	"	1.201	"	0.00085	" "
" " "	1.194	"	1.163	"	0.00071	" "
" " "	1 155	"	1.125	"	0.0005	" "

For instance: An acid of 1.178 spec. grav. at 17.5° C. containing 25 per cent. of anhydrous, or 29.16 per cent. of monohydrated, nitric acid will have, at 20° C., a spec. grav. of (1.178—0.00072×2.5=) 1.762, and at 15° C. a spec. grav. of (1.178+0.00072×2.5=)1.1798.

ACIDUM OXALICUM.

Oxalic Acid.

Colorless, transparent, oblique-rhombic prisms, containing 2 molecules (28 per cent.) of water of crystallization, which they lose in a dry and warm air, the crystals crumbling down to a soft white powder of anhydrous oxalic acid, which may be sublimed, in a great measure, without decomposition. Heated upon platinum-foil, the crystals first fuse, with slight crepitation, and finally are completely dissipated, emitting inflammable fumes.

Oxalic acid is soluble in 8 parts of cold water, and in its own weight, or less, of boiling water; it is also soluble in 7 parts of glycerin, and in about 4 parts of alcohol, but sparingly in ether and chloroform. Its solution has a very sour taste, and a strong acid reaction; it forms with the alkaline metals soluble, with all other bases, for the most part, insoluble, salts, which, however, are soluble in dilute acids.

When a cold saturated aqueous solution of oxalic acid is dropped into strong alcohol, it should not produce a turbidity; when dropped into lime-water, a copious white precipitate must ensue at once, which remains unchanged upon addition of acetic acid, as well as of ammonium chloride, but which is readily dissolved by hydrochloric and nitric acids. Added to a solution of calcium sulphate, a precipitate is also produced after a while.

When heated with concentrated sulphuric acid, oxalic acid is resolved into water and equal volumes of carbonic oxide and carbonic acid, without being charred.

Examination:

Binoxalates and *quadroxalates* of *potassium* (sorrel and lemon salts) are detected by heating about half a drachm of the oxalic acid in a platinum or porcelain capsule, to redness, and until no more fumes are emitted; a white fused residue, turning red litmus-paper blue, and effervescing with a few drops of hydrochloric acid, would indicate potassium or traces of calcium.

Crude commercial acid mostly leaves a very small trace of residue, too insignificant, however, to impair the quality of the acid, or render it unfit for its common technical applications.

Tartaric, *citric*, and *racemic acids*, and their salts, as accidental admixtures in oxalic acid, may be detected by gently heating, in a test-tube, some of the crystals in strong sulphuric acid. The crystals, as well as the acid, must not become dark-colored or blackened, otherwise the presence of the one or other of such admixtures is indicated.

ACIDUM PHOSPHORICUM.

Phosphoric Acid.

Glacial, *monobasic*, or *metaphosphoric acid* forms colorless, transparent, glass-like, fusible masses, deliquescent, and slowly but freely soluble in water and in alcohol, yielding colorless, inodorous acid solutions. The fused acid consists of 142 parts of metaphosphoric acid and 18 parts of water, containing, therefore, 11.2 per cent. of water. Its aqueous solution produces white monobasic precipitates with albumen and with soluble salts of silver, barium, and calcium. When its solution is heated for some time, the monobasic acid is converted into the tri-basic acid, which is contained in the medicinal acidum phosphoricum dilutum. This process is accelerated by the addition of a little nitric acid to the boiling solution of the monobasic acid.

Acidum phosphoricum dilutum forms a colorless, inodorous, acid fluid of 1.056 spec. grav., containing 7.2 per cent. of anhydrous tri-basic phosphoric acid, 100 grains of which saturate 23.4 grains of crystallized potassium bicarbonate, without precipitation. This acid does not coagulate albumen nor precipitate barium chloride; when neutralized with sodium carbonate, it yields a light-yellow precipitate with argentic nitrate, soluble in nitric acid and in ammonium hydrate. It produces, in dilute solution of ammoniated magnesium sulphate, a white crystalline precipitate of ammonio-magnesium phosphate, which, when collected and washed upon a filter, becomes yellow upon being moistened with dilute solution of argentic nitrate.

Examination of Glacial Phosphoric Acid:

Ammonium may be detected by heating a few fragments of the fused acid in a strong solution of potassium hydrate in a test-tube; the odor of ammonium hydrate and the formation of white fumes, when a glass rod, moistened with acetic acid, is held over the orifice of the tube, will indicate ammonium.

For further examination, about two drachms of the glacial acid are dissolved in six fluid-drachms of water, in a small porcelain capsule; when dissolved, 20 drops of concentrated nitric acid are added, and the fluid boiled until reduced to about half its original bulk; then so much water is added as to make the fluid measure one ounce.

Magnesium and *aluminium* are detected in a small portion of this acid, after dilution with water, by the formation of a white precipitate upon over-saturation with strong ammonium hydrate; if this is soluble in acetic acid, magnesium is indicated; if it is insoluble or only partly soluble, aluminium is also indicated, and may be confirmed by heating a little of it on platinum wire before the blow-pipe; the colorless bead will assume a blue color, after being moistened with solution of cobaltous nitrate, and reheated, when aluminium is present.

Calcium and Silicic Acid.—Another part of the above solution of the acid is mixed with about four times its volume of absolute alcohol; if a precipitate is formed, it is collected upon a filter, and, after the alcoholic liquid has passed through the filter, is treated with liquor potassæ; if a residue remains on the filter, the presence of calcium is indicated; if the precipitate, however, be partly or entirely dissolved, the solution is oversaturated with hydrochloric acid and evaporated to dryness at a gentle heat; the residue is dissolved in water acidulated with sulphuric acid; if a white turbid solution results, silicic acid is indicated.

Metals.—The rest of the solution of the acid may be examined for metals in the same mode as described under acidum phosphoricum dilutum.

Examination of Acidum Phosphoricum Dilutum:

Monobasic phosphoric acid may be detected by the formation of a gelatinous white precipitate when tested with solution of albumen.

Phosphorous acid is indicated, if no arsenious acid be present, by decoloration when a little of the acid is heated in a test-tube, with one or two drops of potassium permanganate. It may also be detected in the diluted acid by adding a few drops of solution of mercuric chloride, and gently warming the liquid; an ensuing white turbidity would confirm the presence of phosphorous acid.

Hydrochloric acid is detected by a white precipitate with argentic nitrate in the diluted acid, to which previously have been added a few drops of concentrated nitric acid.

Nitric acid is indicated by decoloration when a little of the acid is gently heated with one drop of indigo-solution. Its presence may be confirmed by mixing with the acid nearly an equal bulk of concentrated solution of ferrous sulphate, and by placing this mixture upon concentrated sulphuric acid, with the precaution that the two fluids do not mix (Fig. 35); a red-brown coloration upon the line of contact between the two fluids will confirm the presence of nitric acid.

Sulphuric acid is detected in the diluted acid, to which a few drops of nitric acid have been added, by a white precipitate with barium nitrate.

Metals are detected by saturating the diluted acid with hydrosulphuric-acid gas, and allowing the liquid to stand for 12 hours in a corked test-tube or flask; the occurrence of a coloration or precipitate will indicate metals; a light-yellow floccu-

Fig. 35.

lent one, arsenic; a brown or black one, copper, lead, or tin. They may be discriminated by the methods described on page 41.

Arsenious as well as *arsenic acids* may be detected in phosphoric acid by the two following equally accurate methods:

1. A mixture of about two drachms of the officinal phosphoric acid with two drachms of concentrated pure hydrochloric acid and about 20 drops of concentrated solution of stannous chloride, is heated in a test-tube; the mixture, at first white and turbid, becomes transparent and clear; a brown coloration or turbidity will indicate arsenic.

2. Three fluid-drachms of the phosphoric acid are warmed in a test-tube; then one drop of solution of potassium permanganate is added; if decoloration takes place, the solution is added drop by drop until decoloration of the reagent ceases; one fluid-drachm of concentrated sulphuric acid (previously tested for arsenic) is added, and the mixture placed in a long

test-tube (Fig. 36), care being taken that the interior upper parts of the tube remain dry; a few pieces of pure zinc are

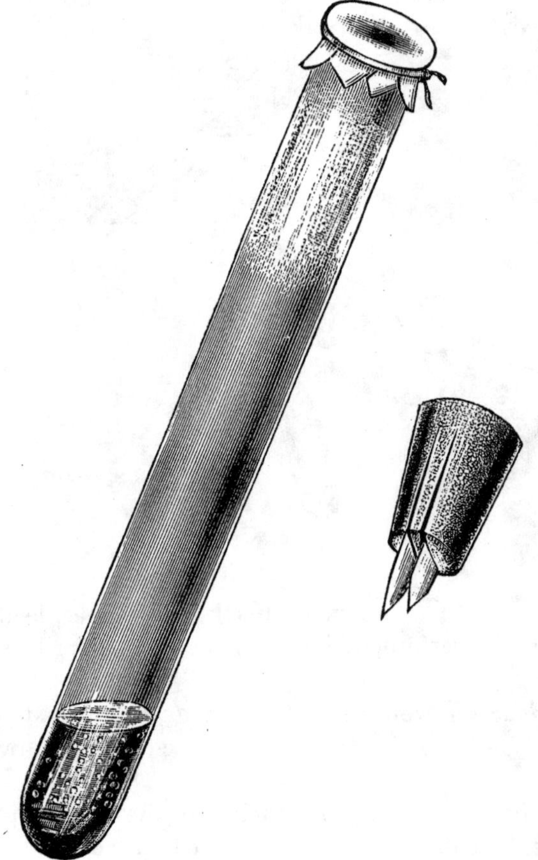

FIG. 36.

then added, and the test conducted in the mode described on page 30. A dark coloration of the solution of argentic nitrate will confirm the presence of arsenic.

TABLE

OF THE QUANTITY BY WEIGHT OF ANHYDROUS AND OF TRI-HYDRATED PHOSPHORIC ACID CONTAINED IN 100 PARTS BY WEIGHT OF AQUEOUS PHOSPHORIC ACID AT DIFFERENT DENSITIES.

TEMPERATURE 17.5° C.

Specific Gravity.	Percent. of PO_5.	Percent. of PO_5+3H_2O	Specific Gravity.	Percent. of PO_5.	Percent. of PO_5+3H_2O	Specific Gravity.	Percent. of PO_5.	Percent. of PO_5+3H_2O
1.809	68	93.67	1.462	46	63.37	1.208	24	33.06
1.800	67.5	92.99	1.455	45.5	62.68	1.203	23.5	32.37
1.792	67	92.30	1.448	45	61.99	1.198	23	31.68
1.783	66.5	91.61	1.441	44.5	61.30	1.193	22.5	30.99
1.775	66	90.92	1.435	44	60.61	1.188	22	30.31
1.766	65.5	90.23	1.428	43.5	59.92	1.183	21.5	29.62
1.758	65	89.54	1.422	43	59.23	1.178	21	28.93
1.750	64.5	88.85	1.415	42.5	58.55	1.174	20.5	28.24
1.741	64	88.16	1.409	42	57.86	1.169	20	27.55
1.733	63.5	87.48	1.402	41.5	57.17	1.164	19.5	26.86
1.725	63	86.79	1.396	41	56.48	1.159	19	26.17
1.717	62.5	86.10	1.389	40.5	55.79	1.155	18.5	25.48
1.709	62	85.41	1.383	40	55.10	1.150	18	24.80
1.701	61.5	84.72	1.377	39.5	54.41	1.145	17.5	24.11
1.693	61	84.03	1.371	39	53.72	1.140	17	23.42
1.685	60.5	83.34	1.365	38.5	53.04	1.135	16.5	22.73
1.677	60	82.65	1.359	38	52.35	1.130	16	22.04
1.669	59.5	81.97	1.354	37.5	51.66	1.126	15.5	21.35
1.661	59	81.28	1.348	37	50.97	1.122	15	20.66
1.653	58.5	80.59	1.342	36.5	50.28	1.118	14.5	19.97
1.645	58	79.90	1.336	36	49.59	1.113	14	19.28
1.637	57.5	79.21	1.330	35.5	48.90	1.109	13.5	18.60
1.629	57	78.52	1.325	35	48.21	1.104	13	17.91
1.621	56.5	77.83	1.319	34.5	47.52	1.100	12.5	17.22
1.613	56	77.14	1.314	34	46.84	1.096	12	16.53
1.605	55.5	76.45	1.308	33.5	46.15	1.091	11.5	15.84
1.597	55	75.77	1.303	33	45.46	1.087	11	15.15
1.589	54.5	75.08	1.298	32.5	44.77	1.083	10.5	14.46
1.581	54	74.39	1.292	32	44.08	1.079	10	13.77
1.574	53.5	73.70	1.287	31.5	43.39	1.074	9.5	13.09
1.566	53	73.01	1.281	31	42.70	1.070	9	12.40
1.559	52.5	72.32	1.276	30.5	42.01	1.066	8.5	11.71
1.551	52	71.63	1.271	30	41.33	1.062	8	11.02
1.543	51.5	70.94	1.265	29.5	40.64	1.058	7.5	10.33
1.536	51	70.26	1.260	29	39.95	1.053	7	9.64
1.528	50.5	69.57	1.255	28.5	39.26	1.049	6.5	8.95
1.521	50	68.88	1.249	28	38.57	1.045	6	8.26
1.513	49.5	68.19	1.244	27.5	37.88	1.041	5.5	7.57
1.505	49	67.50	1.239	27	37.19	1.037	5	6.89
1.498	48.5	66.81	1.233	26.5	36.50	1.033	4.5	6.20
1.491	48	66.12	1.228	26	35.82	1.029	4	5.51
1.484	47.5	65.43	1.223	25.5	35.13	1.025	3.5	4.82
1.476	47	64.75	1.218	25	34.44	1.021	3	4.13
1.469	46.5	64.06	1.213	24.5	33.75	1.017	2.5	3.44

With the decrease or increase of temperature, the density of phosphoric acid suffers a corresponding increase or decrease, amounting for each degree of the centigrade thermometer in either direction—

For acids of a specific gravity of	1.809	to those of 1.613	to about	0.001.
" " "	1.597	" 1.462	"	0.00082.
" " "	1.448	" 1.336	"	0.00068.
" " "	1.325	" 1.228	"	0.00052.
" " "	1.218	" 1.122	"	0.0004.
" " "	1.113	" 1.079	"	0.00035.

For instance: An acid of 1.130 spec. grav. at 17.5° C., containing 16 per cent. of anhydrous or 22.04 per cent. of tri-hydrated phosphoric acid, will have, at 20° C., a spec. grav. of (1.130—0.0004 × 2.5=) 1.129, and at 15° C., a spec. grav. of (1.130+0.0004 × 2.5=) 1.131.

ACIDUM SUCCINICUM.

Succinic Acid.

Colorless, oblique-rhombic prisms, without odor when pure, and with a more or less strong odor when the acid is obtained from amber by sublimation, and is only imperfectly freed from the empyreumatic oils. When heated upon platinum-foil, it fuses at 180° C., and emits irritating fumes, leaving a charred residue, which, however, is entirely dissipated upon heating to red heat.

Pure succinic acid dissolves in from 25 to 28 parts of cold, and about three parts of boiling, water, in 15 parts of cold, and in 1½ part of boiling, alcohol; it is but sparingly soluble in ether and in oil of turpentine (distinction from benzoic acid); its aqueous solution yields, when neutralized with aqua ammoniæ, a copious reddish-brown precipitate with diluted solutions of ferric salts, which is re-dissolved upon addition of hydrochloric acid.

Examination:

Fixed Admixtures.—If a residue remains when the acid is heated and dissipated upon platinum-foil, about two drachms of it are completely incinerated in a porcelain crucible; the residue, when cold, is tested with moist turmeric as well as with red litmus-paper; it is then divided into three parts; one of which is mixed with a little strong alcohol, and this ignited; a green color of the flame, especially toward the termination of the ignition, indicates *boracic acid;* another portion of the residue is moistened with one drop of concentrated hydrochloric acid; effervescence would indicate *potassium bitartrate*, or other *tartrates*, *citrates*, or *oxalates;* the third portion is dissolved in a few drops of water and one drop of nitric acid, and tested with barium nitrate; a white precipitate would indicate *potassium sulphate* or *bisulphate.*

Ammonium salts are detected by the odor of ammonia and by white fumes when a glass rod, moistened with acetic acid, is held in the test-tube, when the acid is heated in solution of potassium hydrate.

Metallic impurities may be detected in the concentrated solution of the acid, by the admixture of an equal volume of hydrosulphuric acid.

Organic acids and *salts* may be detected by agitating four grains of the succinic acid in two drachms of hot water; a complete solution must take place, and continue after cooling; when a white precipitate separates on cooling, it is collected upon a filter, washed with a few drops of cold water, and, when perfectly dry, treated with a few drops of warm oil of turpentine, wherein it will dissolve, if it is *benzoic acid.* A portion of the aqueous solution of the acid is then dropped into strong alcohol; an ensuing turbidity would indicate mineral salts; another part of the solution is dropped into lime-water, taking care that this, with its alkaline reaction, remains prevalent; the formation of a white turbidity would indicate *oxalic* or *tartaric acids*, or *their salts;* if a turbidity is not formed until after warming the liquid, *citric acid* or *citrates* would be indicated; a little solution of ammonium chloride is added and agitated: oxalate will remain unaltered, tartrate or citrate will partly or entirely dissolve.

The following may serve as a general test for the purity of succinic acid: Ten grains of the acid are dissolved in 3 fluid-drachms of strong or absolute alcohol; the solution is aided by dipping the test-tube in hot water; when cold, it is divided into two parts, one of which is mixed with an equal bulk of chloroform, the other with an equal bulk of aqua ammoniæ;

a complete solution must take place in the first test, and a clear mixture in the second, otherwise one or more of the above-mentioned adulterations are present.

When a crude acid, containing empyreumatic substances, has to be examined, it is first agitated and washed with a little ether, and is then dissolved in boiling water, and the solution, when cold, passed through a filter previously moistened with water.

ACIDUM SULPHURICUM.

Sulphuric Acid.

A dense, colorless, inodorous, highly-corrosive liquid, of a spec. grav. of from 1.843 to 1.850, containing from 18 to 20 per cent. of water, and 96.8 per cent. of monohydrated sulphuric acid. At this density one fluidounce of the acid weighs a small fraction over 14 drachms. It has a strong attraction for water, absorbing it from the atmosphere, and withdrawing it or its elements from organic compounds immersed in, or mixed with, the acid; sulphuric acid, therefore, when in contact with organic substances, or with air containing dust, gradually loses its colorless appearance, and becomes more or less brown, and rapidly chars and destroys most organic substances.

Sulphuric acid is miscible with water, glycerin, alcohol, chloroform, carbon bisulphide, and other solvents, with evolution of heat, and produces, with most organic liquids, a more or less vehement decomposition; in its relations to other compounds, it maintains the character of one of the strongest acids, its affinity for bases being so powerful as to withdraw them from most of their compounds, forming sulphates, which, with the exception of those of barium, strontium, lead, and calcium, are soluble in water. By the same powerful affinity, sulphuric acid decomposes water, and deprives it of its oxygen, when in contact with water and certain metals (iron, zinc, tin, magnesium), or part of the acid itself is decomposed by the abstraction of oxygen, with the formation of metallic oxides and of sulphates, when concentrated sulphuric acid acts upon some metals (copper, mercury, silver).

A piece of pine-wood, dipped into concentrated sulphuric acid, will become black, and when a piece of loaf-sugar is placed in a little of the acid, in a test-tube, and gently heated, the acid will become black, and emit the odor of sulphurous acid; when one drop of the acid is mixed with a test-tubeful of water, this will let fall a white precipitate upon the addition of a few drops of barium chloride.

Volumetric Estimation, see page 61.

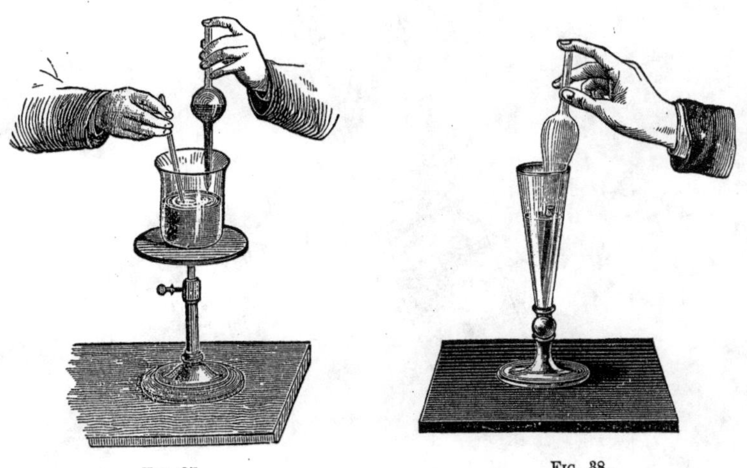

FIG. 37. FIG. 38.

Examination:

Fixed impurities are recognized by a residue after complete evaporation of a few drops of the acid upon platinum-foil.

Lead is indicated by a white turbidity taking place upon the careful admixture of one part of the acid with about four or five times its volume of alcohol (Fig. 37). Another method of readily recognizing the presence of lead in sulphuric acid is, to fill a small conical cylinder about half with concentrated hydrochloric acid, and then to place below the acid, by means of a pipette, a nearly equal volume of the sulphuric acid, with care that the fluids do not mix (Fig. 38); an ensuing white turbidity at the junction of the two fluids would confirm the presence of lead.

The presence of lead and other metallic impurities may be also detected by nearly neutralizing the acid, diluted with two or three times its bulk of water, with aqua ammoniæ or liquor potassæ, and by adding, while yet warm, an equal volume of

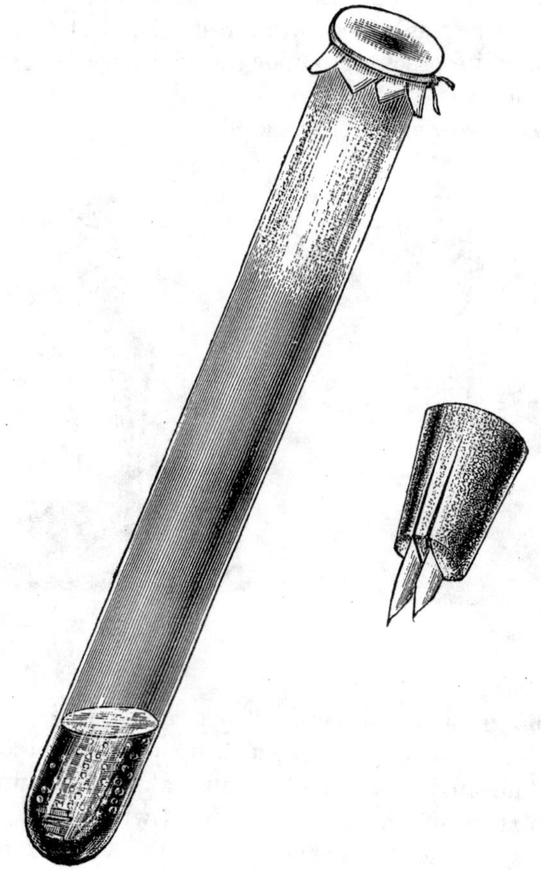

Fig. 89.

hydrosulphuric acid, and allowing the liquid to stand for several hours; a precipitate would indicate lead or other metals, which, when required, may be further examined by the methods described in the course of analytical investigation (page 41).

Arsenic may be detected by either of the two following methods:

1. About half a fluidounce of a mixture of equal parts, by volume, of the concentrated sulphuric acid and water is, when cool, heated in a test-tube, and a solution of potassium permanganate added drop by drop, until decoloration ceases; then about 20 grains of sodium chloride, and subsequently 20 drops of concentrated solution of stannous chloride, are added, and the mixture heated. An ensuing brown turbidity would indicate arsenic.

2. To about one fluid-drachm of the sulphuric acid are added one or two drops of solution of potassium permanganate, or, if decoloration takes place, enough to make this cease. The acid is then added to 4 parts, by volume, of water, and introduced into a long test-tube, of which it should fill not more than about one-tenth, and whose interior above the fluid should remain dry (Fig. 39); a few pieces of pure zinc are then added, and the test performed in the mode described on page 30.

Sulphurous acid and *oxides of nitrogen* are indicated by decoloration upon the addition of one drop of sulphuric-acid indigo-solution, and gentle warming, as also by decoloration of solution of potassium permanganate. A special test for nitrogen oxides consists in dissolving two or three drops of pure anilin in about one drachm of diluted sulphuric acid, and in adding gradually, with gentle stirring with a glass rod, about twice the volume of the concentrated sulphuric acid (Fig. 37). Nitrogen oxides, if present, cause rose-colored lines in the track of the glass rod.

TABLE

OF THE QUANTITY BY WEIGHT OF ANHYDROUS AND OF MONO-HYDRATED SULPHURIC ACID CONTAINED IN 100 PARTS BY WEIGHT OF AQUEOUS SULPHURIC ACIDS AT DIFFERENT DENSITIES.

TEMPERATURE 17.5° C.

Specific Gravity.	Percent. of SO_3.	Percent. of SO_3+H_2O.	Specific Gravity.	Percent. of SO_3.	Percent. of SO_3+H_2O.	Specific Gravity.	Percent. of SO_3.	Percent. of SO_3+H_2O.
1.841	81.6	100	1.559	53.8	66	1.235	26.1	32
1.840	80.8	99	1.547	53.0	65	1.257	25.3	31
1.839	80.0	98	1.536	52.2	64	1.219	24.5	30
1.838	79.2	97	1.525	51.4	63	1.211	23.6	29
1.837	78.3	96	1.514	50.6	62	1.202	22.8	28
1.835	77.5	95	1.503	49.8	61	1.194	22.0	27
1.833	76.7	94	1.493	49.0	60	1.186	21.2	26
1.830	75.9	93	1.482	48.1	59	1.178	20.4	25
1.826	75.1	92	1.471	47.3	58	1.170	19.6	24
1.821	74.3	91	1.461	46.5	57	1.163	18.7	23
1.815	73.4	90	1.450	45.7	56	1.155	17.9	22
1.808	72.6	89	1.440	44.9	55	1.147	17.1	21
1.800	71.8	88	1.430	44.0	54	1.140	16.3	20
1.791	71.0	87	1.420	43.2	53	1.132	15.5	19
1.782	70.1	86	1.411	42.4	52	1.125	14.7	18
1.774	69.4	85	1.401	41.6	51	1.117	13.8	17
1.765	68.5	84	1.392	40.8	50	1.110	13.0	16
1.755	67.7	83	1.382	40.0	49	1.103	12.2	15
1.744	66.9	82	1.373	39.2	48	1.095	11.4	14
1.733	66.1	81	1.364	38.3	47	1.088	10.6	13
1.722	65.3	80	1.354	37.5	46	1.081	9.8	12
1.711	64.4	79	1.345	36.7	45	1.074	9.0	11
1.699	63.6	78	1.336	35.9	44	1.067	8.1	10
1.688	62.8	77	1.328	35.1	43	1.060	7.3	9
1.676	62.0	76	1.319	34.3	42	1.053	6.5	8
1.665	61.2	75	1.310	33.4	41	1.046	5.7	7
1.653	60.4	74	1.302	32.6	40	1.039	4.9	6
1.641	59.6	73	1.293	31.8	39	1.032	4.1	5
1.629	58.7	72	1.285	31.0	38	1.025	3.2	4
1.617	57.9	71	1.276	30.2	37	1.019	2.4	3
1.605	57.1	70	1.268	29.4	36	1.012	1.6	2
1.593	56.3	69	1.260	28.5	35	1.006	0.8	1
1.582	55.5	68	1.251	27.7	34	1.003	0.4	0.5
1.570	54.7	67	1.243	26.9	33	0.000	0	0

With the decrease and increase of temperature, the density of sulphuric acid suffers a corresponding increase or decrease, amounting for each degree of the centigrade thermometer in either direction—

For acids of a specific gravity of			1.841	to those of	1.782	to about	0.0014.
"	"	"	1.774	"	1.665	"	0.0012.
"	"	"	1.653	"	1.302	"	0.001.
"	"	"	1.293	"	1.219	"	0.00075.
"	"	"	1.211	"	1.140	"	0.00045.
"	"	"	1.132	"	1.067	"	0.00047.

ACIDUM TANNICUM.

Tannic Acid. Tannin.

Amorphous, friable, porous, and inodorous masses, or thin shining scales, of a pale greenish-yellow color, and feeble, mild odor* (mostly combined with a faint odor of ether); when heated upon platinum-foil, tannic acid fuses, swells up, and burns away without residue.

Tannic acid is soluble in about two parts of water, in three parts of glycerin, and in less than its own weight of diluted alcohol; but it is less soluble in strong alcohol, only sparingly in ether, and insoluble in the fixed and volatile oils: its aqueous solution reddens litmus-paper, and has an astringent taste without bitterness; it becomes turbid when boiled, and gradually dark-colored and mouldy, when exposed to the air; it suffers precipitation by the alkaline salts, and by the mineral acids, and forms soluble compounds with the alkaline hydrates, sparingly soluble ones with the earthy oxides, and more or less insoluble ones with most of the metallic oxides; its solution coagulates solutions of gluten, albumen, and starch (distinction from gallic acid), and affords white precipitates, soluble in acetic acid, with the alkaloids; it renders no reaction with ferrous salts, if completely free from ferric salts, but it gives a bluish-black precipitate with the latter, which is soluble in oxalic and mineral acids. When solution of tannic acid is dropped into lime-water, it produces a white turbidity, which soon becomes gray and dingy green, and passes through various shades to a dark purple-brown color.

Examination:

The absence of admixtures of *resinous substances*, of *dextrin*, and of *sugar*, may be ascertained by the property of the acid to yield a clear or almost clear solution with about four or five parts of warm water, which should remain so when tested in two portions, the one by addition of twice its volume of strong alcohol, the other by dilution with water; if any such adulterations be present, they may be separated and recognized by making two solutions of the acid, one in strong alcohol

* The color and odor are due to traces of a greenish resin.

when sugar and dextrin will remain behind, and another one in boiling water, when resinous substances will be expelled on cooling.

ACIDUM TARTARICUM.

Tartaric Acid.

Colorless, transparent crystals, which have the form of an oblique-rhombic prism, more or less modified; they are permanent in the air; but when heated, they fuse, become black, and are decomposed and charred, with the evolution of inflammable vapors of a peculiar odor, and are finally wholly dissipated without residue.

Tartaric acid is soluble in an equal weight of cold, and half its weight of boiling, water, and in three parts of alcohol, forming solutions of a strongly acid taste and reaction, which, when dropped into solutions of neutral potassium salts, give rise to the formation of a white granular precipitate, at once in concentrated solutions, and after a time in diluted ones. This reaction, however, does not take place in solutions containing free mineral acids or acid salts thereof. When solution of tartaric acid is dropped into lime-water, so that the alkaline reaction remains prevalent, a white turbidity occurs (distinction from citric acid) which, disappears again upon addition of solution of ammonium chloride (distinction from racemic acid), and also upon addition of acetic acid (distinction from oxalic acid); solution of calcium sulphate remains unchanged upon addition of tartaric acid (additional distinction from oxalic and racemic acids).

Crystals of tartaric acid, when immersed in concentrated sulphuric acid, dissolve gradually without coloration, unless warmed, when they become black.

One drachm of tartaric acid requires, for saturation, 65.70 grains of potassium carbonate, 79.98 grains of crystallized potassium bicarbonate, 114.39 grains of crystallized sodium carbonate, 67.20 grains of sodium bicarbonate, and 38.19 grains of magnesium carbonate.

Volumetric Estimation, *see* page 61.

Examination:

Salts.—An admixture of salts is recognized by an addition of an equal volume of alcohol to a cold saturated aqueous solution of the acid, or by dissolving the powdered acid in 6 parts of strong alcohol; a complete solution must take place and remain in either case.

Sulphates may be detected in the diluted solution, by a white turbidity with barium nitrate.

Oxalates may be detected in the concentrated aqueous solution of the acid, by a white precipitate when tested with solution of calcium sulphate.

Calcium salts may be detected in the diluted solution, neutralized with aqua ammoniæ, by a white precipitate with ammonium oxalate.

Metallic impurities (copper or lead) are detected when a saturated aqueous solution of the acid is mixed with about twice its bulk of hydrosulphuric acid.

ACIDUM VALERIANICUM.

Valerianic Acid.

Pure valerianic acid forms a thin, colorless, or nearly colorless liquid, having the persistent odor of valerian-root and a pungent, acid taste; it reddens litmus, bleaches the skin, and burns with a bright, smoky light. In contact with water, it absorbs about 20 per cent. of its weight without losing its oily consistence, and is itself soluble in about 26 parts of water; it is miscible with aqua ammoniæ, alcohol, and ether, in all proportions. Its spec. grav. is from 0.935 to 0.940, and it boils at 175° C. The commercial acid is generally trihydrated, having a density of 0.950, and the boiling-point at 165° C.

Examination:

Fatty Acids.—Twenty grains of the acid are weighed in a tared flask, and water, of a temperature of from 12° to 15° C., is carefully added, with constant agitation, until the acid is just dissolved. The bottle is weighed again, and the quantity of water required for solution must be not less than 26 times the weight of the acid; in this instance, not less than one ounce and two scruples. If the acid dissolves in less water, it is not pure, containing admixtures (alcohol, acetic acid, and butyric acid), which by their greater solubility increase that of the valerianic acid. On the other hand, the quantity of water required for solution must not exceed thirty times (10 drachms) the weight of the valerianic acid, in which case it would contain less soluble or insoluble admixtures (caproic and similar monatomic acids, valeric aldehyde, etc.).

If the preceding test leaves doubt as to the purity of the acid, or if a more conclusive examination be required, one drachm of the acid is weighed in a beaker-glass, and mixed with two drachms of hot water; then from a burette, or a graduated pipette (Fig. 40), a solution of potassium carbonate, of 1.33 spec. grav. (containing 33⅓ per cent. of anhydrous carbonate), is added drop by drop, until the acid is exactly neutralized. The quantity by weight of the solution of potassium carbonate used must very little, if at all, exceed twice the quantity of the acid; if a greater quantity be required, the presence of butyric, acetic, and similar homologous acids, is evident. When, in this test, oily drops are separated upon the surface of the liquid, the admixture of some neutral oily compound is indicated.

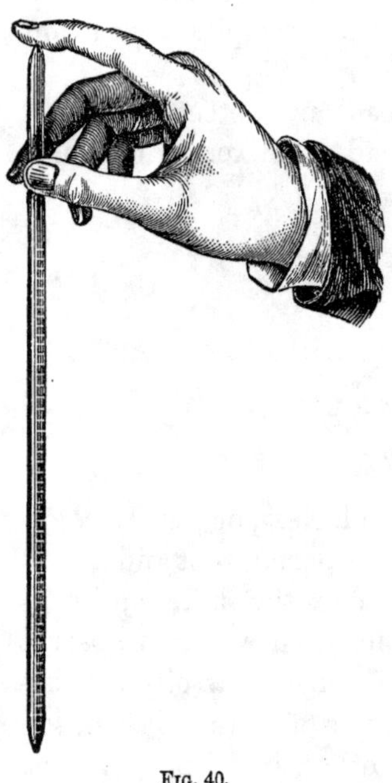

FIG. 40.

Acetic acid may be detected, by adding to the liquid obtained in the preceding test half as much hydrochloric acid, of 1.124 spec. grav. (containing 25 per cent. of anhydrous acid), as was required of the potassium-carbonate solution for saturation; after agitation, the liquid is transferred upon a small glass funnel, provided with a stop-cock. After the separation of the supernatant fatty acid, the lower aqueous layer is drawn off through the stop-cock. A part of this solution is carefully neutralized, in a test-tube, with diluted aqua ammoniæ, and a few drops of a solution of ferric perchloride are then added; a reddish-brown precipitate will ensue, which, upon agitation, acquires a resinous appearance; the supernatant liquid must appear colorless; a reddish color would indicate acetic acid.

Mineral acids may be detected in the aqueous solution of the valerianic acid by adding a few drops of nitric acid, and by subsequently testing portions of it with barium nitrate for sulphuric acid, and with argentic nitrate for hydrochloric.

ACONITIA.

ACONITINUM.

Aconitia. Aconitine.

White, amorphous, pulverulent grains, or a white or yellowish-white powder, or colorless rhombic or six-sided prisms, fusing, when heated upon platinum-foil, at about 80° C., without loss of weight; at about 120° C. it becomes brown, and, at a higher temperature, burns away without a residue.

Aconitia is but sparingly soluble in cold water; with hot water it becomes soft and resin-like, and dissolves gradually, in the proportion of one part of aconitia to 50 parts of hot water; upon cooling, most of the aconitia separates again. The aqueous solution has a persistent bitter and acrid taste, and a feeble alkaline reaction; upon addition of concentrated sulphuric acid, the solution remains unchanged, or becomes slightly yellow; but, when heated by immersing the test-tube in boiling water, it turns purple, and, upon the addition of potassium bichromate, green.

Alcohol, ether, chloroform, carbon bisulphide, benzol, and diluted acids, dissolve aconitia freely; its solution in water, acidulated with hydrochloric or sulphuric acids, yields a white precipitate with solutions of tannic acid, of phosphoric acid, of potassio-mercuric iodide, and of the alkaline hydrates and carbonates, a brown one with solution of iodinized potassium iodide.

Aconitia dissolves in concentrated nitric acid without change; with concentrated sulphuric acid, it forms a coherent mass, which dissolves upon agitation, with a yellowish-brown color, gradually changing to purple or brown. It is also dissolved by phosphoric acid; this solution becomes purple upon evaporation on a porcelain capsule, in a water-bath.

A kind of aconitia is occasionally met with in commerce which differs from that above described in its chemical and therapeutical properties, and which is evidently derived from one or more non-officinal species of Aconitum; it is readily recognized by its slighter solubility in chloroform, requiring about 230 parts for solution, while the officinal aconitia dissolves in 2½ parts of chloroform.

ÆTHER.

Ether. Ethylic Ether. Ethyl Oxide.

A colorless, light, and limpid liquid, of a characteristic fragrant odor, very volatile and inflammable; it does not redden litmus, but becomes slightly acid by the absorption of oxygen and the formation of acetic acid, from contact with the air in imperfectly-stoppered bottles. When pure, its spec. grav., at 15.5° C., is about 0.720.

The United States Pharmacopœia provides two strengths of ether, one of the spec. grav. of 0.750, at 15.5° C., containing about 70 per cent., and æther fortior, of a spec. grav. not exceeding 0.728 at 15.5° C., and containing about 90 per cent. of ethylic ether.

Ether is miscible, in all proportions, with alcohol, carbon bisulphide, chloroform, and benzol; pure ether dissolves about two per cent. of water, increasing thereby its density from 0.720 to 0.723, at 15.5° C. Water, on the other hand, dissolves nearly 10 per cent. of ether. From its solution in ether, the water can wholly be removed again by anhydrous potassium carbonate, provided that the ether be pure and free from alcohol. When completely free from alcohol and water, ether has no action on dry tannic acid, which deliquesces to a thick, syrupy fluid in æther fortior.

Ether dissolves sulphur and phosphorus sparingly, but bromine, iodine, gun-cotton, caoutchouc, essential oils, and most of the fatty and resinous substances, freely; it is also a solvent for a number of alkaloids, and for some metallic salts, e. g., mercuric, auric, platinic, and ferric chlorides, etc.

Examination:

Alcohol and Water.—Shaken with an equal bulk of water, in a small graduated cylinder (Fig. 41), officinal ether should not lose more than from one-fifth to one-fourth, and æther fortior not more than from one-tenth to one-eighth, of its volume; otherwise an excess of one or other of the above is contained in the ether, which fact will also be indicated by a greater density of the ether than that above stated.

A still more accurate result of this test is obtained when pure glycerine is employed instead of water.

Acids.—Blue neutral litmus-paper, when immersed in both strata in the cylinder, should remain unaltered, as also when a small quantity of the ether is evaporated in a porcelain capsule until reduced to a few drops, and then tested with litmus-paper; a slight acid reaction would indicate acetic acid, and, in crude ether, possibly sulphurous or sulphuric acid; the acid reaction may also be caused by traces of ethyl sulphate; traces of this and other ethylic or amylic ethers or alcohols are also indicated when about half an ounce of the ether is evaporated from a flat porcelain capsule by causing the fluid to flow to and fro; when the ether is evaporated, the surfaces of the capsule should be covered with a deposit of moisture, without taste or smell, and without any oily appearance.

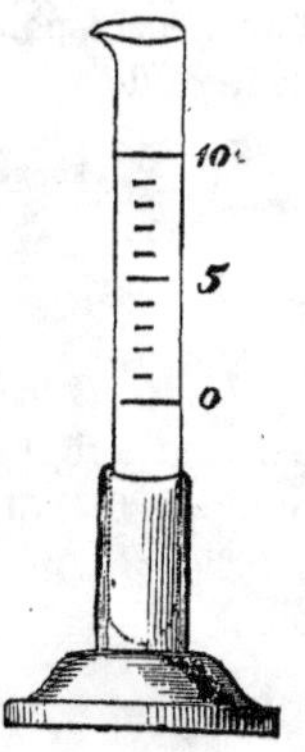

Fig. 41.

TABLE

OF THE QUANTITY BY WEIGHT OF PURE ETHYLIC ETHER CONTAINED IN 100 PARTS BY WEIGHT OF ETHER AT DIFFERENT DENSITIES.

TEMPERATURE 17.5° C.

Specific Gravity.	Percentage of Ethylic Ether.	Specific Gravity.	Percentage of Ethylic Ether.	Specific Gravity.	Percentage of Ethylic Ether.	Specific Gravity.	Percentage of Ethylic Ether.
0.7185	100	0.7310	87	0.7456	74	0.7614	61
0.7198	99	0.7320	86	0.7468	73	0.7627	60
0.7206	98	0.7331	85	0.7480	72	0.7640	59
0.7215	97	0.7342	84	0.7492	71	0.7653	58
0.7224	96	0.7353	83	0.7504	70	0.7666	57
0.7233	95	0.7364	82	0.7516	69	0.7680	56
0.7242	94	0.7375	81	0.7528	68	0.7693	55
0.7251	93	0.7386	80	0.7540	67	0.7707	54
0.7260	92	0.7397	79	0.7552	66	0.7721	53
0.7270	91	0.7408	78	0.7564	65	0.7735	52
0.7280	90	0.7420	77	0.7576	64	0.7750	51
0.7290	89	0.7432	76	0.7588	63	0.7764	50
0.7300	88	0.7444	75	0.7601	62	0.7778	49

With the decrease and increase of temperature, the density of ether suffers a corresponding increase or decrease, amounting for each degree of the centigrade thermometer in either direction—

For ether of a specific gravity of 0.7198 to that of 0.7331, about 0.0013.
" " " 0.7342 " 0.7504, " 0.0011.
" " " 0.7516 " 0.7627, " 0.0009.
" " " 0.7640 " 0.7764, " 0.0008.

For instance: An ether of 0.7206 spec. grav. at 17.5° C., containing 98 per cent. ethyl oxide, will have, at 20° C., a spec. grav. of (0.7206−0.0013 ×2.5=) 0.7173, and, at 15° C., a spec. grav. of (0.7206+0.0013×2.5=) 0.7239.

ÆTHER ACETICUS.

Acetic Ether. Ethyl Acetate.

A colorless, light, limpid liquid, of an agreeable, ethereal, and fruity odor and taste; very volatile and combustible. When pure, its spec. grav. is 0.900 to 0.904 at 15° C., and its boiling-point at 74 to 76° C. Acetic ether is miscible in all proportions with ether, alcohol, chloroform, carbon bisulphide, and benzol; it is miscible with water, and soluble therein, in approximately the same proportion as ether. It absorbs oxygen from the air, especially if it contains some water, forming acetic acid; both the water and the acid can be removed from the ether by shaking it with exsiccated potassium carbonate, which will become more or less liquefied when these fluids are present.

Examination:

Alcohol.—When shaken with an equal volume of water in a graduated glass cylinder (Fig. 41, page 115), the ether, after subsiding, should not have decreased in bulk more than one-tenth to one-eighth; when pure glycerine is employed instead of water, the volume of both liquids should remain almost unaltered. Alcohol and water are also indicated by a greater density, and consequently a lower specific gravity of the ether.

Acids.—Neutral litmus-paper, when immersed in both strata in the cylinder, should remain unaltered, as also in the remainder of a little of the ether when evaporated in a porcelain capsule.

Estimation of the ethyl acetate contained in acetic ether. Fifty grains of pure crystallized barium hydrate are introduced into a 2-ounce vial, with so much water as to leave room for 20 grains of the ether; the vial is then corked and weighed, and 20 grains of acetic ether are added. The flask is then tightly closed, and allowed to stand in a warm place or in tepid water, agitating it occasionally for several hours, until the ethereal odor of the liquid has entirely ceased. The contents of the flask are then emptied into a beaker or flask, the flask being rinsed with a little water to avoid loss. Carbonic-acid gas is now passed into the liquid until this has an acid reaction; it is then warmed, the turbid liquid transferred to a filter, and the precipitate washed by means of a washing-bottle. The entire filtrate is completely precipitated with diluted sulphuric acid; the ensuing precipitate is collected upon a moistened tared filter, washed, dried, and weighed. The weight of the precipitate, divided by 1.32, gives, as a quotient, the quantity of ethyl acetate contained in 20 grains of the acetic ether, which result, multiplied by five, gives the percentage.

ALCOHOL.

SPIRITUS RECTIFICATUS.

Ethyl Alcohol.

Alcohol absolutum, Spiritus alcoholisatus. Spec. grav., 0.795. 100 per cent. of real alcohol by volume.

Alcohol fortius, Spiritus rectificatissimus. Spec. grav., 0.817. 95 per cent. of real alcohol by volume.

Alcohol rectificatum, Spiritus rectificatus. Spec. grav., 0.835. 90 per cent., by volume, of real alcohol.

Alcohol dilutum, Spiritus tenuinor. Spec. grav., 0.941. 46 per cent. of real alcohol by volume.

A colorless, limpid, neutral liquid, inflammable, and burning with a blue flame, without smoke; its spec. grav. is 0.795 at 15° C.; its boiling-point at 78.4° C.; it is miscible in all proportions with most liquid bodies, except the fatty oils, and, next to water, is the most extensive and important solvent; it combines with some salts, forming alcoholates. Anhydrous alcohol has a great attraction for water, absorbing its vapor from the atmosphere, and abstracting the moisture from organic substances immersed in it. In the act of dilution, a contraction of volume and an increase of the temperature of the mixture take place. When 55 volumes of absolute alcohol are mixed with 45 volumes of water, the mixture, after cooling, will occupy only 96.2 volumes, having therefore suffered a contraction of 3.8 per cent.: and, *vice versa*, an expansion of volume takes place when diluted alcohol is mixed with water: e. g., when 100 volumes of alcohol, of a spec. grav. of 0.966, containing 28 per cent., by volume, of real alcohol, are mixed with 50 volumes of water, 153 volumes will be obtained.

The percentage of real alcohol in its aqueous dilutions can be determined approximately, but accurately enough for any practical purpose, by ascertaining its specific gravity at a known temperature. The specific gravity of any sample of alcohol established will, by the aid of the following table, at once indicate the percentage of real alcohol:

TABLE

OF THE QUANTITY OF REAL ALCOHOL CONTAINED IN 100 PARTS OF AQUEOUS ALCOHOL BY WEIGHT AND BY VOLUME AT DIFFERENT DENSITIES.

TEMPERATURE 15° C.

Specific Gravity.	100 Volumes contain:		100 Parts by Weight contain:	Specific Gravity.	100 Volumes contain:		100 Parts by Weight contain:
	Alcohol.	Water.	Alcohol.		Alcohol.	Water.	Alcohol.
0.7951	100	0.00	100.00	0.9348	50	53.72	42.53
0.8000	99	1.28	98.38	0.9366	49	54.70	41.59
0.8046	98	2.54	96.83	0.9385	48	55.68	40.66
0.8089	97	3.77	95.35	0.9403	47	56.66	39.74
0.8130	96	4.97	93.89	0.9421	46	57.64	38.82
0.8169	95	6.16	92.45	0.9439	45	58.61	37.90
0.8206	94	7.32	91.08	0.9456	44	59.54	37.00
0.8242	93	8.48	89.72	0.9473	43	60.58	36.09
0.8277	92	9.62	88.37	0.9490	42	61.50	35.18
0.8311	91	10.76	87.04	0.9506	41	62.46	34.30
0.8344	90	11.88	85.74	0.9522	40	63.42	33.40
0.8377	89	13.01	84.47	0.9538	39	64.37	32.52
0.8409	88	14.12	83.22	0.9553	38	65.32	31.63
0.8440	87	15.23	81.96	0.9568	37	66.26	30.75
0.8470	86	16.32	80.72	0.9582	36	67.20	29.88
0.8500	85	17.42	79.51	0.9595	35	68.12	29.01
0.8530	84	18.52	78.29	0.9607	34	69.04	28.14
0.8559	83	19.61	77.09	0.9620	33	69.96	27.27
0.8588	82	20.68	75.91	0.9633	32	70.89	26.41
0.8616	81	21.76	74.75	0.9645	31	71.80	25.56
0.8644	80	22.82	73.59	0.9657	30	72.72	24.70
0.8671	79	23.90	72.43	0.9668	29	73.62	23.85
0.8698	78	24.96	71.30	0.9679	28	74.53	23.00
0.8725	77	26.03	70.16	0.9690	27	75.43	22.16
0.8752	76	27.09	69.04	0.9700	26	76.33	21.31
0.8778	75	28.15	67.93	0.9711	25	77.23	20.47
0.8804	74	29.20	66.82	0.9721	24	78.13	19.63
0.8830	73	30.26	65.72	0.9731	23	79.09	18.79
0.8855	72	31.30	64.64	0.9741	22	79.92	17.96
0.8880	71	32.35	63.58	0.9751	21	80.81	17.12
0.8905	70	33.39	62.50	0.9761	20	81.71	16.29
0.8930	69	34.44	61.43	0.9771	19	82.60	15.46
0.8954	68	35.47	60.38	0.9781	18	83.50	14.63
0.8978	67	36.51	59.33	0.9791	17	84.39	13.80
0.9002	66	37.54	58.29	0.9801	16	85.29	12.98
0.9026	65	38.58	57.25	0.9812	15	86.19	12.15
0.9049	64	39.60	56.23	0.9822	14	87.09	11.33
0.9072	63	40.63	55.21	0.9833	13	88.00	10.51
0.9095	62	41.65	54.20	0.9844	12	88.90	9.69
0.9117	61	42.67	53.19	0.9855	11	89.80	8.87
0.9139	60	43.68	52.20	0.9867	10	90.72	8.06
0.9161	59	44.70	51.20	0.9878	9	91.62	7.24
0.9183	58	45.72	50.21	0.9890	8	92.54	6.43
0.9205	57	46.73	49.24	0.9902	7	93.45	5.62
0.9226	56	47.73	48.26	0.9915	6	94.38	4.81
0.9247	55	48.74	47.29	0.9928	5	95.30	4.00
0.9267	54	49.74	46.33	0.9942	4	96.24	3.20
0.9288	53	50.74	45.37	0.9956	3	97.17	2.40
0.9308	52	51.74	44.41	0.9970	2	98.11	1.60
0.9328	51	52.73	43.47	0.9985	1	99.05	0.80

Since, however, the temperature exercises a considerable expanding and contracting influence upon alcohol and its dilution with water, it is necessary to ascertain, simultaneously with the specific gravity, also the temperature of the sample; for this reason, the areometers (Alcoholometers) constructed for measuring the density of alcohol are provided with a thermometer, and differences in the temperature of the alcohol under estimation may readily be corrected by calculation based upon this rule: The number of degrees of temperature of the alcohol above or below 15° must be multiplied by four-tenths; the product is then to be added to the percentage of the real alcohol indicated by the specific gravity, when the sample was cooler than 15° C., and subtracted, when it was warmer.

If, e. g., the spec. grav. of a sample is found to be 0.860, at a temperature of 5° C., its percentage of real alcohol would be, according to the above table, 81 per cent., by volume; since, however, the alcohol was weighed at a temperature 10° lower than the standard temperature of the above table, its density was accordingly greater. Therefore, in order to correct this difference, 10 has to be multiplied by four-tenths; the product (= 4) must be added to the percentage of alcohol (81) inferred from the spec. grav., and the sum (= 85) expresses the real quantity of alcohol in 100 parts by volume.

Examination:

Fusel-oil (a mixture of ethyl and amyl alcohols, containing traces of propylic, butylic, and other alcohols) and *aldehyde* may be detected by mixing about half an ounce of the alcohol with an equal volume of pure ether, and by subsequently adding an amount of water equal to the volume of the mixture; the whole is shaken, and, when subsidence has taken place, the ethereal layer is decanted and evaporated at the common, or at a slightly-raised, temperature. After the evaporation of the ether, the remainder will give the characteristic odor of fusel oil, or of any flavors indicative of a previous employment of the alcohol for the extraction of vegetable substances.

Methyl Alcohol.—Among the several methods for the detection of methylic alcohol as an admixture in ethyl alcohol, the following is least open to objection: Thirty grains of powdered potassium bichromate are dissolved, in a little flask, in half an ounce of water; 25 drops of concentrated sulphuric acid, and subsequently 40 drops of the alcohol under examination, are added, and the flask closed with a cork provided with a delivery-tube, leading into a large test-tube, cooled in ice-water (Fig. 42). After allowing the mixture to stand for about a quarter of an hour, the flask is gently warmed until about half a fluidounce of distillate is obtained; the distillate is then slightly over-saturated with a little crystallized sodium carbonate, and evaporated to the bulk of about one-quarter of an ounce; when cool, it is filtered into a small test-tube, and then slightly over-

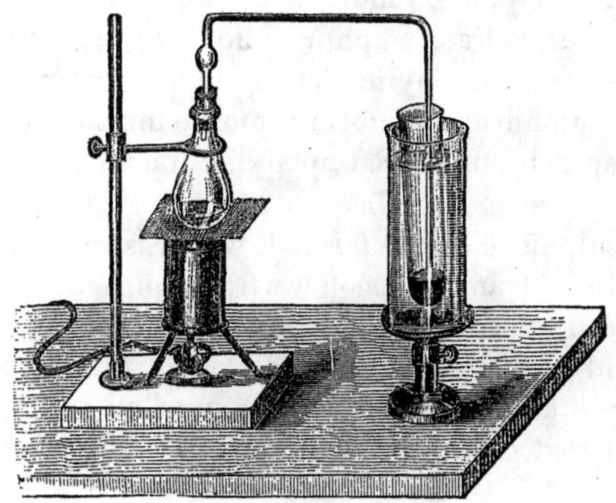

Fig. 42.

saturated with a few drops of acetic acid; 20 drops of solution of argentic nitrate (1:20) are then added, and the mixture gently heated for a few minutes. If the liquid merely darkens a little, but continues quite translucent, the alcohol is free from methylic alcohol; but, if a dark-brown or black precipitate of reduced silver separates, and the test-tube, after being rinsed and filled with water, shows upon its interior a brilliant metallic resplendence, which, when seen against white paper, appears brown by transmitted light, the alcohol is methylated.

ALCOHOL AMYLICUM.

Amylic Alcohol. Fusel-Oil.

A colorless or pale-yellow, mobile, oily liquid, of a strong, offensive odor, and acrid, burning taste; its spec. grav. is 0.818 and its boiling-point between 132° and 133° C.; it solidifies at about —22° C. When dropped upon water, it floats upon the surface like an oil. Fusel-oil requires about 40 parts of water for solution; it is miscible, in all proportions, with alcohol, ether, carbon bisulphide, chloroform, and essential and fatty oils; it dissolves iodine, sulphur, phosphorus, camphor, and resins. Three parts of amylic alcohol mixed with two parts of concentrated sulphuric acid, form a purple mixture, from which the alcohol separates unaltered upon dilution with water; when the mixture, however, has been allowed to stand for several hours, amyl-sulphuric acid is formed, which is soluble in water. When a mixture of amylic alcohol with strong sulphuric acid is heated with a fragment of potassium bichromate, the odor of valerianic acid is evolved. Amylic alcohol does not take fire by contact with a flame, and, when dropped on paper, does not leave a permanent greasy stain.

ALUMEN.

ALUMINII ET POTASSII SULPHAS. ALUMINII ET AMMONII SULPHAS.

Alum.

Colorless, transparent, octahedral crystals, which often exhibit the faces of a cube, permanent in the air; when exposed to heat, they undergo aqueous fusion at about 100° C., and, at a stronger heat, lose their water of crystallization, and swell up to a white porous mass, which, when moistened with a few drops of solution of cobaltous nitrate and reheated, assumes a blue color.

Alum is soluble in 25.3 parts of water at 0° C., and in about 15 parts of cold, and in its own weight of boiling, water; it is also soluble in glycerin, but insoluble in alcohol, ether, and chloroform. Its solution has a sweetish, astringent taste, reddens litmus-paper, and gives a bulky white precipitate with the alkaline hydrates, of which that with potassium or sodium hydrate is soluble in an excess of the reagent, but it is reprecipitated by ammonium chloride; the alkaline carbonates and phosphates, and solutions of barium salts, also throw down white precipitates.

Most commercial alum is a mixture of ammonium and potassium alum, the former being substituted, to a greater or less extent, for the latter; since their properties are almost the same, this admixture is of little consequence in the common uses of alum. Potassium alum consists, in 100 parts, of 18.39 parts of potassium sulphate, 36.14 parts of aluminium sulphate, and 45.47 parts of water of crystallization; while ammonium alum contains, in 100 parts, 14.58 of ammonium sulphate, 37.83 of aluminium sulphate, and 47.59 of water of crystallization. The presence of ammonium alum is recognized by the odor of ammonia, and by its reaction upon moistened red test-paper, and with a glass rod moistened with acetic acid, when a little of the powdered alum is heated in liquor potassæ.

Examination:

Iron is recognized in the solution of alum, after the addition of a few drops of sulphuric acid, by a blue coloration when tested with potassium ferrocyanide: most crude alum contains traces of ferric salts; their quantity, however, should not be so considerable as to produce a purple coloration of a solution of the alum upon the addition of a few drops of solution of tannic acid.

Other *metallic* impurities may be detected in the solution, after addition of a little tartaric acid and subsequent over-saturation with liquor potassæ, by ammonium sulphydrate; a dark coloration or precipitate indicates metallic impurities; a white precipitate, not disappearing upon addition of potassium hydrate, would show *zinc.* If required, the nature of such precipitate may be determined by the method described on pages 41 to 43.

ALUMINII SULPHAS.

ALUMINIUM SULPHURICUM.

Aluminium Sulphate.

A white crystalline powder, or small pearly plates, of a sour-sweet and astringent taste, permanent in the air; when exposed to heat the salt fuses, loses at first the water of crystallization, and, at a strong heat, the sulphuric acid also, leaving behind aluminium oxide (alumina or argilla), which, when moistened with solution of cobaltous nitrate and reheated, assumes a blue color.

Aluminium sulphate is soluble in twice its weight of cold water, forming a clear solution, of an acid reaction and astringent taste, which affords a bulky white precipitate with the alkaline hydrates, of which that with the fixed hydrates is soluble in an excess of the precipitant (evidence of the absence of magnesium), but is precipitated again upon addition of ammonium chloride. The alkaline solution should yield no reaction upon the addition of a few drops of ammonium sulphydrate; the occurrence of a brown or yellowish-red precipitate would indicate *ferric* and *manganic* salts, and of a white one, *zinc.*

AMMONII BENZOAS.

AMMONIUM BENZOICUM.

Benzoate of Ammonium. Ammonium Benzoate.

Small, colorless, shining, thin, four-sided, tabular crystals, deliquescent in the air, and having a feeble odor of benzoic acid, and a saline, somewhat balsamic, and bitterish taste. When heated, the salt sublimes without residue. It is soluble in water, glycerin, and alcohol; its aqueous solution, if not too dilute, emits the odor of ammonia when heated with potassium hydrate, and gives a white precipitate of benzoic acid upon addition of hydrochloric acid, and a copious, pale, reddish-yellow precipitate with ferric salts. The diluted solution of ammonium benzoate must remain clear when mixed with lime-water (evidence of the absence of ammonium oxalate).

AMMONII BROMIDUM.

AMMONIUM BROMATUM.

Bromide of Ammonium. Ammonium Bromide.

Colorless, transparent, anhydrous crystals, or a granular salt, which, by the action of the atmospheric oxygen, gradually becomes yellow; entirely volatile by heat, and subliming unchanged.

Ammonium bromide is soluble in 1½ part of cold water, in about 15 parts of alcohol, and but sparingly in ether. Its solution has a saline, pungent taste, is neutral, and, when dropped into a dilute solution of argentic nitrate, acidulated with a few drops of nitric acid, gives a white, curdy precipitate, soluble upon addition of aqua ammoniæ; when dropped into a very dilute solution of mercuric chloride, no reaction takes place (distinction from alkaline iodides). When the aqueous solution is mixed with a little mucilage of starch, and afterward with a little chlorine-water, a yellowish coloration takes place, which will transfer its color to chloroform, ether, or carbon bisulphide, when added and agitated with the solution. A blue tint upon the addition of the chlorine-water, and a violet or reddish coloration of the chloroform, would indicate iodides. Heated with potassium hydrate, it emits the odor of ammonia.

Examination:

Ammonium bromate is detected in the aqueous solution by an ensuing red coloration upon the addition of concentrated hydrochloric acid.

Ammonium chloride, as an admixture, may approximately be ascertained by completely precipitating a solution of 10 grains of the dry crystallized salt in one ounce of water, acidulated with nitric acid, by a solution of argentic nitrate; the precipitate is collected upon a moist double filter, both being of equal size, is washed, dried, and, when completely dry, weighed, the one filter serving to counterpoise the other one, which contains the precipitate; if the salt was pure ammonium bromide, the obtained argentic bromide must weigh 19.15 grains; if it contained alkaline chloride, the weight will be greater in proportion to the quantity of the latter.

AMMONII CARBONAS.

AMMONIUM CARBONICUM.

Carbonate of Ammonium. Ammonium Sesqui-carbonate.

Colorless, hard, translucent, crystalline masses, of a strong ammoniacal odor, completely volatile, without fusion, when heated; exposed to the air, the salt decomposes rapidly, giving off water and ammonium hydrate, and is converted into a white, opaque, pulverulent mass of ammonium bicarbonate, which requires eight parts of cold water for solution, while the sesqui-carbonate dissolves in four parts of water. Ammonium sesqui-carbonate is soluble in about five parts of glycerin; it is but sparingly soluble in alcohol, and dissolves in acids with effervescence; one drachm of the salt requires for saturation 71.16 grains of citric, and 76.26 grains of tartaric, acid.

Volumetric Estimation, see page 58.

Examination:

Ammonium bicarbonate is already indicated by the change of the hard crystalline state of the salt into the soft one; it is further recognized by agitating one drachm of the triturated salt with from five to six drachms of cold water; in this case, pure sesqui-carbonate renders a clear solution, while the bicarbonate will not dissolve until two or three more drachms of water have been added.

Ammonium sulphate is detected in the aqueous solution, over-saturated with nitric acid, by a white precipitate with barium nitrate.

Ammonium chloride and *hyposulphite* are recognized in the aqueous solution, over-saturated with acetic acid, by testing it with argentic nitrate; a white precipitate, insoluble in diluted nitric acid, will indicate chloride; a white turbidity, which gradually turns black, indicates hyposulphite.

Metals are detected in the aqueous solution, over-saturated with nitric acid, by hydrosulphuric acid.

AMMONII CHLORIDUM.

AMMONII MURIAS. AMMONIUM CHLORATUM. SAL AMMONIACUM.

Ammonium Chloride.

A colorless anhydrous salt, either in translucent, crystalline masses, of a tough fibrous texture, or in small cubic or octahedral crystals, or a granular white powder. When heated, the salt volatilizes without fusion or decomposition, and condenses, upon cooling, in thick white fumes; it yields a white and soft streak when scratched with a knife; and remains white, and evolves the odor of ammonia, when moistened with liquor potassæ.

Ammonium chloride is soluble in three parts of cold, and in its own weight of boiling, water, and in about six parts of glycerin, but only sparingly in alcohol, and not at all in ether or chloroform; its aqueous solution has a sharp, saline taste, reddens blue litmus-paper, emits the odor of ammonia when heated with liquor potassæ, and forms a copious, curdy, white precipitate with argentic nitrate.

Examination:

Sulphates are detected in the diluted solution by a white precipitate, when tested with barium nitrate.

Fixed impurities are indicated by a residue left after complete volatilization of the ammonium chloride, upon platinum-foil or in a porcelain crucible.

Metallic impurities may be detected in the aqueous solution by a coloration or precipitate when tested with hydrosulphuric acid, and by subsequent addition of aqua ammoniæ; if a precipitate be formed with the first reagent, it is collected upon a filter, after several hours' digestion, and the filtrate is then rendered alkaline with ammonium hydrate; if a second precipitate be then formed, it is also collected. The first precipitate may contain tin, copper, and zinc; the second one, iron and manganese; if required, they may be discriminated by the methods described on pages 41 to 43.

Iron may also be recognized at once by a blue coloration, when the solution of the salt is acidulated with hydrochloric acid, and tested with potassium ferrocyanide.

AMMONII IODIDUM.

AMMONIUM IODATUM.

Iodide of Ammonium. Ammonium Iodide.

A white, granular, crystalline salt, which, when exposed to the air, becomes yellowish or reddish brown, from oxidation and consequent liberation of a minute quantity of iodine. When heated, it is completely volatilized with the evolution of purple vapors.

Ammonium iodide is very soluble in water, and freely soluble in alcohol (distinction from ammonium and potassium bromides, which are less soluble in alcohol); its solution has a pungent, saline taste, and emits the odor of ammonia when heated with liquor potassæ; it yields a red precipitate with mercuric bichloride, and a yellowish-white one with argentic nitrate, which latter precipitate remains unchanged upon the addition of dilute nitric acid and of aqua ammoniæ; the solution assumes a blue color upon the addition of mucilage of starch and a little chlorine-water.

Examination:

An admixture of alkaline iodides, bromides, or chlorides, is approximately recognized, when a concentrated aqueous solution of the salt is dropped into strong alcohol; the liquid must remain clear; the separation of a white crystalline powder would indicate such an admixture.

Chlorides and *bromides* are detected by completely precipitating the solution of the salt with argentic nitrate, and by subsequent addition of aqua ammoniæ; the filtrate is then oversaturated with nitric acid; a slight turbidity may take place; a white precipitate would indicate chlorides and bromides. In this case, and in order to distinguish argentic chloride or bromide, the precipitate is collected and washed upon a filter, and is then rinsed through the pierced filter into a test-tube; the supernatant water is decanted as far as practicable, and good chlorine-water is poured upon and agitated with the silver salt. This will remain unchanged if it consists of argentic chloride, but, if it contains argentic bromide, the chlorine-water assumes a yellowish or reddish color, which will also be absorbed by chloroform, when agitated with the liquid.

Sulphate may be detected in the diluted solution, acidulated with tartaric acid, by a white precipitate upon the addition of a few drops of barium nitrate.

AMMONII NITRAS.

AMMONIUM NITRICUM.

Nitrate of Ammonium. Ammonium Nitrate.

Long, flexible needles, or a fibrous crystalline mass, when obtained by crystallization above 40° C., and large six-sided prisms, terminated by six-sided pyramids, when crystallized at a temperature below 38° C.; they contain four molecules of water of crystallization, which evaporates slowly at common temperatures. Exposed to the air, ammonium nitrate deliquesces slightly, loses ammonium hydrate, and becomes acid; it fuses at 108° C., commences boiling at 180° C., and between 210° and 240° C. it is resolved and entirely volatilized into watery vapor and nitrous-oxide gas. When heated rapidly, ammonium hydrate, nitric oxide, and ammonium nitrite, are also formed; when thrown upon a red-hot surface, it burns with a slight noise and a pale-yellow flame; in contact with heated combustible matter, it deflagrates, like all nitrates.

Ammonium nitrate dissolves in about half its weight of cold water, producing considerable depression of temperature; it is also soluble in twice its weight of alcohol, so that its concentrated aqueous solution remains perfectly limpid upon addition of strong alcohol. Its aqueous solution emits the odor of ammonium hydrate, when heated with liquor potassæ; and when mixed with a few drops of a solution of ferrous sulphate, and carefully poured upon concentrated sulphuric acid, it affords a dark zone at the junction of the liquids, characteristic of the oxides of nitrogen.

Examination:

Ammonium chloride and *sulphate* may be detected by white precipitates, when the dilute aqueous solution of the salt is acidulated with nitric acid and tested in separate portions, with argentic nitrate for the former salt, and with barium nitrate for the latter.

AMMONII PHOSPHAS.

AMMONIUM PHOSPHORICUM.

Phosphate of Ammonium. Tri-basic Ammonium Phosphate.

Transparent, colorless prisms, efflorescent in the air; when heated upon platinum-foil, ammonium hydrate, and subsequently thick, white vapors of phosphoric acid, are evolved, and the salt is wholly dissipated at a red heat. When treated with liquor potassæ, ammonium hydrate is given off.

Ammonium phosphate is readily and freely soluble in water, but insoluble in alcohol; its solution is slightly alkaline, but, when the salt has been exposed to the air or is old, the solution is neutral, or even acid, from the loss of ammonium hydrate; this reaction, with the formation of an acidulous ammonium phosphate, takes place when the alkaline solution of the salt is boiled.

With solution of argentic nitrate, the diluted solution of ammonium phosphate gives a yellow precipitate, soluble upon addition of ammonium hydrate, as well as of nitric acid. It must give no reaction with ammonium sulphydrate, and, after acidulation with diluted hydrochloric acid, none with hydrosulphuric acid or with barium chloride. A precipitate with the first two reagents would indicate *metals*, a white precipitate with the latter reagent, insoluble in diluted nitric acid, would indicate *sulphate*.

If a solution of twenty grains of ammonium phosphate is completely precipitated with solution of ammoniated magnesium sulphate, and the precipitate collected and washed upon a filter with diluted aqua ammoniæ, dried, and subsequently heated to redness in a tared porcelain crucible, the residue obtained must weigh 16.8 grains.

AMMONII VALERIANAS.

AMMONIUM VALERIANICUM.

Valerianate of Ammonium. Ammonium Valerianate.

Colorless, transparent, quadrangular plates, or a white, translucent, crystalline mass, with the odor of valerianic acid. When heated, the salt melts and emits vapors of the odor of ammonium hydrate and of valerianic acid; at a stronger heat it becomes black, with the evolution of pungent, inflammable vapors, and is at last wholly dissipated. It is decomposed, and emits the odor of ammonium hydrate, when heated in liquor potassæ.

Ammonium valerianate is deliquescent in moist air, and is freely soluble in water, glycerin, and alcohol; its aqueous solution, if not very dilute, separates, upon super-saturation with acids, an oily layer of valerianic acid. The underlying aqueous liquid, when nearly saturated with aqua ammoniæ, must not become red upon the addition of one drop of dilute solution of ferric chloride, for in this case *acetic acid* (admixture of potassium or sodium acetate) would be indicated, which may also be recognized by a fixed residue upon complete dissipation of the salt upon platinum-foil, which residue will effervesce when moistened with one drop of concentrated hydrochloric acid.

ANTIMONII ET POTASSII TARTRAS.

ANTIMONIUM TARTARATUM. ANTIMONIUM ET POTASSIUM TARTARICUM. TARTARUS STIBIATUS.

Tartar Emetic. Potassio-antimonious Tartrate.

Colorless, transparent, triangular-faced crystals, or a white powder. The crystals contain one molecule (2.7 per cent.) of water of crystallization; they effloresce slightly when exposed to the air, and lose their water of crystallization completely at 110° C., becoming white and opaque; when heated, they decrepitate and blacken; when powdered and heated in a dry test-tube, tartar emetic emits sour empyreumatic vapors, and leaves a charred residue which, when cool, turns moist turmeric-paper brown; when the residue is placed upon charcoal and heated before the blow-pipe, white fumes are evolved, coating the coal, and brittle globules of antimony are formed.

Tartar emetic dissolves in about 15 parts of cold, and 2 to 2½ parts of boiling, water; it is almost insoluble in alcohol. Its solution has a nauseous metallic taste, a slightly acid reaction upon blue litmus-paper, and gradually decomposes if not containing a small addition of alcohol; it is decomposed both by acids (except acetic, tartaric, and citric acids) and by alkaline hydrates; an excess of the latter redissolves the precipitate; it is also precipitated by all soluble carbonates, but not by bicarbonates. Hydrosulphuric acid produces an orange-red coloration in concentrated solutions of tartar emetic, and gradually a precipitate of the same color; in very dilute solutions, only a coloration takes place; but, upon warming, or upon the addition of an acid, or when the tartar emetic is contaminated with potassium bitartrate, a turbidity ensues immediately.

Solution of tartar emetic reduces a solution of mercuric chloride to mercurous chloride, gradually at common temperatures, and quickly at elevated ones.

Examination:

Arsenic is indicated by the garlic-like odor when about 10 grains of the powdered tartar emetic are charred in an iron spoon, or in a porcelain crucible, and subsequently heated to redness.

If the result of this test be doubtful, or confirmatory evidence be required, about 10 grains of the finely-powdered tartar emetic are dissolved in a test-tube in about three fluid-drachms of concentrated hydrochloric acid, to which from 30 to 50 drops of water have been added; then a few pieces of tin-foil (real tin), or, if this be not at hand, about 25 drops of concentrated solution of stannous chloride, are added to the solution, and this heated to boiling; the liquid must remain clear and colorless on cooling; a brown turbidity or precipitate would indicate *arsenic*.

Chlorides and *sulphates* may be detected in the aqueous solution of tartar emetic, to which a little tartaric acid has been added, by testing it in separate portions, with argentic nitrate for the former, and barium nitrate for the latter. A white precipitate in either instance would indicate the respective impurity.

Potassium Bitartrate.—An admixture of this salt may be approximately recognized by the difference of the solubility of tartar emetic (1.15) and of cream of tartar (1.184) in cold water; when, therefore, one part of the tartar emetic is agitated with 16 to 18 parts of warm water (27 grains of the former in one ounce of the latter), a complete solution must take place, and remain unchanged after cooling. If cream of tartar be present, it will separate in small crystals.

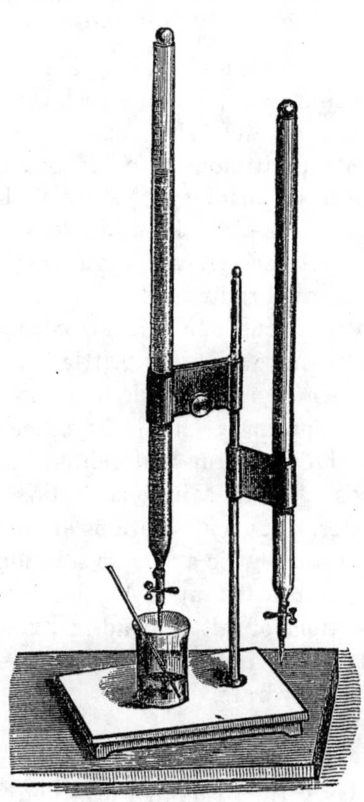

Fig. 43.

The absence of any admixture of potassium bitartrate or other salts may be established either by completely precipitating an aqueous solution of 20 grains of tartar emetic with hydrosulphuric-acid gas, whereby a precipitate is obtained which, when collected, washed, and completely dried upon a tared filter, must weigh 9.91 grains; or by volumetric estimation of a solution of one gramme of tartar emetic in 30 grammes of water, to which have been added 5 grammes of a saturated solution of potassium bicarbonate, and subsequently a little mucilage of starch. To this solution, test-solution of iodine is added from a burette (Fig. 43), with constant gentle agitation of the liquid, until this just acquires a fixed blue tint. The quantity of iodine required, multiplied by 1.315, gives as product the quantity of potassio-antimonious tartrate contained in one gramme of the tartar emetic. If the tartar emetic was pure, 60 centimetres of the test-solution must have been required for the reaction.

ANTIMONII OXIDUM.

ANTIMONIUM SEU STIBIUM OXYDATUM.

Oxide of Antimony. Antimonious Oxide.

A grayish-white or pale-buff powder, when obtained by precipitation; or small, colorless, transparent, brilliant needles, when obtained by sublimation (Flores Antimonii). When heated, antimonious oxide becomes yellow, and fuses at a dull-red heat, forming a yellowish liquid, which solidifies, on cooling, into a crystalline mass of a pearl color; at a stronger heat, it volatilizes in white vapors; when mixed and heated with exsiccated sodium carbonate on charcoal before the blow-pipe, antimonious oxide is reduced, forming globules of metallic antimony which are brittle when cold.

Antimonious oxide is insoluble in water and in aqua ammoniæ, sparingly soluble in nitric and in acetic acid, but readily soluble in warm hydrochloric acid, in warm liquor potassæ or sodæ, and in solutions of tartaric acid, and the alkaline tartrates. Its acid solutions afford an orange-red precipitate with hydrosulphuric acid; its solutions in the fixed alkaline hydrates are not acted upon by this reagent (distinction from alkaline solutions of salts of lead and zinc), but, with argentic nitrate, a black precipitate is formed, insoluble upon addition of ammonium hydrate.

Examination:

Antimonic oxide would be indicated by a turbidity on dissolving the oxide in warm hydrochloric acid.

Antimonious oxy-chloride (Algaroth's Powder) is indicated by a white precipitate, when a diluted solution of the oxide in an excess of tartaric acid is tested with argentic nitrate; its presence may be confirmed by digesting a little of the oxide for about one hour with a dilute solution of sodium carbonate, and by testing the subsequent filtrate, after over-saturation with nitric acid, with argentic nitrate.

Arsenic is recognized by the garlic-like odor, when a little of the oxide is mixed and subsequently fused and reduced upon charcoal before the blow-pipe. Its presence may be confirmed by dissolving about 10 grains of the oxide in about 3 fluid-drachms of concentrated hydrochloric acid, and by heating the solution either with a small piece of real tin-foil, or with about 20 drops of a concentrated solution of stannous chloride. A brown turbidity would indicate arsenic.

ANTIMONII OXY-SULPHURETUM.

ANTIMONIUM OXY-SULPHURATUM. STIBIUM SULPHURATUM RUBEUM.

Mineral Kermes. Sulphurated Antimony. Antimonious Oxy-sulphide.

An insipid powder of a dark brick-red color, becoming gradually lighter by the action of air and light. It is a mixture of antimonious sulphide with a small and variable amount of antimonious oxide, the former appearing under the microscope in amorphous globules, or laminæ, and the latter in small, colorless crystals, or fragments of such. When heated upon charcoal before the blow-pipe, mineral kermes fuses and burns away, with the evolution of white fumes, and the odor of sulphurous acid; heated with the addition of a little dried sodium carbonate, brittle globules of antimony are obtained.

Antimonious oxy-sulphide is insoluble in water and alcohol, but readily and wholly soluble in hydrochloric acid, which solution, after the hydrosulphuric acid has been completely expelled by heat, gives a white precipitate when dropped into water, which, however, is redissolved upon the addition of tartaric acid; this solution yields an orange-red precipitate with hydrosulphuric acid.

Antimonious oxy-sulphide is only slightly soluble in ammonium hydrate, but warm liquor potassæ dissolves the greater part of it, leaving behind the antimonious oxide, which may be washed in water and dissolved in solution of tartaric acid. If these solvents are employed in the reversed order, the oxide may be extracted first, leaving behind the sulphide, now completely soluble in liquor potassæ.

Examination:

About half a drachm of kermes is agitated with about two drachms of water; the filtrate must not affect either blue or red litmus-paper, nor leave a residue when evaporated upon platinum-foil (though, when potassium carbonate has been employed instead of sodium carbonate in the preparation of the kermes, the solution generally contains traces of potassium salt, and leaves a small fixed residue).

Antimonious Oxide.—The remaining kermes of the preceding test is rinsed, with a little tepid water, through the pierced filter into a test-tube; about 10 grains of tartaric acid are added, and the mixture agitated for a few minutes; it is then filtered, and hydrosulphuric acid added to the filtrate. The ensuing turbidity will be in proportion to the quantity of antimonious oxide contained in the kermes.

Arsenic may be detected by dissolving about 10 grains of the kermes in about 3 fluid-drachms of hot hydrochloric acid; the solution is boiled until the odor of hydrosulphuric acid entirely ceases, and is then filtered; one fluid-drachm of hydrochloric acid and a piece of tin-foil (real), or about 25 drops of a concentrated solution of stannous chloride, are now added, and the mixture heated; it must remain clear, as a brown turbidity would indicate the presence of arsenic.

The following confirmatory test may also be employed: Half a drachm of the antimonious oxy-sulphide and six fluid-drachms of hydrochloric acid of 1.12 to 1.15 spec. grav. are introduced into a small flask, which is provided, by means of a twice-perforated cork, with a funnel-tube reaching nearly to the bottom of the flask, and with a long delivery-tube, the end of which dips into a little hydrochloric acid in a long test-tube cooled in ice-water (Fig. 44). Heat is applied to the flask until gentle boiling ensues, and this is continued until nearly half of the liquid has distilled over, the orifice of the delivery-tube being constantly kept below the level of the acid. If arsenic be present, a yellow deposit within the delivery-tube above the recipient, and a turbidity of the hydrochloric acid and distillate in the receiving-tube, will occur, which latter appears at first white, but gradually, when allowed to stand in the corked test-tube, becomes a yellow, flocculent precipitate.

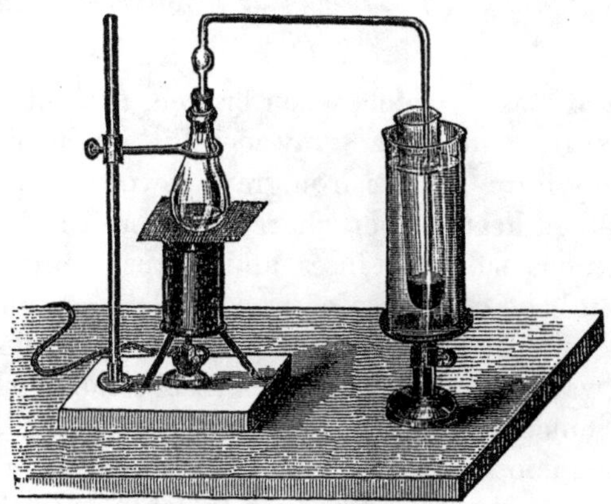

FIG. 44.

If the precipitate be considerable, it should be collected upon a filter, washed with a few drops of water, dried, and transferred, with a little potassium cyanide, into a dry reduction-tube; when the humidity is expelled by a gentle heat, stronger heat is applied to produce the arsenic mirror.

Admixtures of *ferric* or *ferrous* oxides are readily detected by an infusible brown residue when the mineral kermes is fused upon charcoal before the blow-pipe; and in the solution of the kermes in hydrochloric acid, if, when diluted and filtered, it yields a deep-blue turbidity upon the addition of potassium ferrocyanide. Admixtures of powdered silicates (brick-stone, etc.) remain unfused before the blow-pipe, and undissolved in hydrochloric acid.

ANTIMONII SULPHURETUM NIGRUM.

ANTIMONIUM SULPHURATUM NIGRUM. STIBIUM SULPHURATUM CRUDUM.

Native Sulphide of Antimony. Tri-sulphide of Antimony. Antimonious Sulphide.

Heavy fused masses which, when broken, present a striated crystalline texture, and a lead-gray metallic brilliancy; when pulverized, they form a dark iron-gray powder. Spec. grav. about 4.6. When heated upon charcoal before the blow-pipe, black antimonious sulphide fuses and burns, emitting thick white fumes and the odor of sulphurous acid; when mixed with some dried sodium carbonate and potassium cyanide, and heated in the same way, metallic globules are obtained, which are brittle when cooled.

Black antimonious sulphide, when reduced to a fine powder, is soluble in concentrated boiling hydrochloric, nitric, and sulphuric acids, and in a strong boiling solution of potassium hydrate. The solution in hydrochloric acid, when dropped into water, produces a copious white turbidity, which turns orange-red upon addition of hydrosulphuric acid (a brown or black color of the precipitate would indicate the presence of lead or other metals).

The native antimonious sulphide generally contains sulphides of iron, lead, copper, and arsenic; and there are also found, especially in the commercial black powder, silicates and mineral admixtures.

Examination:

Arsenic may be detected by dissolving about 15 grains of the black antimonious sulphide in 3 fluid-drachms of boiling concentrated hydrochloric acid; when the hydrosulphuric acid is expelled by continued boiling, the solution is filtered into a test-tube, and the filtrate, after the addition of 1 fluid-drachm of concentrated hydrochloric acid and about 20 drops of a concentrated solution of stannous chloride, is heated to boiling; a brown turbidity would indicate arsenic.

Arsenic and Lead.—Another test for arsenic may be combined with that for lead. Half a drachm of the black antimonious sulphide is treated with 6 drachms of hydrochloric acid in exactly the same mode as described on page 136, the operation, as well as the process, being the same. If the sulphide contains any arsenic, a yellow deposit will be formed in the delivery-tube, and the hydrochloric acid of the receiver will show a turbidity, which is at first white, but becomes gradually a yellow flocculent precipitate. If the sulphide contains lead, this will be indicated in the remainder in the flask, by the formation of small white needles after cooling, which will increase in quantity upon the addition of a dilute solution of tartaric acid. They may be collected upon a filter, washed with cold diluted hydrochloric acid, and subsequently dissolved upon the filter in boiling water; the obtained filtrate is, if necessary, refiltered until clear, and is then tested for lead in separate portions, by sulphuric acid, by potassium iodide, and by hydrosulphuric acid.

Metals may be detected in black antimonious sulphide by dissolving a little of it in boiling nitric acid; the solution is diluted with about four times its bulk of water, is filtered until clear, and is then tested in separate portions with one drop of sulphuric acid for *lead*, with potassium ferrocyanide for *iron*, and by over-saturation with ammonium hydrate for *copper;* in the latter test the liquid has to be filtered, if necessary, and the filtrate will have a bluish appearance, if copper is present.

Admixtures of black manganic peroxide, of pyrites, and of other crude minerals, are recognized by dissolving the black powder in boiling hydrochloric acid; the first-named gives rise to the evolution of chlorine, the latter remain mostly undissolved.

The artificially-prepared black antimonious sulphide contains frequently more or less metallic antimony, which may be recognized by its insolubility in warm hydrochloric acid, remaining behind in minute brilliant iron-gray particles, which, however, dissolve upon the addition of potassium chlorate.

ANTIMONII SULPHURETUM AURANTIACUM.

ANTIMONIUM SULPHURATUM. ANTIMONIUM SEU STIBIUM SULPHURATUM AURANTIACUM. SULPHUR AURATUM ANTIMONII.

Golden Sulphur. Penta-sulphide of Antimony. Antimonic Sulphide.

A fine orange-red powder,* nearly odorless and tasteless, becoming gradually paler colored by the action of air and light; when heated in a dry test-tube, it gives off sulphur, leaving behind black antimonious sulphide; when heated upon charcoal before the blow-pipe, it burns away with a pale, bluish flame, emitting the odor of sulphurous acid, and causing a white incrustation of the coal.

Antimonic sulphide is insoluble in water and alcohol, and in diluted mineral acids. When treated with ten to fifteen times its weight of warm concentrated hydrochloric acid, it dissolves for the most part with effervescent evolution of hydrosulphuric acid, leaving behind a scanty residue of red-colored sulphur; the solution, when deprived of the hydrosulphuric acid by heat, produces, when dropped into water, a white turbidity, disappearing upon the addition of tartaric acid, which solution is precipitated orange-red by ammonium sulphydrate.

Antimonic sulphide is completely soluble in potassium hydrate; triturated with a little water, and added to solutions of ammonium hydrate or alkaline carbonates and digested for some time, it is nearly or completely dissolved; if a small residue is left, it will dissolve in solution of tartaric acid.

Examination:

Half a drachm of the antimonic sulphide is triturated with one ounce of tepid water and agitated for about ten minutes; after subsidence, the supernatant water should not act upon neutral litmus-paper, nor leave a residue upon evaporation, nor, when acidulated with a few drops of nitric acid, yield a white turbidity, either with barium or with argentic nitrate.

Antimonious oxide may be detected in the sulphide left in the preceding test, by adding a few grains of tartaric acid and agitating the mixture for a few minutes; the subsequent filtrate

* The color is the brighter and lighter, the more dilute the solution was from which it is precipitated, and the darker, the more concentrated it was.

is then tested with hydrosulphuric acid; the occurrence of an orange-red turbidity or precipitate would indicate antimonious oxide.

Antimonious Sulphide.—A small portion of the golden sulphur, treated with solution of tartaric acid, and left in the flask or test-tube, is washed, and then repeatedly treated with ammonium hydrate; complete solution must ensue, as a brown remainder would indicate antimonious sulphide.

Arsenic may be detected by triturating about 10 grains of the antimonic sulphide and 10 grains of sodium bicarbonate with half a fluidounce of water; the mixture is macerated in a

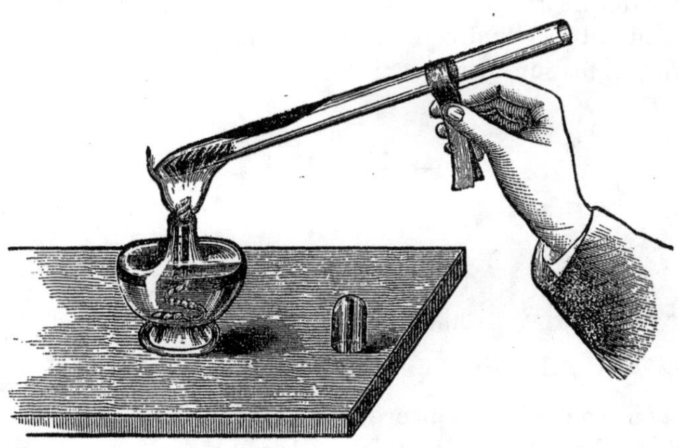

FIG. 45.

corked flask for about half an hour, with occasional agitation, is then filtered, and the filtrate over-saturated with hydrochloric acid; a lemon-yellow precipitate, occurring at once or after a short time, would indicate arsenic.

Another test for arsenic is to dissolve 10 grains of the antimonic sulphide in about 3 fluid-drachms of concentrated hydrochloric acid, diluted with 30 drops of water; the solution is heated until the hydrosulphuric acid is completely expelled; it is then filtered, and the filtrate, after the addition of 1 fluid-drachm of hydrochloric acid, and about ten drops of concentrated solution of stannous chloride, is heated to boiling; an ensuing brown turbidity would indicate arsenic.

If a confirmatory test is required, some of the antimonic sulphide may be mixed and agitated for a few minutes with a cold saturated solution of crystallized ammonium carbonate; the subsequent filtrate is evaporated to dryness in a porcelain capsule; the residue is scraped off by means of a pestle and a little powdered magnesite or pumice-stone; after the addition of a little potassium cyanide, the powder is once more warmed, and, when dry, is introduced into a warm reduction-tube, and heated to redness to produce the arsenic mirror (Fig. 45).

Admixtures of *ferric oxides* and of *silicates* are indicated by their insolubility in potassium hydrate, as well as by their stability at a red heat upon charcoal. Iron is recognized in the diluted and filtered hydrochloric-acid solution, by a blue turbidity with potassium ferrocyanide.

AQUA AMMONIÆ.

LIQUOR AMMONIÆ. AQUA SEU LIQUOR AMMONII CAUSTICI.

Solution of Ammonia. Solution of Ammonium Hydrate.

Aqua ammoniæ is an aqueous solution of the gaseous hydrogen nitride called ammonia; this gas is soluble in water to an extraordinary degree, one volume of water at 0° C. absorbing 1149, and, at 15° C., 783 volumes of the gas; the quantity of ammonium hydrate contained in the commercial and officinal solutions varies from 32 to 10 per cent. by weight of gas, the latter strength corresponding with a spec. grav. of 0.958—0.960 at 15° C., being the average strength of the aqua ammoniæ of the majority of the pharmacopœias. The United States and the British pharmacopœias include also an almost-saturated solution, Aqua Ammoniæ fortior, the former of 0.900 spec. grav., containing 29 per cent., the latter of 0.891 spec. grav., containing 32.5 per cent. of the gas.

This gas is also soluble in alcohol, which solution is officinal as Spiritus Ammoniæ, or Spiritus Ammonii caustici Dzondii, generally of a strength containing 10 to 12 per cent. of the gas.

These solutions have the properties of the gaseous ammonium hydrate, its pungent odor, sharp burning taste, and caustic action upon animal membranes; they have strong alkaline reactions, and form white fumes when brought into contact with the vapors of chlorine or acids, however diluted with atmospheric air they may be. Solution of ammonium hydrate is miscible with water, glycerin, and alcohol, and is saturated by all acids; it decomposes and precipitates most of the earthy and metallic oxides from their compounds; several of these precipitates are redissolved in an excess of the precipitant. It also precipitates most of the alkaloids.

The purity of commercial aqua ammoniæ depends upon the mode of preparation, the materials employed, and the water used for the absorption of the gas. For medicinal use, distilled water ought to be employed, while this precaution is not required for solutions used in the arts and trades. The strength of solutions of ammonia may be determined by ascertaining their specific gravity; this method, however, is reliable and accurate only when the water contains ammonium hydrate alone, and is free from other, and especially from fixed, substances, which would increase the density of the solution.

For *Volumetric Estimation*, *see* page 58.

Examination:

Fixed substances are recognized by remaining behind upon evaporation of a little of the aqua ammoniæ upon platinum-foil or a watch-glass.

Ammonium carbonate is detected by mixing equal volumes of aqua ammoniæ and lime-water; a turbidity would indicate carbonate.

Empyreuma, if not recognized by the odor, will be brought out distinctly by over-saturating in a beaker with diluted nitric acid a mixture of equal parts of aqua ammoniæ and water.

Calcium hydrate is detected by testing the aqua ammoniæ with ammonium oxalate.

Chloride, *Cyanide*, and *Sulphate*.—About two ounces of aqua ammoniæ are slightly over-saturated with nitric acid, and the solution evaporated to dryness. The residue is dissolved in a little water; small portions of this solution are tested severally with argentic nitrate for chloride and cyanide, and with barium nitrate for sulphate; when a precipitate has been formed with argentic nitrate, its nature may be ascertained by slightly over-saturating a little of the aqua ammoniæ with hydrochloric acid, and then testing it with a few drops of solution of ferric chloride; a greenish-blue coloration or blue precipitate would indicate the presence of cyanide; if such reaction does not occur, the silver precipitate, if insoluble in diluted nitric acid, consists of argentic chloride.

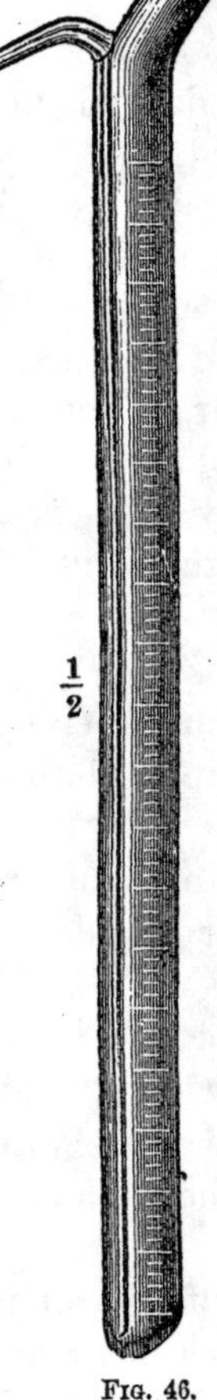

FIG. 46.

Metals are detected in aqua ammoniæ, neutralized with nitric acid, by hydrosulphuric acid, and by subsequent addition of the same aqua ammoniæ. A dark reaction in either instance would indicate metals.

The *ammonia-strength* of impure aqua ammoniæ, or of crude ammoniacal liquors (free of other alkaline hydrates or carbonates), may be estimated by the following simple method:

One hundred cubic centimetres of oxalic acid test-solution (page 56) are poured into a beaker, and reddened with a few drops of litmus-solution. Then the ammoniacal liquor is added from a burette (Fig. 46), until the red color commences to pass into a violet one. The number of cubic centimetres required for neutralization is then read off, and is used as divisor of the number 170; the quotient indicates at once the percentage of ammonia contained in the ammoniacal liquor. For instance, if 40 cubic centimetres of the latter have been used, it contains 4.25 per cent.; if 29 cubic centimetres, it contains 5.86 per cent.; if 17 cubic centimetres, it contains 10 per cent. of ammonia.

TABLE

OF THE QUANTITY BY WEIGHT OF AMMONIA CONTAINED IN 100 PARTS BY WEIGHT OF AQUA AMMONIÆ AT DIFFERENT DENSITIES.

TEMPERATURE 14° C.

Specific Gravity.	Percent. of Ammonia.	Specific Gravity.	Percent. of Ammonia.	Specific Gravity.	Percent. of Ammonia.	Specific Gravity.	Percent. of Ammonia.
0.8907	33.0	0.9116	24.6	0.9373	16.2	0.9677	7.8
0.8911	32.8	0.9122	24.4	0.9380	16.0	0.9685	7.6
0.8916	32.6	0.9127	24.2	0.9386	15.8	0.9693	7.4
0.8920	32.4	0.9133	24.0	0.9393	15.6	0.9701	7.2
0.8925	32.2	0.9139	23.8	0.9400	15.4	0.9709	7.0
0.8929	32.0	0.9145	23.6	0.9407	15.2	0.9717	6.8
0.8934	31.8	0.9150	23.4	0.9414	15.0	0.9725	6.6
0.8938	31.6	0.9156	23.2	0.9420	14.8	0.9733	6.4
0.8943	31.4	0.9162	23.0	0.9427	14.6	0.9741	6.2
0.8948	31.2	0.9168	22.8	0.9434	14.4	0.9749	6.0
0.8953	31.0	0.9174	22.6	0.9441	14.2	0.9757	5.8
0.8957	30.8	0.9180	22.4	0.9449	14.0	0.9765	5.6
0.8962	30.6	0.9185	22.2	0.9456	13.8	0.9773	5.4
0.8967	30.4	0.9191	22.0	0.9463	13.6	0.9781	5.2
0.8971	30.2	0.9197	21.8	0.9470	13.4	0.9790	5.0
0.8976	30.0	0.9203	21.6	0.9477	13.2	0.9799	4.8
0.8981	29.8	0.9209	21.4	0.9484	13.0	0.9807	4.6
0.8986	29.6	0.9215	21.2	0.9491	12.8	0.9815	4.4
0.8991	29.4	0.9221	21.0	0.9498	12.6	0.9823	4.2
0.8996	29.2	0.9227	20.8	0.9505	12.4	0.9831	4.0
0.9001	29.0	0.9233	20.6	0.9512	12.2	0.9839	3.8
0.9006	28.8	0.9239	20.4	0.9520	12.0	0.9847	3.6
0.9011	28.6	0.9245	20.2	0.9527	11.8	0.9855	3.4
0.9016	28.4	0.9251	20.0	0.9534	11.6	0.9863	3.2
0.9021	28.2	0.9257	19.8	0.9542	11.4	0.9873	3.0
0.9026	28.0	0.9264	19.6	0.9549	11.2	0.9882	2.8
0.9031	27.8	0.9271	19.4	0.9556	11.0	0.9890	2.6
0.9036	27.6	0.9277	19.2	0.9563	10.8	0.9899	2.4
0.9041	27.4	0.9283	19.0	0.9571	10.6	0.9907	2.2
0.9047	27.2	0.9289	18.8	0.9578	10.4	0.9915	2.0
0.9052	27.0	0.9296	18.6	0.9586	10.2	0.9924	1.8
0.9057	26.8	0.9302	18.4	0.9593	10.0	0.9932	1.6
0.9063	26.6	0.9308	18.2	0.9601	9.8	0.9941	1.4
0.9068	26.4	0.9314	18.0	0.9608	9.6	0.9950	1.2
0.9073	26.2	0.9321	17.8	0.9616	9.4	0.9959	1.0
0.9078	26.0	0.9327	17.6	0.9623	9.2	0.9967	0.8
0.9083	25.8	0.9333	17.4	0.9631	9.0	0.9975	0.6
0.9089	25.6	0.9340	17.2	0.9639	8.8	0.9983	0.4
0.9094	25.4	0.9347	17.0	0.9647	8.6	0.9991	0.2
0.9100	25.2	0.9353	16.8	0.9654	8.4		
0.9106	25.0	0.9360	16.6	0.9662	8.2		
0.9111	24.8	0.9366	16.4	0.9670	8.0		

With the decrease and increase of temperature, the density of aqua ammoniæ suffers a corresponding increase or decrease, amounting for each degree of the centigrade thermometer in either direction—

For aqua ammoniæ of a specific gravity of 0.9001	to that of 0.9221	to about 0.00055.
" " " " 0.9251	" 0.9414	" 0.0004.
" " " " 0.9520	" 0.9670	" 0.0003.
" " " " 0.9709	" 0.9831	" 0.0002.

For instance: An aqua ammoniæ of 0.9593 spec. grav., at 14° C., containing 10 per cent. of ammonia, will have, at 18° C., a spec. grav. of (0.9593−0.0003×4=) 0.9581, and, at 12° C., a spec. grav. of (0.9593+0.0003+2=) 0.9599.

AQUA AMYGDALARUM AMARARUM.

Bitter-Almond Water.

Bitter-almond water, when prepared from the essential oil of bitter almonds, is a clear, when derived by distillation from bitter almonds, mostly a turbid, colorless liquid, with the odor of oil of bitter almonds, which odor must not disappear after the elimination of the hydrocyanic acid by argentic nitrate. It should be of such strength, that 1,000 parts contain one part of anhydrous hydrocyanic acid, which is associated in the water with benzoic aldehyde.

Ferrous salts render no reaction with bitter-almond water; but a slight blue precipitate will take place when, after the addition of the reagent, first a few drops of liquor potassæ are added, and subsequently a slight excess of diluted hydrochloric acid.

Argentic nitrate and mercurous nitrate produce but a slight turbidity in bitter-almond water; but, when a few drops of aqua ammoniæ have been added previously, and the liquid is then over-saturated with diluted nitric acid, argentic nitrate will yield a white, and mercurous nitrate, upon warming, a dark-gray precipitate.

The quantity of hydrocyanic acid in bitter-almond water may be determined by weighing 1,000 grains of the water into a vial, and adding and agitating it with a solution of 8 grains of argentic nitrate in 60 grains of water and 30 grains of aqu ammoniæ; 2 drachms of diluted nitric acid are then added, and the whole is gently warmed by immersing the flask in hot water. The precipitate of argentic cyanide is collected upon a tared and moist filter, washed, and dried, and should weigh not less than 5 grains, indicating one part by weight of anhydrous hydrocyanic acid in 1,000 parts of the water.

AQUA CHLORI.

AQUA CHLORINII. LIQUOR CHLORI.

Chlorine-Water.

A saturated solution of chlorine gas in distilled water, containing about twice its bulk, and 0.4 per cent. by weight, of the gas. Chlorine-water has the pale, greenish-yellow color, the irritating and suffocating odor, and the chemical properties, of the gas. When cooled to near the freezing-point, it forms yellow crystalline plates of chlorine hydrate, which dissolve again upon increase of the temperature. When heated, it loses the chlorine by evaporation; when exposed to sunlight, it is gradually changed, with the participation of the elements of water, into hydrochloric acid and free oxygen.

Chlorine-water destroys instantaneously the color of dilute indigo-solution, and all vegetable colors. By the strong affinity of chlorine for all the elements except oxygen, nitrogen, and carbon, and for many compound radicals, it is a powerful chemical agent, and, especially by its property of abstracting or displacing hydrogen, bromine, and iodine, from almost all their combinations by equivalent substitution, a most energetic oxidizer.

Examination:

Hydrochloric acid, and consequently decomposition, is indicated when blue litmus-paper, upon being immersed in chlorine-water, is reddened before it is bleached; such decomposition may further be ascertained by filling a one-ounce vial, in which a few globules of mercury have been placed, with the chlorine-water, and agitating the corked bottle until the odor of chlorine has entirely disappeared. The water should now leave blue litmus-paper unchanged, and should form no precipitate with argentic nitrate, although all chlorine-water, however freshly prepared, will yield a slight turbidity with the latter reagent. Loss of the strong odor of the gas, an acid reaction upon litmus, and the formation of a white precipitate with argentic nitrate, would indicate a degree of decomposition which renders the chlorine-water unfit for use.

Mineral salts, as an evidence of the employment of spring-water for the absorption of the gas instead of distilled water, may be ascertained by the fixed residue remaining upon evaporation of a little of the water on platinum-foil, or on a watch-glass; as well as by testing it, after addition of a little aqua ammoniæ, with oxalic acid; a white precipitate would prove the presence of calcium salts, and would be indicative of spring-water.

Estimation of the strength of chlorine-water.

I. Approximate estimation:

1. One hundred parts by weight of chlorine-water are agitated with a solution of 3 parts of crystallized or granulated ferrous sulphate (free from peroxide) in 10 parts of water acidulated with hydrochloric acid. When, now, a few drops of diluted test-solution of potassium permanganate are added, no discharge of its color should take place.

2. One fluidounce of chlorine-water is mixed with a solution of 10 grains of ferrous sulphate (free from peroxide) in 2 drachms of water. This mixture must yield no blue precipitate upon the addition of potassium ferricyanide.

II. Quantitative volumetric estimation:

One ounce by weight of chlorine-water is mixed with a solution of 10 grains of potassium iodide in 2 drachms of water; to this is added from a burette (Fig. 47), a test-solution of sodium hyposulphite, containing 10 weight-parts of the reagent in 100 volume-parts of the solution, until the brown color has just disappeared. The number of volume-units required, divided by seventy, gives as quotient the quantity by weight of the chlorine contained in one ounce of the solution.

For volumetric estimation, with the standard test-solution of sodium hyposulphite, *see* page 64.

AQUA DESTILLATA.

Distilled Water.

Distilled water must not leave a fixed residue upon evaporation. When reduced by evaporation to one-fourth or one-sixth of its volume, and then tested in separate portions with lime-water for *carbonates*, with argentic nitrate and a few drops of nitric acid for *chlorides*, with barium nitrate for *sulphates*, with ammonium oxalate for *calcium*, and with hydrosulphuric acid and by subsequent acidulation with hydrochloric acid for *metals*, it must in no instance yield any reaction.

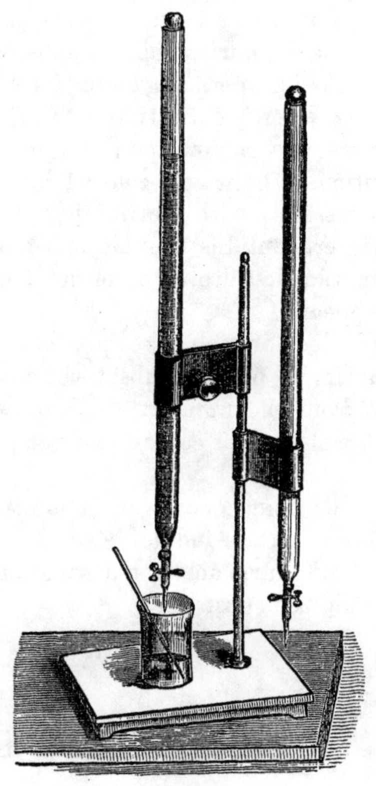

FIG. 47.

Ammonium salts may be detected by a white turbidity, occurring after fifteen to thirty minutes, when one drop of a strong solution of potassium carbonate, and subsequently 3 or 4 drops of solution of mercuric chloride, are added to about one ounce of the water.

Nitrous acid may be detected by mixing in a beaker a little sulphuric-acid mucilage of starch and one drop of a solution of potassium iodide; the mixture must remain colorless; the water under examination is then added, stirring it with a glass rod; if the liquid assumes a bluish tint, traces of nitrous acid are indicated.

Nitric acid may be detected by reducing about 2 ounces of the water by evaporation to about 2 drachms; to these are added, in a test-tube, a few drops of solution of aniline sulphate, and, afterward, about 10 drops concentrated sulphuric acid; the liquid is slowly stirred with a glass rod; if nitric acid be present, rose-colored lines will appear after a while, and the whole liquid also will gradually assume this tint.

Organic substances may be recognized in the water, if it is free from nitrous acid, by warming to near 60° C. about six ounces of it in a beaker placed upon white paper; then a few drops of sulphuric acid, and subsequently a very dilute solution of potassium permanganate (1:1000), are added drop by drop. If the water is entirely free from organic substances, it should assume and retain, upon the addition of the first drop of the solution, a slight rose-colored hue, which increases in intensity progressively with the number of drops added. If the water, however, contains organic substances, the coloration received from the first drop will either not appear at all, or will soon disappear.

If decoloration takes place, an approximate estimate of the quantity of organic substances contained in the water may be had from the number of drops which it is necessary to use before this effect ceases, and the permanent color begins to appear.

This same decoloration is also produced when the water contains certain inorganic substances, as nitrous, sulphurous, or hydrosulphuric acid, ferrous and other sub-salts, and other readily-oxidized substances.

ARGENTI NITRAS.

ARGENTUM NITRICUM.

Nitrate of Silver. Argentic Nitrate.

Anhydrous, colorless, transparent, rhombic plates, or, when fused and cast into moulds, thin, white, transparent, cylindrical sticks (lunar caustic). Permanent in the air, but decomposed by the combined action of organic substances and solar light. Argentic nitrate fuses at 219–220° C.; at about 320° C. it is decomposed; when fused upon charcoal before the blow-pipe, it deflagrates, emitting yellow vapors and sparks, while a reticular coating of metallic silver remains behind.

Argentic nitrate is soluble in an equal weight of cold, and in half that quantity of boiling, water; it is also soluble in alcohol, but only sparingly in ether and chloroform; its strong aqueous solution, therefore, when dropped into alcohol, suffers no precipitation; it is, however, precipitated by a solution of ferrous sulphate acidulated with nitric acid. When the supernatant liquid is decanted from the precipitate and placed upon strong sulphuric acid (Fig. 48), it yields the dark-brown reac-

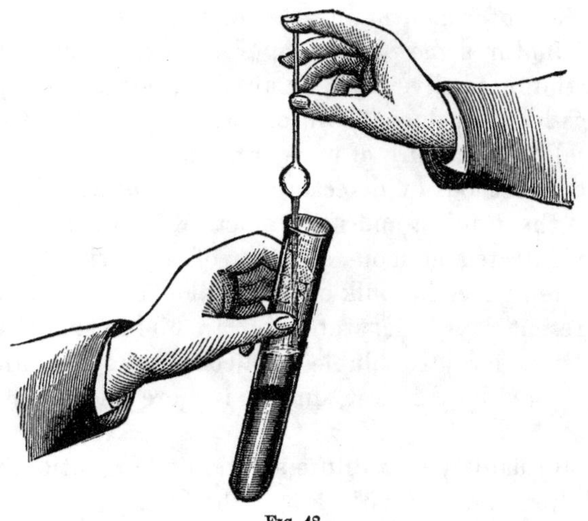

FIG. 48.

tion of the nitrogen oxides on the line of contact between the two fluids.

The aqueous solution of argentic nitrate must be clear; a white turbidity would indicate argentic chloride or nitrite; a bluish color, cupric nitrate; a grayish-black turbidity in the solution of the fused salt, a partial reduction by an excess of heat in the fusion, or cupric chloride or oxide.

The solution gives a white, curdy precipitate with hydrochloric acid and with soluble chlorides, a black one with hydrosulphuric acid, a brown one with the alkaline hydrates, a white one with the carbonates and oxalates, and a yellow one with tri-basic phosphoric acid and its soluble salts.

Argentic nitrate consists, in 100 parts, of 68.23 parts of argentic oxide (containing 63.53 parts of silver), and 31.77 parts of nitric acid; it yields, when completely precipitated by hydrochloric acid or chlorides, 84.39 parts of dry argentic chloride.

Examination:

Potassium nitrate is used to adulterate both the fused and the crystallized argentic nitrate; sodium nitrate cannot well be employed for such adulteration, on account of its hygroscopic character; argentic nitrate fuses with either of these alkaline nitrates in all proportions, and such a mixture in the proportion of one part of argentic nitrate and 2 parts of potassium nitrate is officinal in some pharmacopœias. Such admixture is indicated in the fused silver salt, by an alteration of its appearance, which is less translucent, whiter, and without the distinct radiate crystalline structure of pure argentic nitrate.

Among the methods of detecting such an adulteration, the following are the readiest and most practicable ones:

1. A concentrated aqueous solution of argentic nitrate is dropped into ten times its bulk of strong alcohol; if potassium nitrate be present, it will separate after a while in small, white granules, as it is far less soluble in alcohol. Sodium nitrate cannot be detected by this test, since it is more soluble in alcohol.

2. A small quantity of a dilute solution of argentic nitrate is completely precipitated with diluted hydrochloric acid; the liquid is then warmed, and must, when filtered, leave no fixed residue upon evaporation; such a residue would indicate alkaline nitrates or other impurities.

3. A number of larger and smaller crystals are mixed and broken in a mortar; a small portion of the coarse powder, or fused argentic nitrate, if this is to be tested, is fused and completely reduced on charcoal before the blow-pipe (Fig. 49); a slight reticular metallic coating will remain behind, and some alkaline carbonate, if potassium or sodium nitrates were present; they will be recognized by the alkaline reaction, when moist, red litmus-paper is pressed upon the spot of the coal where the fusion took place.

Argentic chloride is indicated, as stated, by a white turbidity of the solution of argentic nitrate; its identity may be ascertained by its dissolving upon the addition of ammonium hydrate, but remaining insoluble in nitric acid.

Copper and *iron* may be detected by completely precipitating the aqueous solution of the salt with hydrochloric acid, and by subsequent approximate neutralization of the filtrate with aqua ammoniæ; this solution is then tested with a few drops of a solution of potassium ferrocyanide; a red precipitate would indicate copper, a blue one, iron; copper may also be detected, or its presence confirmed, by dissolving about 5 grains of the argentic nitrate in 10 drops of water, and by

FIG. 49.

dropping this solution into a small test-tube containing about half a drachm of aqua ammoniæ; an ensuing blue coloration would indicate copper; a white turbidity, lead or zinc.

Nitrous acid, traces of which are frequently met with in the fused argentic nitrate, causing slight turbidity upon solution, may be detected by completely precipitating the solution with potassium chloride. The filtrate is then tested with a few drops of sulphuric-acid mucilage of starch, to which previously has been added one drop of solution of potassium iodide; if nitrous acid, free or combined, be present, a blue coloration takes place.

ARGENTI OXIDUM.

ARGENTUM OXYDATUM.

Oxide of Silver. Argentic Oxide.

An olive-brown, odorless powder, becoming black when heated, and decomposed at a red heat, leaving behind spongy metallic silver; it is also gradually reduced by the solar light.

Argentic oxide is sparingly soluble in water, but freely dissolved in ammonium hydrate as well as in boiling nitric acid; it is insoluble in the fixed alkaline hydrates. Its aqueous solution has an alkaline reaction and a metallic taste, and is rendered turbid by a small quantity of carbonic acid, but becomes clear again by an excess of it.

The purity of argentic oxide is ascertained by its solubility in aqua ammoniæ, and also in hot nitric acid, without effervescence, and by the fact that this solution, when completely precipitated with hydrochloric acid, gives a filtrate which leaves no residue upon evaporation, and yields no reaction with hydrosulphuric acid and subsequent over-saturation with aqua ammoniæ.

ARSENICI IODIDUM.

ARSENICUM IODATUM.

Iodide of Arsenic. Arsenious Iodide.

An orange-red or purple, crystalline solid, fused and volatilized by heat; it is soluble in $3\frac{1}{2}$ parts of water, and also in glycerin, alcohol, ether, and carbon bisulphide; its aqueous solution is yellow, and gradually and partly decomposes into arsenious and hydriodic acids; it forms a yellow precipitate with hydrosulphuric acid, and emits violet vapors of iodine when heated with nitric acid.

ATROPIA.

ATROPINUM.

Atropia. Atropine.

Colorless, silky, acicular crystals, or a yellowish-white crystalline powder, without odor, and of a bitter and acrid taste. Heated upon platinum-foil, atropia fuses at about 90° C.; at 140° C. it is partly volatilized and decomposed, swelling and emitting white fumes which burn with a white flame, leaving some coal, which is wholly dissipated at a red heat.

Atropia is soluble in about 300 parts of cold, and in a less amount of boiling, water, in 2½ parts of cold alcohol, in about 35 parts of ether, and freely in amylic alcohol and in chloroform. The aqueous solution has an alkaline reaction, and powerfully dilates the pupil of the eye.

Atropia dissolves in concentrated nitric acid, imparting to it a yellowish color, and in concentrated sulphuric acid, without color (sometimes a transient purple tint appears), but the latter solution becomes yellow after some time; it remains, however, unchanged upon addition of nitric acid, but gradually assumes a green color upon addition of potassium bichromate. When some peroxide of manganese is added to the solution of atropia in strong sulphuric acid, the odor of oil of bitter almonds is evolved, and afterward that of benzoic acid.

The aqueous solution of atropia must remain clear upon addition of a few drops of solution of sodium carbonate; a white turbidity would indicate a contamination with *belladonnia.* This is also indicated in atropia by its lesser solubility in water and a greater solubility in ether.

ATROPIÆ SULPHAS.

ATROPINUM SULPHURICUM.

Sulphate of Atropia or Atropine. Atropia Sulphate.

A colorless, slightly-crystalline powder, which fuses when heated upon platinum-foil, assumes a transient red coloration, and is finally wholly dissipated by heat. It gives the same reactions as atropia with concentrated nitric and sulphuric acids, and, in the latter solution, with peroxide of manganese. Atropia sulphate is freely soluble in water and alcohol, but insoluble in ether and chloroform. Its aqueous solution is neutral, gives a white precipitate with solutions of barium salts, and also dilates the pupil of the eye.

A solution of one grain of atropia sulphate in half an ounce of water must remain unchanged when a portion of it is tested with solution of sodium carbonate; a white turbidity would indicate a contamination with *belladonnia;* another portion of it may be heated with a few drops of Fehling's solution; the occurrence of a reddish-brown deposit would indicate the presence of *glucoside.*

BARII CHLORIDUM.

BARYUM CHLORATUM. BARYTA MURIATICA.

Chloride of Barium. Barium Chloride.

Colorless and transparent, flat, four-sided plates, containing two molecules (six per cent.) of water of crystallization; they are permanent in the air, but lose their water at 100° C., leaving the anhydrous salt as a white mass, which, when heated, fuses without decomposition, and imparts a yellowish-green color to the flame.

Barium chloride is soluble in about $2\frac{1}{2}$ parts of cold, and $1\frac{1}{2}$ part of boiling, water, and but sparingly in alcohol; it is less soluble in diluted hydrochloric and nitric acids, and is therefore partly precipitated from its aqueous solution, if not very dilute, upon the addition of concentrated hydrochloric or nitric acids; the salt is, however, redissolved upon dilution with water. The aqueous solution has a bitter, nauseous, saline taste, does not act upon litmus-paper, and yields copious white precipitates with sulphuric acid and sulphates, and with argentic nitrate, insoluble in diluted nitric acid, but the last one soluble in ammonium hydrate; it forms precipitates also with phosphoric acids and the phosphates, and with the alkaline carbonates; they are soluble in hydrochloric and nitric acids.

Examination:

Alumina may be detected, in the dilute aqueous solution, by a white turbidity with aqua ammoniæ; a bluish coloration of the liquid would indicate *copper*.

Metals will be detected by a dark precipitate, or, if only traces of iron are present, by a dark-greenish coloration, upon addition of ammonium sulphydrate to the aqueous solution; if a precipitate be formed, it is collected upon a filter, and dissolved in a few drops of nitric acid, and the solution over-saturated with aqua ammoniæ; a blue coloration would confirm the presence of *copper;* a brown precipitate, that of *iron.*

Calcium, potassium, and *sodium chlorides*, are detected, in the aqueous solution, by completely precipitating the same with diluted sulphuric acid, and by subsequent examination of the filtrate in separate portions; *calcium* is recognized by a white precipitate when one of these portions is slightly over-saturated with aqua ammoniæ, and tested with ammonium oxalate. *Potassium* and *sodium chlorides* will be indicated by a fixed residue upon complete evaporation of another part of the filtrate; they may be distinguished by dissolving the residue in a few drops of water, and testing the solution with potassium antimoniate; a white turbidity would indicate sodium salt.

Strontium chloride is detected by agitating some of the powdered salt with an equal weight of strong alcohol, and by igniting the filtrate; the presence of strontium will be indicated by a red color of the flame, especially apparent toward the end of the combustion.

BISMUTHI ET AMMONII CITRAS.

BISMUTHUM ET AMMONIUM CITRICUM.

Citrate of Bismuth and Ammonium. Bismuth and Ammonium Citrate.

White, glossy, translucent scales, of a slightly acidulous and somewhat metallic taste. When heated upon charcoal, before the blow-pipe, they yield a black fuse, with a yellow coating of the coal; when heated in a dry test-tube, they are charred, with the evolution of moisture and of ammoniacal and empyreumatic vapors; and, at a red heat, a black fuse remains, which, upon cooling, acquires a lemon-yellow color on the surfaces, and which is readily soludle in warm concentrated nitric acid; this solution, when dropped into a quantity of water, produces a white turbidity. Heated with liquor potassæ, it emits the odor of ammonia.

Citrate of bismuth and ammonium is readily soluble in water, sparingly so in alcohol, and insoluble in ether and chloroform. By exposure to the air, it loses its transparency, and becomes gradually more or less insoluble; it becomes, however, soluble again upon addition of a little ammonium hydrate. Its aqueous solution reddens blue litmus-paper slightly, gives white precipitates with dilute hydrochloric acid (soluble in an excess of the acid), with potassium hydrate, and with the alkaline carbonates, the latter precipitate being insoluble in an excess of the precipitants; it is not acted upon by ammonium hydrate; with potassium bichromate, it forms a yellow precipitate, soluble in dilute nitric acid, and, with hydrosulphuric acid, a brownish-black precipitate, insoluble in dilute acids or alkaline hydrates.

BISMUTHI SUBCARBONAS.

BISMUTHUM CARBONICUM.

Carbonate, Subcarbonate, or Oxy-Carbonate of Bismuth. Bismuthous Carbonate.

A white or yellowish-white, odorless, and tasteless powder, which is blackened when in contact with gaseous or aqueous hydrosulphuric acid. When exposed to heat, it gives off moisture and carbonic acid (amounting to 9½ per cent.), while yellow bismuthous oxide remains behind; when heated with exsiccated sodium carbonate, on charcoal before the blow-pipe, it yields metallic globules of bismuth and an incrustation on the coal, which is orange when hot and yellow when cold.

Bismuthous carbonate is insoluble in water, but slightly soluble in water saturated with carbonic acid; it is readily soluble, with effervescence, in acids, forming solutions which, when nearly neutralized by the bismuthous oxide, produce white precipitates of basic salts when poured into a quantity of water.

Examination:

Nitrate may be detected, in a solution of the carbonate in acetic acid, by imparting to it a faint bluish tinge; upon adding one or two drops of neutral indigo-solution, and warming, the blue color will disappear, if any nitrate be present.

Ammonium salts may be detected by the odor of ammonia, and by white fumes from a glass rod moistened with acetic acid, and held in the orifice of the test-tube, when the bismuthous carbonate is heated with liquor potassæ.

The examination for other admixtures or impurities is the same as described with bismuthous nitrate, on pages 160–163.

BISMUTHI SUBNITRAS.

BISMUTHUM SUBNITRICUM. BISMUTHUM ALBUM.

Subnitrate, or Oxy-Nitrate of Bismuth. Basic Bismuthous Nitrate.

A heavy, white powder, in minute crystalline scales; it reddens moistened blue litmus-paper, and becomes black in contact with hydrosulphuric acid. When heated in a dry test-tube, it first emits moisture, and afterward reddish-yellow, acid vapors, leaving a residue which is readily soluble in warm nitric or hydrochloric acid, forming a solution which, when poured into a quantity of water, produces a white precipitate; when heated to redness, a straw-yellow powder of bismuthous oxide (amounting to eighty per cent. of the bismuthous nitrate) remains, which is fusible at a high temperature; when heated with exsiccated sodium carbonate, on charcoal before the blow-pipe, brittle globules of bismuth are obtained, and the charcoal becomes covered with a slight incrustation, which is orange when hot and yellow when cold.

Bismuthous nitrate is nearly insoluble in water; upon a

continued digestion, however, it suffers an alteration in its composition, whereby its solubility in water is increased; it is readily soluble in nitric and hydrochloric acids, and these solutions, when poured into a large amount of water, form white precipitates of basic bismuthous salts. Bismuthous nitrate is but sparingly soluble in potassium and sodium hydrates, and but little more so in ammonium hydrate.

Examination:

Carbonates and *insoluble admixtures* are detected, the former by effervescence in the cold, the latter by remaining behind, when about 10 grains of the bismuthous nitrate are mixed and dissolved, with the aid of heat, in about three fluid-drachms of a mixture consisting of equal parts of concentrated nitric acid and water.

Chlorides and *sulphates* may be detected, in the solution obtained in the preceding test, diluted with an equal bulk of water, and filtered if necessary, by testing it in two separate portions, with a few drops of solution of argentic nitrate for chlorides, and with a few drops of barium nitrate for sulphates.

The test for chlorides and sulphates may also be made by adding about 10 grains of the bismuthous nitrate to about three drachms of a boiling, diluted solution of sodium carbonate; after a few minutes' boiling, the mixed fluid, when nearly cool, is filtered, and the filtrate over-saturated with nitric acid, and then tested, in separate portions, with argentic nitrate for chlorides and with barium nitrate for sulphates.

Salts of Calcium, Magnesium, and Zinc.—About three fluid-drachms of a mixture of equal parts of concentrated hydrochloric acid and water are heated, and so much of the bismuthous nitrate as will be dissolved is added, little by little; about 20 drops of concentrated hydrochloric acid, and subsequently six drachms of water, are then added, and, after a while, the solution is filtered; the filtrate is now repeatedly saturated, and completely precipitated, with hydrosulphuric-acid gas, and again filtered; the filtrate is then warmed, and slightly over-saturated with sodium carbonate; an ensuing white precipitate would indicate salts of calcium, magnesium, or zinc; if it be comparatively considerable, and sufficient in amount to ascertain its nature, it is collected and washed upon a filter, and afterward agitated with aqua ammoniæ; if a residue remains, it is likewise collected and washed upon a filter, and the filtrate tested with hydrosulphuric acid; a white precipitate indicates *zinc* salts; the residue upon the filter is dissolved in a few drops of diluted hydrochloric acid, and the solution neutralized by a few drops of aqua ammoniæ, and then divided into two portions, one of which is tested with ammonium oxalate, for *calcium* salts, the other with sodium phosphate, for *magnesium* salts.

Calcium phosphates may be detected, in the bismuthous nitrate, by adding to a solution of one part of it, in a sufficient quantity of a mixture of equal parts of nitric acid and water, two parts of citric acid dissolved in a little water, and then adding an excess of aqua ammoniæ; the occurrence of a white, bulky precipitate, either at once or after a while, will indicate calcium phosphates.

The presence or absence of earthy admixtures in bismuthous nitrate may also be ascertained, by boiling about 10 grains of it, for 10 minutes, in about two fluid-drachms of acetic acid; the liquid is then filtered, and completely precipitated with hydrosulphuric acid; the filtrate must leave no fixed residue upon evaporation; if any such residue remains, earthy oxides are present.

Metallic Impurities:

Lead.—Two drachms of a mixture consisting of equal volumes of concentrated nitric acid and water are heated, and so much of the bismuthous nitrate is added, in small portions, as will be dissolved; the solution is then decanted, and diluted with four times its volume of water; after about a quarter of an hour, the liquid is filtered, if necessary, and a portion of it tested with one or two drops of concentrated sulphuric acid; if a white precipitate, indicating the presence of lead, occurs, the larger part of the liquid, or the whole of it, is precipitated with sulphuric acid; the precipitate is washed with water by careful decantation, and then agitated for 10 minutes with tepid liquor potassæ; the liquid is then passed through a moist filter, and the filtrate mixed with an equal volume of hydrosulphuric acid; a black precipitate will confirm the presence of lead.

Copper salts are detected, in the diluted nitric-acid solution of the bismuthous nitrate, by a reddish-brown precipitate with potassium ferrocyanide.

Arsenic may surely enough be recognized by its characteristic odor, when a mixture of five grains of the bismuthous nitrate with 10 grains of powdered potassium bitartrate is incinerated, either in an iron spoon or in a small cavity upon charcoal, before the blow-pipe (Fig. 50).

FIG. 50.

As an additional or confirmatory evidence of the absence or presence of arsenic, either of the two following tests may be applied: About 20 grains of the bismuthous nitrate are added to about two fluid-drachms of concentrated sulphuric acid, in a test-tube or in a small flask, and the mixture boiled; the tube or flask is held, as much as practicable, in an inclined position, so as to allow the nitric and nitrous acid vapors to escape; when the evolution of such vapors ceases, two fluid-drachms of concentrated hydrochloric acid are added, and the mixture is allowed to cool, and may serve, in two separate portions, for the two following tests:

1. One portion of the mixture is added to about an equal volume of concentrated hydrochloric acid, and a piece of tin-foil (real tin), or about 20 drops of concentrated solution of stannous chloride, are added, and heat applied; a brown turbidity of the mixture, either at once or after a while, and a grayish-brown precipitate after subsiding, would indicate *arsenic.*

2. The other portion of the sulphuric-acid mixture is carefully added to five times its volume of a mixture of equal parts of hydrochloric acid and water, and the mixture carefully

poured upon granulated zinc, in a long test-tube, only one-tenth of which should be filled by the liquid, taking care that the upper interior walls of the tube do not get wet; a bunch of cotton, moistened with solution of plumbic acetate, is then introduced into the orifice of the tube, and this loosely closed,

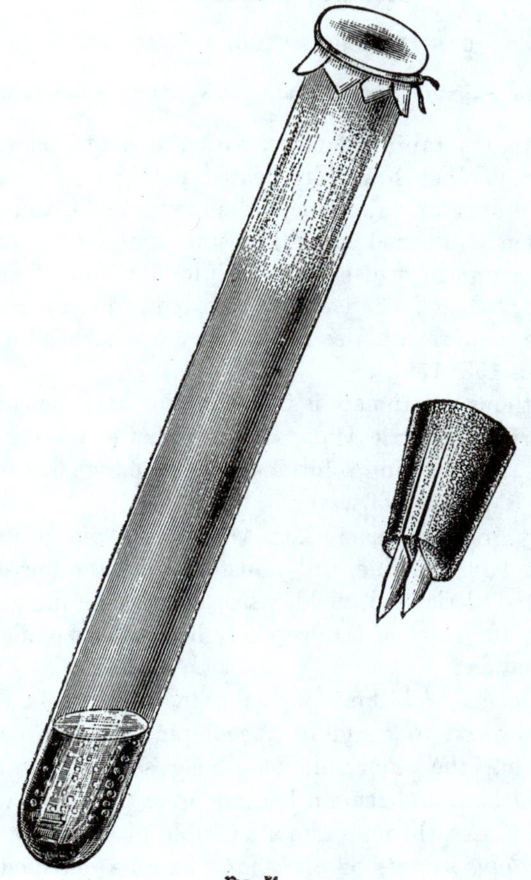

Fig. 51.

either by paper moistened with one drop of solution of argentic nitrate, or by a cork provided with two strips of paper, saturated with solutions of plumbic acetate and of argentic nitrate respectively (Fig. 51). The test is then conducted as described on page 30.

A dark coloration of the argentic-nitrate solution would indicate *arsenic.*

BISMUTHI VALERIANAS.

BISMUTHUM VALERIANICUM.

Basic Valerianate of Bismuth. Bismuthous Valerianate.

A white, crystalline powder, with the strong odor of valerianic acid. When gradually heated in a porcelain crucible, it emits the odor of valerianic and butyric acids, and becomes black; upon continued heating, bismuthous oxide is left behind, appearing brownish-yellow while hot and straw-yellow on cooling; when heated with exsiccated sodium carbonate, it affords the same reaction as the bismuthous carbonate and nitrate (pages 158, 159).

Bismuthous valerianate is insoluble in water, but soluble in hydrochloric and nitric acids, separating an oily layer of valerianic acid, and forming solutions which produce white precipitates with a quantity of water.

Twenty grains of bismuthous valerianate, placed in a small tared porcelain crucible, and moistened with a few drops of concentrated nitric acid, yield, when completely incinerated at a red heat, 16 grains of straw-yellow bismuthous oxide.

Examination:

The solution of bismuthous valerianate in acids, for tests, must be allowed to stand in a cool place for one or several hours, so that the valerianic acid may separate and collect upon the surface, and remain behind upon the paper when the solution is passed through a moist double filter.

Bismuthous nitrate or *carbonates* may be detected by dissolving a few grains of the valerianate in concentrated hydrochloric acid; effervescence indicates carbonates, and, in this case, a solution of 10 grains of the salt, in a sufficient quantity of concentrated hydrochloric acid, is completely precipitated with hydrosulphuric acid, and the filtrate slightly over-saturated with sodium carbonate; a white precipitate would indicate salts of *calcium*, *magnesium*, or *zinc;* they may be discriminated by the same method as described on page 160.

Another part of the solution of the valerianate in hydrochloric acid is diluted with about four times its volume of water, and is faintly tinted with one drop of solution of indigo, and is gently warmed; if decoloration takes place, *nitrate* is indicated.

The examination of bismuthous valerianate, i. e., of its solutions in nitric or hydrochloric acid, for *chlorides* and *sulphates*, for *calcium phosphate* and salts of *calcium*, *magnesium*, and *zinc*, and for *metallic* impurities, is performed in the same way as with the corresponding solutions of bismuthous nitrate, described on pages 160–163.

BROMUM.

BROMINIUM.

Bromine.

A heavy, dark-red, very volatile liquid, of an intense and suffocating odor, somewhat resembling that of chlorine; its spec. grav. is 2.95 to 3.00. Bromine solidifies at —7° C., forming, at temperatures below that, a dark lead-gray, brittle, crystalline mass, of a semi-metallic lustre; from —6° to 63° C., it is liquid, and volatile at common temperature; at 63° it boils, forming yellowish-red vapors.

Bromine is soluble in 32 parts of water, yielding an orange-yellow solution, which has the odor of bromine, bleaches vegetable colors and solution of indigo, and colors starch orange-yellow, and which, when cooled down to 0° C., forms small yellow scales of solid hydrate; the aqueous solution of bromine is deprived of its bromine and of its color when agitated with ether, chloroform, or carbon bisulphide; these ethereal solutions, however, are themselves decolorized when agitated with potassium hydrate; but either of them, with the exception of carbon bisulphide, will form a new solution of the bromine, and consequently regain the color, upon the addition of an excess of any mineral acid.

Bromine is freely soluble in alcohol, chloroform, benzol, carbon bisulphide, and the alkaline hydrates, and is miscible with ether in all proportions. In its chemical relations, it resembles chlorine, having a powerful affinity for hydrogen, though not quite so strong, and hence it acts with energy on many organic compounds, abstracting hydrogen with equivalent substitution.

Examination:

Chlorine may be detected by shaking a little of the bromine with three times its volume of water; the liquid is then poured into a conical glass, and the supernatant aqueous solution decanted into a test-tube, and shaken with an equal volume of ether; the lower or aqueous layer is then transferred, by means of a pipette, to another test-tube, and boiled until the ether, dissolved, is volatilized; it is then, after the addition of a few drops of diluted nitric acid, completely precipitated with argentic nitrate; the precipitate is washed, by repeated decantation, and subsequently dissolved, by the aid of heat, in a solution of ammonium sesqui-carbonate. After cooling, the filtered solution is reduced by evaporation to a small volume, and finally over-saturated with nitric acid, when argentic chloride is separated, if the bromine is contaminated with chlorine.*

This test may be made at once quantitatively, by performing it with a known weight of bromine, and by collecting, drying, and weighing the precipitate of argentic chloride, and calculating the amount of chlorine contained in the quantity of the latter, on the basis of the fact that argentic chloride contains 24.73 per cent. of chlorine.

Iodine is indicated by a purple coloration of carbon bisulphide, when this is shaken with a solution of the bromine in a strong solution of sodium hydrate, which has been over-saturated with fuming nitric acid previous to the addition of the carbon bisulphide.

* This test depends upon the solubility of chloro-bromine in water, and the deportment of ether, which decomposes the chloro-bromine, abstracting the bromine from the aqueous solution, and leaving the chlorine as hydrochloric acid in the water, and finally upon the solubility of argentic chloride in ammonium sesqui-carbonate.

CADMII SULPHAS.

CADMIUM SULPHURICUM.

Sulphate of Cadmium. Cadmium Sulphate.

Colorless, transparent, oblique prisms, containing three molecules of water of crystallization; they are efflorescent in the air, and freely soluble in water, but insoluble in alcohol. The solution has an astringent, acidulous, and slightly austere taste; it yields a white precipitate with aqua ammoniæ, soluble in an excess of the reagent; on subsequent addition of hydrosulphuric acid, a yellow precipitate is formed, which is soluble in strong, warm hydrochloric and nitric acids, but insoluble in diluted acids and in the alkaline hydrates (distinction from arsenious sulphide); it gives a white precipitate with the alkaline carbonates, insoluble in an excess of the precipitant.

The absence of other metallic oxides may be ascertained by completely precipitating the aqueous solution of the cadmium sulphate with hydrosulphuric acid, and by subsequent complete evaporation of the filtrate; if a residue remains, admixtures of other salts are indicated. In such case, the residue should be dissolved in diluted hydrochloric or nitric acid, and examined for metallic, earthy, and alkaline oxides by the systematic method of analysis as described on pages 41–44.

CALCII CARBONAS PRÆCIPITATA.

CALCIUM CARBONICUM PRECIPITATUM. CALCARIA CARBONICA PRECIPITATA.

Precipitated Calcium Carbonate.

A white, light powder, consisting, when magnified, of minute rhombohedral crystals; it does not change moistened turmeric-paper, but turns it brown after having been exposed to a red heat before the blow-pipe.

Calcium carbonate is almost insoluble in water, but somewhat soluble if the water is saturated with carbonic-acid gas; the solution reddens litmus, but changes the yellow color of turmeric-paper to brown; by boiling or exposure to the air, the carbonic acid is evolved, and the calcium carbonate deposited. It is readily soluble, with effervescence, in dilute hydrochloric, nitric, and acetic acids. The acid solution is precipitated by oxalic acid, but not by a solution of calcium sulphate (distinction from barium and strontium carbonates), nor by ammonium hydrate (distinction from aluminium carbonate), nor by potassium hydrate (distinction from magnesium carbonate). Nor must the solution yield any reaction with hydrosulphuric acid, and subsequent over-saturation with aqua ammoniæ (free from carbonate).

Examination:

An insufficient washing in the manufacture, or a fraudulent or accidental admixture of calcium sulphate, may be detected by agitating some of the carbonate with water, and by testing the filtrate, acidulated with one or two drops of nitric acid, in separate portions, with argentic nitrate for chloride, and with barium nitrate for sulphate.

The crude kinds of calcium carbonate—chalk, prepared oyster-shells, and others derived from animal organisms—contain more or less of other bases (magnesium, iron, potassium, sodium, etc.) and acids (phosphoric, silicic, and sulphuric), and always, also, traces of organic substances; they do not render a complete solution in dilute acetic or hydrochloric acid, and the filtered solution usually gives a slight coloration or turbidity with hydrosulphuric acid, and a precipitate when over-saturated with aqua ammoniæ, indicating phosphates.

CALCII HYPOPHOSPHIS.

CALCIUM HYPOPHOSPHOROSUM. CALCIS HYPOPHOSPHIS.

Hypophosphite of Lime. Calcium Hypophosphite.

Small, colorless, transparent, six-sided prisms, or a white crystalline powder, of a pearly lustre; heated in a dry test-tube, the salt deflagrates, emitting inflammable phosphorus vapors, and leaving a residue, which appears, after cooling, slightly reddish, being calcium pyrophosphate, with a little red phosphorus.

Calcium hypophosphite dissolves in six parts of cold water, but is insoluble in alcohol (distinction from sodium hypophosphite); the aqueous solution has a slightly bitter taste, and, when greatly diluted with water, suffers no change upon the addition of diluted sulphuric acid, nor with solutions of barium and calcium chlorides, nor of plumbic acetate (distinction from soluble phosphates and phosphites); it forms, however, white precipitates with the soluble carbonates, with oxalic acid and oxalates, and with argentic nitrate, which latter precipitate soon becomes black.

CALCII PHOSPHAS PRÆCIPITATA.

CALCIS PHOSPHAS. CALCIUM PHOSPHORICUM. CALCARIA PHOSPHORICA.

Precipitated Phosphate of Lime. Tri-basic Calcium Phosphate.

A light, white, inodorous, and tasteless powder, emitting vapor of water when heated in a dry test-tube, but otherwise remaining unaltered; it fuses, without decomposition, at an intense heat. When the powder is moistened with diluted solution of argentic nitrate, it assumes a straw-yellow color.

Calcium phosphate is insoluble in water, but a little soluble in water saturated with carbonic acid; it dissolves, to some extent, in acetic acid, and quite readily in diluted hydrochloric and nitric acids; the latter solution remains clear, when an excess of sodium acetate is added, and gives a copious white precipitate on the subsequent addition of oxalic acid.

Examination:

Carbonates are indicated by effervescence when a little of the calcium phosphate is first thoroughly mixed with a little water, and concentrated nitric acid added afterward.

Sulphates are detected in the diluted nitric-acid solution by a white precipitate with a few drops of barium nitrate.

Iron and *metallic salts* are detected by adding hydrosulphuric acid to the nitric-acid solution, and by subsequent over-saturation with ammonium hydrate. The ensuing precipitate with the latter reagent must be perfectly white; a brown color would indicate iron; a dark one, other metals besides.

CALX CHLORINATA.

CALX CHLORATA. CALCARIA CHLORATA. CALCIUM HYPOCHLOROSUM. CALCARIA HYPOCHLOROSA.

Chlorinated Lime. Bleaching-Powder. Calcium Hypochlorite.

A dull-white powder, with a feeble odor of chlorine, being a mixture mainly of calcium hypochlorite, hydrate, and chloride. Exposed to heat, it gives off oxygen and chlorine, and is finally converted into calcium hydrate or oxide and calcium chlorate and chloride. Mixed with ten or more parts of water, its soluble constituents enter into solution, leaving behind calcium hydrate, and the insoluble impurities of the lime employed in the manufacture of bleaching-powder; the filtered solution is colorless, and of an acrid, nauseous taste, changes red litmus for a moment into blue, and decolorizes it almost at once, and completely; it emits the odor of chlorine with acids, and forms a white precipitate with sulphuric and oxalic acids.

Chlorinated lime, exposed to the carbonic acid and moisture of the air, to acids, or to acid salts, evolves hypochlorous acid, which, when free, readily breaks up into water, chlorine, and chloric acid; the latter is also soon resolved into oxygen, water, chlorine, and perchloric acid; a deliquescent residue, consisting of calcium hydrate, carbonate, and chloride, is left of the chlorinated lime. Upon this decomposition depends the energetic chemical action of chlorinated lime as an oxidizing agent, which, therefore, is proportionate to the percentage of calcium hypochlorite, or, in other words, of the available chlorine. In order to estimate this, and to determine the value of commercial bleaching-powder, several methods of testing are employed, among which the following two are simple and reliable:

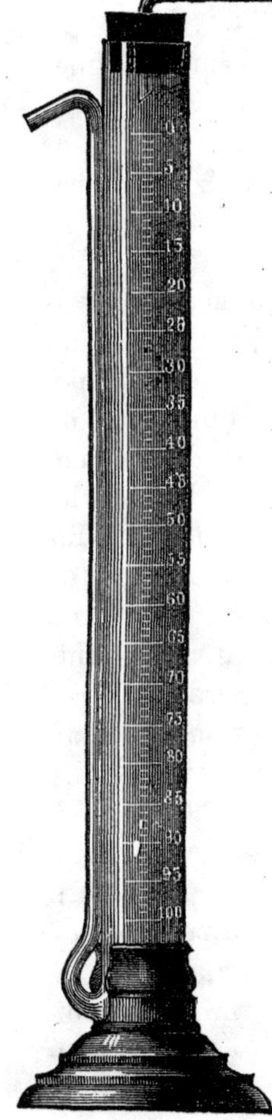

FIG. 52.

1. Seventy-eight grains of ferrous sulphate (free of ferric oxide) are dissolved in about two ounces of water, previously boiled, and the solution is acidulated with a few drops of hydrochloric acid; it will require about 10 grains of chlorine to transform the ferrous oxide, contained in that quantity of the salt, into ferric oxide.

Fifty grains of the chlorinated lime to be examined are next rubbed up with a little tepid water, and the whole transferred into a burette (Fig. 52) divided into 100 degrees, which is then filled up to 0 with water, washing the mortar; after which, the contents are mixed by agitating the burette. Each degree of the mixture, therefore, contains half a grain of chlorinated lime. This liquid is allowed to flow, little by little, into the ferrous solution in a beaker, with constant stirring with a glass rod, until the ferrous oxide is oxidized, which may be recognized upon a porcelain dish, or a plate of glass placed upon white paper, and provided with a number of drops of a dilute, freshly-prepared solution of potassium ferricyanide; the oxidation is accomplished as soon as a part of a drop of the liquid under examination, when brought in contact with one of the drops of the ferricyanide solution on the plate, ceases to cause a deep-blue coloration.

The number of degrees of the solution of chlorinated lime employed is then read off: since 78 grains of ferrous sulphate require for oxidation 10 grains of chlorine, the quantity of the latter in 50 grains of the chlorinated lime may easily be calculated; thus, suppose 56 such degrees have been used for oxidation; then—

Measures.		Grains of Chlorine.		Measures.		Grains of Chlorine.
56	:	10	=	100	:	17.67

Each degree containing only half a grain of chlorinated lime, this, therefore, contains 35.34 per cent. of available chlorine.

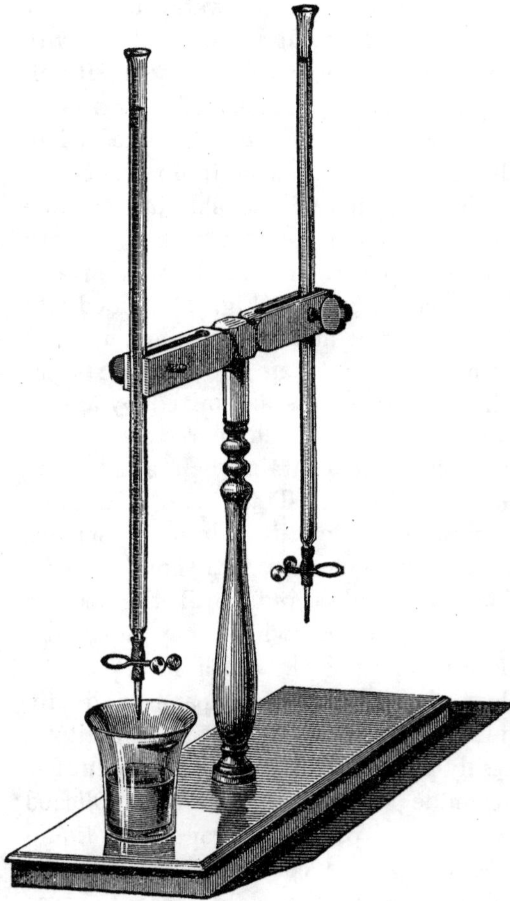

FIG. 53.

2. 1.17 grammes of the chlorinated lime are mixed with about 6 ounces of water, to which 3.5 grammes of potassium iodide, 60 drops of concentrated hydrochloric acid, and a little mucilage of starch, have been added. To this mixture the test-solution of sodium hyposulphite (page 63) is added, from a burette (Fig. 53), until the appearance of a bluish tint of the mixture indicates the termination of the test.

If the sample contains 30 per cent. of obtainable chlorine, 100 cubic centimetres of the test-solution will be required. For instance, as 248 units (in this instance, centigrammes) of sodium hyposulphite indicate the presence of 355 units (centigrammes) of chlorine, 2.48 grammes of the former, the quantity contained in 100 cubic centimetres of the test-solution, indicate 0.355 grammes of chlorine.

If 100 cubic centimetres of the test-solution have been used, therefore, 0.355 grammes of chlorine is obtainable from 1.17 grammes of the chlorinated lime; and if 1.17 yield 0.355 of chlorine, 100 grammes will yield 30.35 grammes.

Samples of the best commercial bleaching-powder contain, on an average, about 35 to 36 per cent. of available chlorine.

CARBONEI SULPHURETUM.

CARBONEUM SULPHURATUM. ALCOHOL SULPHURIS.

Bisulphuret of Carbon. Carbon Bisulphide.

A transparent, colorless, very volatile liquid, of great refractive and dispersive power, of a pungent, somewhat aromatic taste, and a peculiar odor, which, when pure, slightly resembles that of chloroform. Its spec. grav. is 1.27 at 15° C., and it boils at 46.5° C.; it burns with a blue flame, yielding, as the products of combustion, carbonic and sulphurous acids; its vapor, mixed with atmospheric air, forms an explosive gas.

Carbon bisulphide is not miscible with water, and sinks in it; when agitated with iodine-water, it absorbs the minute quantity of iodine dissolved in the water, and acquires a faint-purple color.

Carbon bisulphide is remarkable and important on account of its extensive solvent powers; it is miscible, in all proportions, with absolute alcohol (the solubility decreasing with the decrease of strength of the alcohol), with ether, chloroform, benzol, essential and fatty oils; it dissolves readily and freely, among other substances, sulphur, phosphorus, bromine, iodine, iodoform, camphor, caoutchouc, gutta-percha, resins, wax, paraffin, stearin, chloral hydrate, and those alkaloids which are soluble in ether and alcohol.

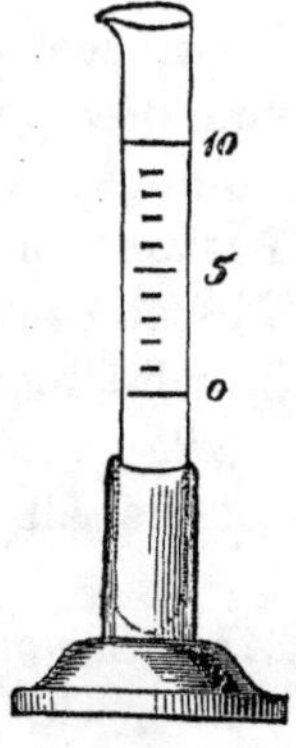

FIG. 54.

Examination:

The odor of carbon bisulphide must not be repulsive, nor

fetid; it should not cause a dark turbidity or precipitate in a solution of plumbic acetate, when agitated with it, nor change the color of moist litmus-paper.

An admixture of ethyl or methyl alcohol may readily be detected by the lesser density of the liquid, by its impaired property of dissolving fatty oils, and by its losing in volume when shaken, in a graduated measure, with an equal volume of water or glycerin (Fig. 54).

CERII OXALAS.

CERIUM OXALATUM.

Oxalate of Cerium. Cerium Oxalate.

A white, granular powder, without odor or taste, insoluble in water, glycerin, alcohol, ether, or chloroform, but soluble in sulphuric and hydrochloric acids. Exposed to heat, the salt is decomposed, and leaves, at a red heat, a reddish-yellow or brown residue of oxide, which is soluble without effervescence in boiling hydrochloric acid; this solution gives, with a saturated solution of potassium sulphate, a crystalline precipitate of cereo-potassium sulphate. Cerium oxalate dissolves in boiling liquor potassæ, which solution, when filtered and neutralized with acetic acid, gives, with solution of calcium chloride, a white precipitate of calcium oxalate, insoluble in acetic acid, but soluble in hydrochloric acid.

Examination:

Earthy carbonates are indicated by effervescence of the salt with hydrochloric acid.

Earthy oxalates are indicated by effervescence of the residue of the salt when reduced, at a red heat, with hydrochloric acid.

Aluminium salts may be detected by boiling the oxalate of cerium with a strong solution of potassium hydrate, filtering, and adding an excess of solution of ammonium chloride, when a white, flocculent precipitate of aluminium hydrate will be formed, if such be present.

CHLORALI HYDRAS.

CHLORALUM HYDRATUM CRYSTALLISATUM.

Hydrate of Chloral. Chloral Hydrate.

Colorless, semi-transparent, crystalline plates, or crystals, of a peculiar ethereal odor and pungent taste. Exposed in a dry test-tube to a gentle heat, by dipping the tube into hot water, chloral hydrate fuses at about 58° C., and solidifies again when cooled down to 15° C.; at about 95° C., it boils, and is partly resolved into chloral and water, which, however, combine again, and form a crystalline deposit in the cooler parts of the tube; at a higher temperature, it is wholly volatilized without combustion.

Chloral hydrate is soluble in about half its weight of cold water, and freely in both alcohol and ether, but only sparingly soluble in *cold* chloroform, in carbon bisulphide, or in oil of turpentine. Its aqueous solution is neutral, and gives no reaction, when slightly acidulated with diluted nitric acid, with diluted solution of argentic nitrate, nor upon subsequent addition of aqua ammoniæ; but, upon heating this mixture, decomposition takes place with effervescence, and with the formation of argentic chloride and metallic silver, the latter coating the walls of the tube. When the aqueous solution is acidulated with diluted sulphuric acid, and faintly tinged with a few drops of solution of potassium permanganate, no decoloration should take place within a few hours.

Concentrated sulphuric, nitric, and hydrochloric acids dissolve chloral hydrate with decomposition, but without color, and without the evolution of colored vapors. Solutions of the alkaline hydrates decompose it, when heated, into soluble formiates and chloroform. Ammonium sulphydrate dissolves chloral hydrate, with the evolution of heat, forming a turbid, reddish-brown liquid; the same reagent produces, in concentrated as well as in diluted solutions of chloral hydrate, a yellow coloration, which becomes dark brown, forming, with the separation of sulphur, a reddish-brown compound, gradually when cold, immediately upon warming.

Examination:

Decomposition of chloral hydrate is indicated by the issue of vapors, and by a pungent odor upon opening the vial, by the reddening of moistened blue litmus-paper, when immersed in it, and by the formation of white fumes when a glass rod, moistened with acetic acid, is held over the mouth of the vessel. It is further indicated, in the aqueous solution of chloral hydrate, acidulated with a few drops of diluted nitric acid, by a white precipitate with argentic nitrate, and in another portion, acidulated with sulphuric acid, by decoloration of solution of potassium permanganate.

Chloral alcoholate is distinguished from the hydrate by the property of being readily and freely soluble in *cold* chloroform, in carbon bisulphide, or in oil of turpentine, but less soluble in *cold* water than is the hydrate, and by its yielding a reddish-brown or brown solution with warm concentrated sulphuric acid, and by the evolution of red nitrous vapors with concentrated nitric acid.

An admixture of the alcoholate with the hydrate may be detected by dissolving about half a drachm of the sample in two drachms of water; 10 drops of liquor potassæ are added, and the liquid warmed to about 50° C., by dipping the tube into water of that temperature; then, one or more drops of solution of iodinized potassium iodide are added, so as to impart to the liquid, after gentle agitation, a yellowish-brown color; it is subsequently carefully decolorized by a few drops of liquor potassæ; an ensuing slight turbidity will disappear upon gentle agitation, if the chloral hydrate be pure; but it will remain, and small yellow crystals, of iodoform (Fig. 55), although somewhat soluble in solutions of chloral hydrate, will subside, and be recognized under the microscope, if chloral alcoholate be present.

FIG. 55.

The following method of approximate estimation of the purity of chloral hydrate depends upon the volumetric determination of the quantity of chloroform produced by the decomposition of a known quantity of chloral hydrate:

Five hundred grains of the sample are dissolved in about one ounce of water, and the solution poured into a graduated cylinder, divided into 1,000 grain-measures (Fig. 56); the vessel wherein the solution is effected is rinsed with so much water as to make up the solution to 700 grain-measures; 300 grain-measures of aqua ammoniæ, of 0.891 spec. grav., are then added, so that the whole measures 1,000 grains. The cylinder is then closed, agitated, and immersed in tepid water, of about 50° C., for half an hour, the stopper being tightly fitted as soon as the expansion of the liquid, caused by the warmth, allows it; it is then set aside for 12 hours, when the fluid will have separated into two layers, a lower one of chloroform and an upper, more or less red-colored, ammoniacal aqueous solution of ammonium formiate.

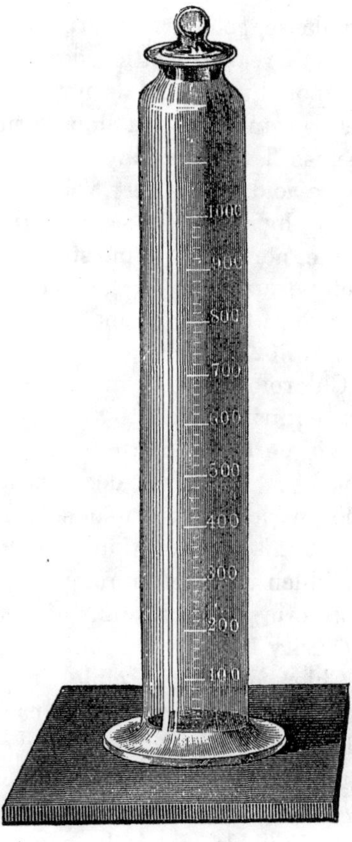

Fig. 56.

If the sample was pure hydrate, the chloroform should measure not less than 235 grains, equal to 351.7 grains by weight, and 70.3 per cent. of the chloral hydrate employed; if it was chloral alcoholate, the chloroform will measure about 200 grains, equal to 299 grains by weight, and to 59.8 per cent. of the chloral alcoholate.

An admixture of the alcoholate with the chlorate will therefore be indicated, proportionately, by the decrease of the quantity of chloroform, ranging, in the above test, between 235 and 200 grain-measures by volume, and 351.7 and 299 grains by weight.

CHLOROFORMUM.

CHLOROFORMIUM.

Chloroform.

A dense, colorless, volatile, and limpid liquid, of an agreeable, ethereal, aromatic odor and sweetish taste; it does not act upon litmus, and is not readily inflammable, but, when a wick is saturated with chloroform, and ignited, it burns with a greenish flame, emitting pungent vapors which contain hydrochloric acid. It is very volatile at common temperatures, producing, by rapid evaporation, great cold, and leaving neither a residue, nor a film of moisture, nor any unpleasant odor, when wholly evaporated by the warmth of the hand, by causing the chloroform to flow to and fro, in a porcelain capsule. It boils at 61° to 62° C.

Chloroform sinks in water, being but slightly soluble, one part requiring about 100 parts of water for solution. The spec. grav. of pure chloroform is about 1.50, at 15.5° C.; in this state of purity, it is liable to decomposition by the combined action of atmospheric oxygen and of solar light; it is, however, protected against this deterioration by a slight percentage of ethylic alcohol, which is therefore retained in the preparation of medicinal chloroform, to the amount of two or three per cent., whereby its density is decreased to a spec. grav. of from 1.496 to 1.480.

Chloroform is miscible, in all proportions, with absolute alcohol, with ether, benzol, carbon bisulphide, and essential and fatty oils, and is an extensive solvent for resins, caoutchouc, gutta-percha, camphor, paraffin, chloral hydrate, etc.; it also dissolves iodine, bromine, and, more or less completely, most vegetable alkaloids, which latter it almost completely withdraws from their aqueous, alkaline solutions.

Chloroform is not miscible with glycerin, and is insoluble in the concentrated mineral acids; when shaken with them, even at an elevated temperature, it undergoes no perceptible change; nor is it acted upon by the alkaline hydrates, iodides, or bromides, nor by argentic nitrate.

Examination of Commercial Impure, and of Purified Chloroform:

As a *preliminary test* for the indication of a partial decomposition of chloroform, a test-tube may be rinsed with aqua ammoniæ, and, subsequently, one or two drops of the chloroform may be allowed to fall to the bottom of the tube; the appearance of white fumes would indicate such decomposition. In another test-tube, equal volumes of the chloroform and of water, the latter slightly blued with neutral litmus-solution, are shaken; a decoloration or a red appearance of the water, after subsiding, would likewise show decomposition.

The result of these tests should also be negative, if the chloroform has been previously exposed, in a white glass bottle, to direct sunlight, for about 10 hours.

Chlorine and Hydrochloric Acid.—Two volumes of chloroform are shaken in a graduated cylinder (Fig. 57) with one volume of water. A perceptible diminution of the volume of the chloroform, after subsiding, would indicate an objectionable percentage of alcohol. The supernatant water must neither appear turbid, nor redden blue litmus-paper, nor render a precipitate when tested with dilute solution of argentic nitrate. An acid reaction upon litmus, and the occurrence of a precipitate with the latter reagent, would indicate *chlorine* and *hydrochloric acid.*

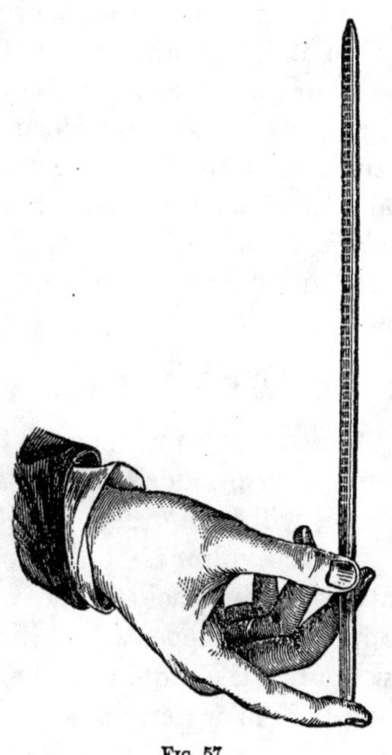

Fig. 57.

Chlorine may also be detected by adding the chloroform, drop by drop, to a solution of potassium iodide (free from iodate) in a test-tube. When agitated, the chloroform, after subsiding, will appear rose-colored, and the aqueous solution yellow, if even traces of chlorine be contained in the chloroform; when this is the case, and the addition of chloroform, in drops, is continued, each drop, falling through the aqueous solution, will assume a slight purplish tint.

Ethene Chloride (Dutch Liquid).—A little fused potassium hydrate is dissolved, in a dry test-tube, in some absolute alcohol; when the solution has subsided, the clear part is decanted into a dry test-tube, and a little chloroform added. No reaction will take place in the clear fluid, unless the chloroform contains Dutch liquid, in which case an elevation of temperature will appear perceptible by a small thermometer immersed in the liquid; a slight evolution of gas from the liquid will also occur, and a crystalline precipitate of potassium chloride.

Alcohol.—Since medicinal chloroform, as stated above, contains about two or three per cent. of alcohol, an examination for an admixture of alcohol by one of the following sensitive tests would obviously be a *contradictio in adjecto.* The specific gravity, the volumetric test in the preceding examination, and the property of chloroform to form a perfectly clear and transparent mixture with sweet oil of almonds, which it will not do if it contains more than five or six per cent. of alcohol, afford a sufficient evidence of the quality of chloroform in this respect. A chloroform which has a spec. grav. of less than 1.48, at 15.5° C., and which renders oil of almonds turbid, and causes a perceptible rise of temperature when suddenly shaken, in a dry test-tube, with an equal volume of concentrated sulphuric acid, cannot be considered as being of officinal strength.

Tests for the Detection of Alcohol in Chloroform.

1. Strong sulphuric acid, to which a little potassium bichromate has been added, when shaken with an equal bulk of chloroform, will turn green, if the latter contains alcohol.

2. Two volumes of chloroform and one volume of concentrated sulphuric acid are mixed in a bottle closed with a glass stopper; after repeated agitation, the bottle is set aside for a few hours; the liquid is then carefully diluted with about an equal bulk of water, the supernatant aqueous liquid is decanted into a beaker, and so much of a mixture of pure barium carbonate in water is added, with constant stirring with a glass rod, as completely to neutralize the acid, so that, after gentle warming, the cooled liquid does not change blue litmus-paper; it is then passed through a moist filter, and the filtrate tested with diluted sulphuric acid. If the chloroform contained traces of alcohol, this would have given rise to the formation of ethyl-

sulphuric acid (sulphovinic acid), and subsequently to soluble barium ethyl-sulphate, contained in the filtered solution, and which is precipitated by sulphuric acid as barium sulphate. Consequently, the occurrence of a white precipitate will be evidence of the presence of alcohol.

FIG. 58.

3. A mixture of two volumes of the chloroform with five volumes of water is warmed, in a test-tube, to about from 30° to 40° C.; after violent agitation for a few minutes, the liquid is passed through a moist filter, and to the filtrate is added a little solution of iodinized potassium iodide; liquor potassæ is then gradually added, until the color of the liquid disappears. After 12 hours' standing in a conical glass, a crystalline deposit of iodoform (Fig. 58) will have taken place, if alcohol be present; the crystals may be recognized under the microscope, when the deposit is carefully removed, by means of a small pipette (Fig. 59), from the lowest point of the conical glass, and transferred to a glass slip.

FIG. 59.

CINCHONIÆ SULPHAS.

CINCHONINUM SEU CINCHONIUM SULPHURICUM.

Sulphate of Cinchonia or Cinchonine. Cinchonia Sulphate.

Colorless, shining, oblique prisms, with dihedral summits, permanent in the air; when gently heated, in a dry test-tube, two molecules of water of crystallization are expelled, and, upon a stronger heat, the salt fuses to a red, resinoid mass, becomes charred, and is finally completely dissipated.

Cinchonia sulphate dissolves in about 56 parts of cold, and in much less boiling water, in seven parts of alcohol and in 35 parts of chloroform, and scarcely at all in ether (distinction from quinia sulphate), but it is readily soluble in dilute acids. Its aqueous solution has a very bitter taste, and exhibits no fluorescence; it is precipitated by solutions of tannic acid, potassio-mercuric iodide, potassium bicarbonate (distinction from morphia and strychnia), of barium and calcium chlorides, and of ammonium hydrate; the precipitate with the latter reagent is not dissolved by ether, when added and agitated with it; nor does aqua ammoniæ cause a green reaction, when added to the solution of cinchonia sulphate in chlorine-water (distinction from quinia).

When concentrated sulphuric acid is poured upon cinchonia sulphate, in a dry test-tube, it dissolves it without color (distinction from salicin and brucia), and the subsequent addition of one drop of a solution of potassium bichromate produces a green coloration (distinction from strychnia). It is also dissolved without color by strong nitric acid (distinction from morphia).

CODEIA.

CODEINUM.

Codeia. Codeine.

Colorless, transparent, rectangular octahedrons, which fuse at a temperature below the boiling-point of water, and, when heated upon platinum-foil, burn away without a fixed residue.

Codeia is soluble in 80 parts of cold, and in 17 parts of boiling, water, and freely in acidulated water and diluted acids, and in alcohol, ether, and chloroform. When heated with less water than required for solution, or when dropped into boiling water, it sinks to the bottom, and melts. The aqueous solution of codeia has an alkaline reaction and a very bitter taste, remains unchanged upon the addition of aqua ammoniæ or liquor potassæ, and forms precipitates with tannic acid, with iodinized potassium iodide, and with potassio-mercuric iodide.

When a small portion of a cold, saturated, aqueous solution of codeia, in a test-tube, is saturated with one or two drops of diluted sulphuric acid, and warmed, it becomes turbid upon the addition of one or a few drops of liquor potassæ, but is rendered transparent again upon dilution with water. Solution of codeia gives no blue reaction with ferric salts, nor does it reduce iodic acid upon addition of an iodate, nor does it suffer any coloration upon the gradual addition of concentrated sulphuric acid, and subsequent addition of a trace of potassium bichromate.

Concentrated sulphuric or nitric acid dissolves codeia without coloration, and the solution in the former acid acquires a bluish tint upon the addition of one drop, or part of a drop, of a solution of a ferric salt.

These characteristics of codeia are sufficient to ascertain its identity and purity. Fraudulent admixtures, like sugar-candy and crystallized salts, are at once indicated by their ready solubility in cold water, and by their insolubility in alcohol and ether, and the latter also by a fixed residue upon incineration on platinum-foil.

COFFEIA.

COFFEINUM.

Coffeia. Coffeine. Theine. Guaranine.

White, slender, silky needles, which contain one molecule (8.4 per cent.) of water of crystallization, and which fuse at a gentle heat, and sublime in feathery needles. When added to concentrated sulphuric acid, in a dry test-tube, coffeia dissolves, forming a colorless solution, which turns yellow upon the addition of one drop of a solution of potassium bichromate. Concentrated nitric acid oxidizes coffeia rapidly, forming compounds which assume a purple color when moistened with, or exposed to the vapors of, ammonia. When coffeia is added to chlorine-water, in a small porcelain capsule, and evaporated to dryness, a purple residue is left, which becomes yellow upon further heating, and red again, when moistened with, or exposed to the vapors of, ammonia.

Coffeia dissolves in about 100 parts of cold, and 10 parts or less of boiling, water; the warm saturated solution separates, on cooling, most of the alkaloid; addition of dilute acids increases its solubility in water. It is freely soluble in chloroform, in 160 parts of absolute alcohol, and in about 300 parts of ether. Its aqueous solution has a bitter taste, and remains unaltered, and does not assume a purple color when it is exposed to the air, after the addition of a little aqua ammoniæ (distinction from phlorizin).

COLCHICIA.

COLCHICINUM.

Colchicia. Colchicine.

An amorphous, yellowish-white powder; when heated upon platinum-foil, it melts and burns away, with intumescence. Added to concentrated sulphuric acid, in a dry test-tube, it agglomerates, and, upon agitation, dissolves, with an orange-yellowish color, which soon turns yellowish-brown, but not violet (distinction from veratria); strong nitric acid, or sulphuric acid to which has been added a little nitric acid, colors it deep violet, which color passes into indigo-blue, and quickly becomes, next, olive-green, and then yellow.

Colchicia is freely soluble in water (additional distinction from veratria), alcohol, and chloroform, and less so in ether. Its aqueous solution has a very bitter but not an acrid taste, becomes turbid with chlorine-water, and assumes, on subsequent addition of aqua ammoniæ, a yellowish-red color.

CONIA.

CONIINUM.

Conia. Conine.

A colorless, transparent, oily-looking, volatile fluid, becoming brown and darker upon exposure to warmth and air; it has a strong, penetrating odor, resembling that of a combination of the odors of tobacco and mice; its taste is acrid, somewhat like that of oil of tobacco. When dropped upon paper, conia produces, like an essential oil, only a transient stain, which by a gentle warmth entirely disappears. It burns with a bright, smoky flame.

The spec. grav. of conia is from 0.88 to 0.89; when dropped upon water, it floats (distinction from nicotia).

Conia combines with one-fourth of its weight of water, forming a hydrate; it is but sparingly soluble in water, one part requiring about 100 parts of cold water; the solution has an alkaline reaction, becomes turbid when warmed, turns brown when exposed to the air, and gradually deposits a brown resinous mass; it forms white precipitates with tannic acid and with potassium mercuric iodide.

Conia dissolves readily in water acidulated with hydrochloric acid, and is miscible with alcohol, ether, chloroform, carbon bisulphide, and fixed and volatile oils.

Conia produces white fumes with the vapors of nitric, hydrochloric, and acetic acids; when dropped into strong sulphuric acid, it dissolves without color; upon addition of one drop of a solution of potassium bichromate, a reddish-brown color ensues, which, however, quickly turns to an olive-green. Concentrated nitric acid turns conia deep purple, and an excess of the acid produces a vehement reaction, with the evolution of nitrous acid. The alkaline hydrates do not act upon conia.

Examination:

The admixture of *volatile* or *fixed oils*, or of *ammonium hydrate* (which may also have resulted from gradual decomposition), may be detected by mixing one drop of conia with 10 drops of water, upon a watch-glass or in a test-tube, and by subsequent addition of one drop of strong hydrochloric acid;

the conia should readily and wholly dissolve to a clear homogeneous solution; any turbidity or oily appearance would indicate such an admixture. If, now, about half a drop of solution of platinum chloride is added, a yellow precipitate will take place if ammonium hydrate be present.

CREASOTUM.

CREOSOTUM. KREOSOTUM.

Creasote. Wood-Tar Creasote.

A distinction has to be made with commercial creasote between the creasote obtained from wood-tar and that derived from coal-tar; the latter is principally a mixture of impure phenol, cresol, and similar homologous alcohols, or only an impure carbolic acid, and exhibits the properties and reactions of carbolic acid (*see* page 78).

Wood-tar creasote is a colorless, somewhat oily liquid, of a peculiar, persistent odor, resembling that of smoked meat, and of a caustic, pungent taste; it does not crystallize, nor become solid, when its temperature is lowered down to —27° C. (distinction from coal-tar creasote); it boils at about 200° C.

Wood-tar creasote is heavier than water, and less readily dissolved therein than carbolic acid, requiring about 80 parts of cold, and 24 parts of boiling, water for solution; its solubility, specific gravity, and boiling-point, however, all vary, since creasote is not an homogeneous substance. It is miscible with alcohol, ether, chloroform, benzol, and carbon bisulphide, with anhydrous glycerin, and with fatty and essential oils; it dissolves resins, camphor, and fats, and is soluble in acetic acid; it coagulates albumen, but not collodion, mixes with concentrated sulphuric and nitric acids with decomposition, and with liquor potassæ without decomposition, but is not miscible with aqua ammoniæ.

A slip of pine-wood immersed in wood-tar creasote, and afterward in hydrochloric acid, acquires a greenish color, on exposure for a short time to the air. Creasote is a powerful refractor of light; when exposed to the air and light, it does not decompose, but absorbs moisture, and becomes in time yellowish or reddish. It is combustible, and burns with a sooty flame.

Examination:

The admixture of *volatile* or *fixed oils* and *oily impurities* may be detected when about one drachm of the creasote is agitated with three drachms of strong acetic acid; a clear solution must take place; a remaining oily layer or oily appearance would indicate such admixtures.

The *distinctions between wood-tar creasote and coal-tar creasote, or wood-tar creasote with an admixture of the latter, are* the following:

Wood-tar creasote remains liquid when cooled in a mixture of broken ice and common salt; coal-tar creasote and carbolic acid either solidify or deposit crystals at such temperatures.

Wood-tar creasote, when mixed and shaken with collodion, produces a clear liquid; carbolic acid and coal-tar creasote form a kind of jelly.

The aqueous solution of wood-tar creasote forms, with a few drops of neutral solution of ferric chloride or sulphate, a yellowish or yellowish-green turbidity; the solution of coal-tar creasote gives, with the same reagent, a blue or bluish-violet coloration. The same reagent, with an alcoholic solution of wood-tar creasote, gives a green coloration, and with that of coal-tar creasote, a brown one.

Wood-tar creasote is not miscible with aqua ammoniæ fortior, while coal-tar creasote and carbolic acid are.

Wood-tar creasote may also be distinguished from carbolic acid, or from coal-tar creasote containing it, by the following comparative tests:

Nine drops of each sample of the creasote or carbolic acid to be tested are placed separately in as many small test-tubes as there are samples; to each tube is then added one drop of liquor ferric chloride, of a spec. grav. of 1.35; creasote will yield a colorless, carbolic acid a yellowish, mixture; five drops of strong alcohol are then added; creasote forms a green solution, carbolic acid a brown one; and when, finally, 60 drops of water are added, the creasote solution becomes turbid, and of a dingy-brownish color, and separates drops of creasote; while that of carbolic acid turns a beautiful blue and remains transparent, or if, at first, a few oily drops are separated, they will redissolve on agitation.

CUPRI ACETAS.

CUPRUM ACETICUM.

Acetate of Copper. Cupric Acetate.

Transparent, dark-green, rhomboidal prisms, slightly efflorescent on the surface, and emitting the odor of acetic acid when triturated or warmed with strong nitric or sulphuric acid.

Cupric acetate dissolves in 14 parts of cold, and in 5 parts of boiling, water, and is also soluble in alcohol, especially when acidulated with acetic acid; its solutions have a bluish-green color, a nauseous, styptic taste, and assume, when much diluted with water, an azure-blue color upon addition of aqua ammoniæ or solution of ammonium carbonate, remaining clear upon the subsequent addition of liquor potassæ.

The absence of iron and other impurities may be sufficiently ascertained by heating and precipitating the aqueous solution of cupric acetate with an excess of liquor sodæ; when cool, the liquid is filtered; the filtrate must render no reaction with hydrosulphuric acid. On the other hand, the aqueous solution of the salt, when acidulated with hydrochloric acid, and completely precipitated with hydrosulphuric acid, must yield a filtrate which leaves no residue upon evaporation upon platinum-foil, and does not become turbid with sodium carbonate.

Cupri Subacetas.—*Aerugo* or *verdigris* is a mixture of several basic cupric acetates with various impurities; it occurs in masses of a pale-green or bluish color, and burns away when heated upon charcoal before the blow-pipe, leaving behind metallic copper. When heated in a test-tube, with concentrated sulphuric acid, it emits acetic-acid vapors. Water resolves verdigris into a soluble acetate and an insoluble basic acetate; it is soluble in diluted acetic, hydrochloric, nitric, and sulphuric acids, and in an excess of solutions of ammonium hydrate or sesqui-carbonate; the insoluble remainder consists mainly of impurities, among which calcium carbonate (crude chalk) is recognized by effervescence of the verdigris with acids. For further examination, part of the solution in acid is completely precipitated with hydrosulphuric acid; the obtained filtrate must leave no residue upon evaporation, nor afford a precipitate when over-saturated with sodium carbonate; a residue and a brown precipitate would indicate iron, and also earthy admixtures.

CUPRI OXIDUM.

CUPRUM OXYDATUM.

Black Oxide of Copper. Cupric Oxide.

A dense, black powder, when prepared by ignition of cupric nitrate, or a less dense, bluish-black, soft powder, when obtained by precipitation. It remains unaltered when heated to redness, is insoluble in water and alcohol, but slightly soluble in saliva and in the gastric juice, and readily soluble in acids; its solutions have a blue or greenish-blue color, and they assume, when so much diluted as to appear almost colorless, an azure-blue color upon addition of aqua ammoniæ, and a red color with potassium ferrocyanide.

Examination:

Cupric nitrate and *nitrite* are recognized, in the oxide, by the evolution of acid nitrous vapors, when heated, in a test-tube, either dry, or with concentrated sulphuric acid; if they are not distinctly recognized by the odor, they are by their action upon moist blue litmus-paper, when held in the orifice of the tube.

Metallic Impurities.—About 20 grains of the cupric oxide are dissolved in three drachms of warm concentrated hydrochloric acid. A small portion of this solution is warmed in a test-tube, and completely precipitated with an excess of hydrosulphuric acid; the filtrate is over-saturated with sodium carbonate, and allowed to stand for several hours; an ensuing precipitate would indicate metallic (ferric) or earthy oxides.

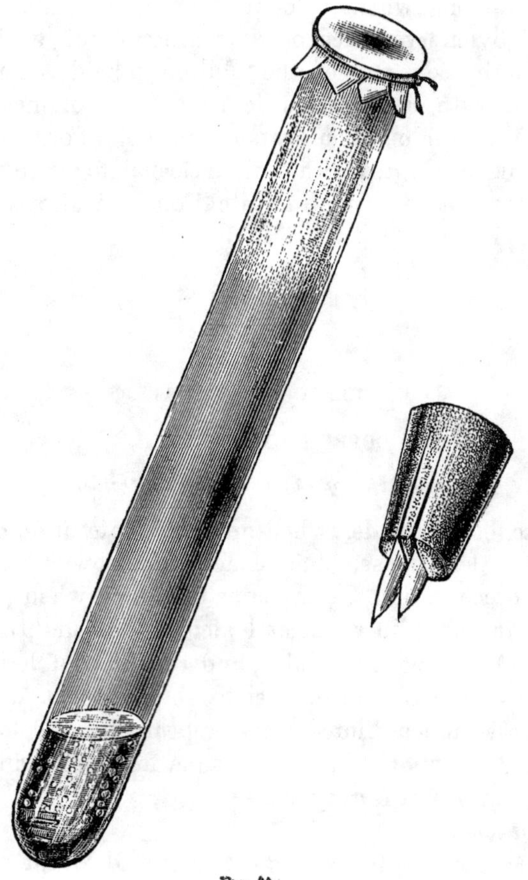

FIG. 60.

The remaining greater portion of the hydrochloric-acid solution may be tested for arsenic by placing it upon granulated zinc, in a large test-tube, which is then closed with a bunch of cotton, moistened with solution of plumbic acetate, and with a paper cover or a cork, provided with a strip of paper moistened with solution of argentic nitrate (Fig. 60). The test is then conducted as described on pages 30 to 32. A dark coloration of the argentic-nitrate solution would indicate *arsenic*.

CUPRI SULPHAS.

CUPRUM SULPHURICUM.

Sulphate of Copper. Cupric Sulphate.

Large, transparent, oblique, rhomboidal prisms of a deep-blue color, containing five molecules (36.08 per cent.) of water of crystallization; they are slightly efflorescent in the air, and lose, when heated, four molecules of water at 100° C., the fifth one being retained until a red heat, when decomposition takes place, with the evolution of sulphurous acid and of oxygen, leaving behind black oxide of copper.

Cupric sulphate is soluble in four parts of cold, and in less than its own weight of boiling, water, but insoluble in alcohol; its solution has an acid reaction, a strong metallic, styptic taste, and, when diluted with so much water as to make it appear almost colorless, reassumes a blue color upon addition of aqua ammoniæ; it gives a white precipitate with barium chloride, a reddish-brown one with potassium ferrocyanide, and a green one with arsenious and arsenic acids, and subsequent neutralization with liquor potassæ or sodæ.

Examination:

Ferrous sulphate is recognized by dissolving the sulphate in diluted aqua ammoniæ, or, in solutions, by adding an excess of aqua ammoniæ; the ferrous hydrate is precipitated along with the cupric hydrate, without being redissolved in an excess of the reagent.

Traces of iron may be detected by mixing an aqueous solution of the salt with twice its volume of chlorine-water, and by the subsequent addition of aqua ammoniæ; the precipitate, formed by the first addition of the ammonium hydrate, will be dissolved by a sufficient addition of the reagent, yielding a complete violet-blue solution. This is then filtered, and, when all the liquid has passed through the filter, the latter is washed with a little dilute aqua ammoniæ; a brown coating, remaining upon the filter, would indicate traces of iron salts.

Magnesium, Zinc, Potassium, and Sodium Sulphates.—About one drachm of the cupric sulphate, taken from a number of triturated crystals, is dissolved in two ounces of hot water, the solution is acidulated with a few drops of diluted hydrochloric acid, and, while warm, completely precipitated with hydrosulphuric-acid gas; after a few hours, it is filtered, and the filtrate evaporated to dryness; if any residue remains, it is dissolved in a little water, and divided into two portions, one of which is tested with ammonium sulphydrate, the other with a few drops of aqua ammoniæ and of solution of sodium phosphate; a white precipitate with the first reagent would indicate *zinc*, with the second one, *magnesium*. If neither of these be present, the residue can only be an *alkaline* sulphate.

Estimation of Commercial Crude Cupric Sulphate:

The following is a simple and ready mode of ascertaining the percentage of cupric sulphate contained in crude blue vitriol, some inferior kinds of which are largely crystallized together with ferrous sulphate: One hundred grains of the salt, taken from a portion of the mixed and triturated crystals, are dissolved, in a small tared beaker-glass, in about one ounce and a half of water; when necessary, the solution is filtered, and the filter washed with one or two drachms of water; the filtrate is returned to the beaker, is acidulated with about one fluid-drachm of concentrated hydrochloric acid, and then a piece of a thin zinc rod, about one inch long, is suspended in the solution by a very thin platinum-wire (Fig. 61); the beaker is then allowed to stand perfectly quiet for 24 hours. After that time, the copper will have precipitated as a bright, spongy mass, around the zinc rod. In order to ascertain if the precipitation has been complete, a few drops of the solution are taken by a glass rod or a pipette, and dropped into a little aqua ammoniæ, or into a little hydro-

Fig. 61.

sulphuric acid acidulated with a few drops of hydrochloric acid; they will produce a blue coloration in the first instance, and a black turbidity with the second reagent, if any copper is left in solution. The copper is then carefully and completely removed from the zinc rod by means of a camel's-hair brush, and, if necessary, the apparatus is allowed, after the addition of a little diluted hydrochloric acid, to stand for 24 hours more; then, when the copper is completely abstracted from the solution, it is brushed down into the liquid, and washed from the zinc by means of a washing-bottle; the zinc is now removed, a little diluted hydrochloric acid added, and the copper allowed to deposit; when this has taken place, the supernatant liquid is carefully removed by decantation, or by means of a pipette, and water is added and removed in the same manner as soon as the copper has subsided; this washing is repeated several times, until the water ceases to redden blue litmus-paper. Then the beaker with the copper is completely dried, and is finally weighed. The weight gives the quantity of metallic copper, and, multiplied by 3.933, the corresponding quantity of crystallized cupric sulphate, contained in 100 grains of the blue vitriol.

CUPRUM AMMONIATUM.

CUPRUM SULPHURICUM AMMONIATUM.

Ammonio-Sulphate of Copper. Ammoniated Cupric Sulphate.

A deep azure-blue, crystalline powder, of an ammoniacal odor; exposed to the air, it loses gradually ammonium hydrate, and assumes a greenish appearance; this same decomposition takes place rapidly upon heating, and leaves behind basic cupric sulphate.

Ammoniated cupric sulphate is soluble in one and a half part of cold water; incomplete solubility indicates partial decomposition. When the solution is diluted largely with water, it becomes turbid, and separates basic cupric sulphate, which, however, is redissolved upon the addition of aqua ammoniæ or of acids.

DIGITALINUM.

Digitalin.

Commercial digitalin varies somewhat in its physical properties, in consequence of different modes of preparation and different grades of purity.

German digitalin, which is mostly used in the United States, forms yellowish-white or yellowish-brown porous scales or powder, inodorous, and of an intensely bitter taste; heated upon platinum-foil, it burns off slowly, with intumescence. It is soluble in water, forming a turbid, neutral liquid, which foams upon agitation; it is also soluble in alcohol, partly so in chloroform, and insoluble in ether. Its aqueous solution, when slightly acidulated with hydrochloric acid, becomes first turbid, and a flocculent white precipitate soon ensues, especially upon gentle warming. When the liquid, after several hours, is filtered off, and over-saturated with sodium carbonate, it turns blue upon the addition of one drop of dilute solution of cupric sulphate, and, when set aside, in a warm place, deposits, after a while, red cuprous oxide.

When about two drops of the aqueous solution of digitalin are mixed, in a test-tube, with four or five drops of strong hydrochloric or sulphuric acid, the liquid remains at first clear, but, when immersed in boiling water, it turns successively yellow, yellowish green, and then yellowish brown, and a precipitate is formed, which, upon addition of water to the liquid, appears white, with a slightly greenish tint; the supernatant liquid shows the same color. After some time, this tint disappears, and the precipitate as well as the liquid becomes colorless. When the same test is performed with a dilute solution of sugar and with hydrochloric acid, a similar reaction takes place, but without the formation of any precipitate.

Digitalin prepared by the process of Homolle and Quevenne forms either a yellowish-white powder or a white, porous, mammillated mass, or small scales, almost insoluble in cold and warm water and in ether, but readily soluble in alcohol and in acids; it is also soluble in chloroform. If not already purified by solution in chloroform, and subsequent evaporation, as it now occurs in commerce, this digitalin, when treated with chloroform, leaves an insoluble residue behind, and the solution yields, upon evaporation, the crystallizable digitalin.* Its solution in hydrochloric acid is of a faint-yellowish color, but soon changes to green; upon dilution with water, it is decolorized, and digitalin separates in a resinous state. Its solution in nitric acid is at first colorless, but becomes yellow and remains so after subsequent dilution with water. Sulphuric acid dissolves it with a green color, disappearing upon dilution with water. Moistened with sulphuric acid and afterward exposed to the vapor of bromine, it assumes a violet color.

EMETIA.

EMETINUM.

Emetine. Emetia.

A whitish or yellowish-white, inodorous powder; heated upon platinum-foil, it melts, and burns off without residue; it is sparingly soluble in cold water and in ether, more soluble in boiling water, and freely in alcohol, chloroform, and diluted acids. The latter solutions have a slightly yellowish color and a bitter taste, and form white precipitates with tannic and gallic acids, with potassio-mercuric iodide, and with solutions of potassium carbonate, bicarbonate, and hydrate, the precipitate with the latter reagent being soluble in an excess of the precipitant. The solution of emetia, in water acidulated with

* Recently a mode of preparing this glucoside has been devised whereby it is obtained from its solution in chloroform in fine, colorless, shining needles, intensely bitter, and, as is claimed, of a far greater physiological action; it gives an intense emerald-green coloration with hydrochloric acid, is almost insoluble in benzol and in pure ether, only sparingly soluble in water, soluble in 12 parts of alcohol, and abundantly soluble in chloroform.

hydrochloric acid, remains colorless upon the addition of diluted solution of ferric chloride (distinction from morphia); nor does it render any reaction with concentrated sulphuric acid, either alone or after the addition of a little potassium bichromate; concentrated nitric acid colors the solution yellowish, and, upon warming, yellowish red.

FERRI ARSENIAS.

FERRUM ARSENICUM.

Arseniate of Iron. Ferrous-Ferric Arseniate.

When freshly prepared, an amorphous white powder, which quickly becomes green or greenish blue; when heated in a dry test-tube, it emits water-vapors, and, afterward, a crystalline sublimate of arsenious acid, leaving behind a dark residue,

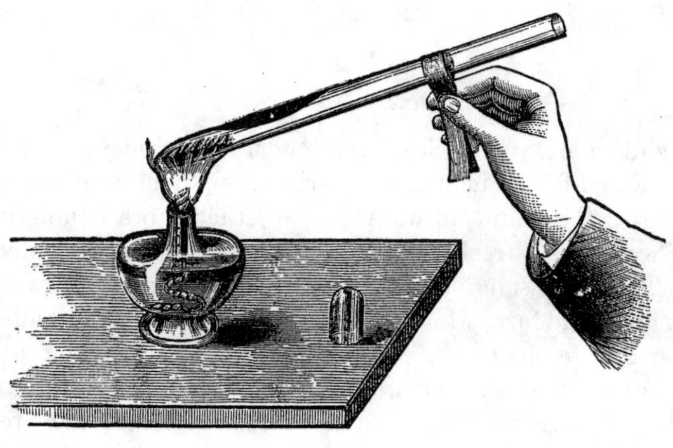

FIG. 62.

which evolves the odor of arsenic when heated upon charcoal before the blow-pipe, and gives rise to the formation of an arsenic-mirror, when heated with a little potassium cyanide, in a narrow tube (Fig. 62).

Arseniate of iron is insoluble in water, but readily soluble in warm hydrochloric acid, forming a yellow solution, which, when highly diluted, gives a blue reaction with both potassium ferrocyanide and ferricyanide; when the solution is mixed with several times its bulk of hydrosulphuric acid, and set aside in a corked vial, in a warm place, first a white turbidity (sulphur) appears, and, after some time, a yellow precipitate. Ferrous-ferric arseniate, when boiled in a solution of sodium carbonate, yields a filtrate which, when exactly neutralized with nitric acid, gives a brick-red precipitate with argentic nitrate, and a white one with ammoniated magnesium sulphate.

Examination:

About one drachm of the powder is shaken with a little tepid water, and the filtrate is tested by evaporation on platinum-foil, as well as with barium chloride; neither should a fixed residue remain, nor a white precipitate be formed, as thereby an insufficient washing of the ferrous-ferric arseniate would be indicated.

FERRI CARBONAS SACCHARATA.

FERRUM CARBONICUM SACCHARATUM.

Saccharated Carbonate of Iron, or Ferrous Carbonate.

A greenish-gray powder, permanent in the air, having a sweet, feebly-chalybeate taste. Heated in a dry test-tube, it is charred, with the evolution of the vapors and odor of burning caramel. When shaken with cold water, this dissolves the sugar, and a little of the ferrous carbonate, which may be precipitated, for the most part, as ferrous hydrate, by boiling the solution; the powder is wholly and readily soluble, with effervescence, in hydrochloric acid, forming a yellow solution which gives, with reagents, the reactions of ferrous salts.

A saccharated carbonate of iron which has a reddish color, and affords no brisk effervescence with acids, should be rejected.

Examination:

Sulphates may be detected by shaking a little of the powder with warm water, in a test-tube, and by testing the filtrate, acidulated with a few drops of diluted nitric acid, with barium nitrate; a white precipitate would indicate sulphate, left from insufficient washing in the preparation.

Copper and Zinc.—The remainder upon the filter of the preceding test is digested in a test-tube, with a little solution of ammonium sesqui-carbonate, for about one hour. The liquid is then filtered; a bluish color of the filtrate would indicate the presence of *copper*, and the formation of a white precipitate, upon the addition of a few drops of ammonium sulphydrate, that of *zinc*.

FERRI CHLORIDUM.

FERRI PERCHLORIDUM. FERRUM SESQUI-CHLORATUM.

Sesqui-Chloride of Iron. Perchloride of Iron. Ferric Chloride.

Orange-yellow, crystalline masses, or, when obtained in the dry way, fragments having a radiate, crystalline structure, and brown color, with a metallic lustre, the former containing 12 molecules (40 per cent.), the latter, 5 molecules (22 per cent.), of water. Ferric chloride is deliquescent; when heated in a retort, it fuses at a gentle heat, giving off water; upon stronger heat, much of the chloride is decomposed, yielding ferric oxide and hydrochloric acid; part of the chloride sublimes, and condenses in the form of small, brilliant, red needles of anhydrous ferric chloride, which is very deliquescent.

Ferric chloride is freely soluble in water, alcohol, and glycerin, and also, but less readily, in ether and chloroform. A strong aqueous solution, of a spec. grav. of 1.355, forms the officinal *Liquor ferri chloridi.* This, as well as the solution of the salt, has an acid and strongly styptic taste, and an acid reaction on test-paper; when diluted with water, they give a blue precipitate with potassium ferrocyanide, a white one with argentic nitrate, and a bulky reddish-brown one with the alkaline hydrates.

Examination:

Ferric chloride must yield a complete and clear solution in water and in alcohol; if a reddish, insoluble residue remains behind, the chloride has undergone partial decomposition.

Ferrous chloride is detected, in the greatly diluted solution, by the formation of a blue precipitate with potassium ferricyanide.

Fixed Impurities, other Metallic Chlorides, and Nitric and Sulphuric Acids.—About 20 grains of the ferric chloride, or one drachm of the officinal liquor ferri chloridi, are dissolved in, or mixed with, about one ounce of water. To this solution, about half an ounce of aqua ammoniæ is added, or so much as completely to precipitate the iron, which may be ascertained by allowing a drop of the liquid to fall upon a glass pane, on which several drops of a solution of potassium ferrocyanide have been placed. The precipitation is accomplished when the blue reaction with this reagent ceases. The liquid is then warmed to near boiling, is subsequently filtered, and evaporated to about one-fourth of its volume. It is then tested, in separate portions, for *fixed impurities*, by evaporation, and heating to redness, upon platinum-foil; for *metallic* impurities (copper and zinc), by mixing a portion of the solution with an equal volume of hydrosulphuric acid; for *nitric acid*, by mixing another part with an equal volume of diluted sulphuric acid, and by testing it, in two portions, with solution of indigo and solution of potassium permanganate; a decoloration of the faintly-colored solutions, upon gently warming, would indicate nitric acid; for *sulphuric acid*, by testing another part of the solution, acidulated with a few drops of hydrochloric acid, with barium chloride.

The presence of nitric acid may be confirmed by mixing a portion of the solution, or of the diluted liquor ferri chloridi, with a strong solution of ferrous sulphate, and by subsequent addition of concentrated sulphuric acid, in single drops; they will cause a dark coloration, if oxides of nitrogen be present.

FERRI CITRAS.

FERRUM CITRICUM.

Citrate of Iron. Ferric Citrate.

Thin, transparent scales, of a garnet-red color, permanent in the air; when heated on platinum-foil, they are charred without fusing, and without the evolution of an ammoniacal odor (distinction from ammonio-ferric citrate); when completely incinerated, red ferric oxide is left, which, when cool, should have no alkaline reaction upon moist turmeric or litmus paper (distinction from potassio-ferric tartrate).

Ferric citrate is slowly soluble in cold, and readily in hot, water, and insoluble in alcohol; its aqueous solution has a yellow color, a mild chalybeate taste, and a neutral reaction; it is not precipitated by ammonium hydrate, and causes a deep-blue turbidity when dropped into water to which a few drops of hydrochloric acid and solution of potassium ferrocyanide have been added.

Examination:

Ferric tartrate may be detected by completely precipitating a solution of ferric citrate with liquor potassæ, and by testing the colorless filtrate by slightly over-saturating a portion of it with acetic acid; when the solution is very dilute, it is first reduced by evaporation, and, when cold, tested with a few drops of concentrated alcoholic solution of potassium acetate; a white granular precipitate, taking place at once or after some time, would indicate tartrate. Another portion of the filtrate is precipitated with calcium chloride, and filtered; the filtrate, when heated to boiling, should yield a white, granular precipitate of calcium citrate, which redissolves on cooling, being a confirmatory evidence of the identity of a citrate.

FERRI ET AMMONII CHLORIDUM.

AMMONIUM CHLORATUM FERRATUM. AMMONIUM MURIATICUM MARTIATUM.

Ammonio-Chloride of Iron. Ammonio-Ferric Chloride.

An orange-yellow, crystalline powder, somewhat deliquescent, readily soluble in water or glycerin, and to some extent in alcohol, forming a yellow, transparent solution, which gives a copious rust-brown precipitate with alkaline hydrates, and, when very dilute, a deep-blue one with potassium ferrocyanide, and a white, curdy one with argentic nitrate.

Examination:

One part of the salt should render a complete and transparent solution with five parts of water; a reddish-brown insoluble residue would indicate decomposition of the ferric chloride by exposure to a too strong heat while drying the salt.

Zinc and Copper.—The diluted aqueous solution is tested with potassium ferricyanide for ferrous chloride, and, when found free from it, is completely precipitated with ammonium hydrate; the filtrate is then tested with ammonium sulphydrate, which will cause a white turbidity, if zinc be present, and a brownish-black one, if copper.

FERRI ET AMMONII CITRAS.

FERRUM ET AMMONIUM CITRICUM. FERRUM CITRICUM AMMONIATUM.

Citrate of Iron and Ammonium. Ammonio-Ferric Citrate.

Thin, transparent, garnet-red scales, of a slightly sweetish and astringent taste; they evolve, when heated, water and ammonia, and, when completely incinerated upon platinum-foil, leave behind not less than 27 per cent. of ferric oxide, which should not change the color of moist red litmus-paper (evidence of the absence of potassio-ferric salts). Heated with liquor potassæ, it evolves ammonia (distinction from ferric citrate), and deposits ferric hydrate.

Ammonio-ferric citrate is readily soluble in water, glycerin, and diluted alcohol, but not in strong alcohol or ether; its solution is neutral or has a slightly alkaline reaction, remains unaltered with aqua ammoniæ, and with solution of potassium ferrocyanide, but produces a deep-blue turbidity in very dilute solutions of the latter reagent, upon addition of mineral acids.

Examination:

Ammonio-ferric tartrate may be recognized, as an incidental or fraudulent admixture or substitution, by completely precipitating a not too dilute aqueous solution of the salt with potassium hydrate; the liquid is heated nearly to boiling, and, when cool, filtered; one portion of the colorless filtrate is examined by slight over-saturation with acetic acid, and by subsequent addition of a little alcoholic solution of potassium acetate, and by allowing the liquid to stand for some hours; the formation of a white, crystalline deposit would indicate *tartrate.*

Another portion of the filtrate is precipitated with calcium chloride, filtered, and the filtrate heated to boiling. A white precipitate of calcium citrate, disappearing again on cooling, will bear evidence of the identity of a citrate.

FERRI ET AMMONII SULPHAS.

FERRUM ET AMMONIUM SULPHURICUM. FERRUM SULPHURICUM OXYDATUM AMMONIATUM.

Sulphate of Iron and Ammonium. Iron Alum. Ammonio-Ferric Sulphate.

Pale-violet, octahedral crystals, containing 24 molecules (44.8 per cent.) of water of crystallization; effervescent in the air. Exposed to heat, they undergo aqueous fusion, lose the water of crystallization, swell up, and leave a pale-brown residue. When the crystals are heated with liquor potassæ, the odor of ammonia is strongly evolved.

Ammonio-ferric sulphate is soluble in four parts of cold, and in less than its weight of boiling, water; it is less soluble in glycerin, and insoluble in alcohol, ether, and chloroform. Its aqueous solution has a slightly acid reaction, a sour, astringent taste, and gives a blue precipitate with potassium ferrocyanide, a brown one with the alkaline hydrates, and a white one, insoluble in acids, with barium nitrate.

When the solution of ammonio-ferric sulphate is completely precipitated with liquor potassæ, the filtrate, when over-saturated with hydrochloric acid, should not render a white precipitate of alumina upon the subsequent addition of an excess of ammonium sesqui-carbonate.

FERRI ET AMMONII TARTRAS.

FERRUM ET AMMONIUM TARTARICUM. FERRUM TARTARICUM AMMONIATUM.

Tartrate of Iron and Ammonium. Ammonio-Ferric Tartrate.

Transparent, deep-red scales, of a sweet taste, and of a rust-brown color when reduced to powder; when heated in a test-tube, the salt emits vapors of water and ammonia, and, when exposed to red heat, upon platinum-foil, it is incinerated, leaving behind 29 per cent. of red ferric oxide. Heated with potassium hydrate, it evolves the odor of ammonia.

Ammonio-ferric tartrate is slowly but freely soluble in water and in glycerin, but insoluble in alcohol and ether; its solution is neutral or slightly alkaline, and, when cold, is not precipitated by the alkaline hydrates or carbonates, but is so upon boiling it with either of these reagents. Its solution is not rendered blue by potassium ferrocyanide, unless acidulated. When completely precipitated by potassium hydrate, the filtrate, if not too dilute, gradually yields, after over-saturation with acetic acid, a white, crystalline deposit of potassium bitartrate. Another part of the filtrate may serve for examination for metallic impurities, by acidulating it with hydrochloric acid, and by subsequent addition of hydrosulphuric acid; a dark turbidity would indicate other metals (copper); if required, the nature of the precipitate of the sulphides may be ascertained, and the metals contained therein recognized, by the method described on page 41.

FERRI ET POTASSII TARTRAS.

FERRUM ET POTASSIUM TARTARICUM.

Tartrate of Iron and Potassium. Potassio-Ferric Tartrate.

Transparent, ruby-red scales, of a sweetish and slightly astringent taste; when heated, they emit at first the odor of burning tartaric acid, and leave, upon incineration at a red heat, a residue which, when cold, blues moist red litmus-paper, and effervesces when moistened with a drop of hydrochloric acid.

Potassio-ferric tartrate is freely soluble in water and in glycerin, but scarcely in alcohol; its solution is neutral or slightly alkaline, and gives, at common temperatures, no precipitate with alkaline hydrates or carbonates, nor with potassium ferrocyanide; but, upon boiling, the former, if they are added in sufficient quantity, cause a complete precipitation of the ferric oxide, and the latter gives a blue precipitate upon acidulation of the solution with mineral acids. When the filtered solution, from the precipitation with alkaline hydrate, is slightly oversaturated with acetic acid, it gives, on cooling, if not too dilute, a crystalline deposit of potassium bitartrate.

FERRI ET QUINIÆ CITRAS.

FERRUM ET CHININUM CITRICUM.

Citrate of Iron and Quinia or Quinine. Quinia Ferric Citrate.

Thin, transparent scales, varying in their color from a yellowish-brown, with a tint of green, to a reddish-brown, according to the thickness of the scales. When heated, they burn away with white fumes, and leave, upon incineration, a residue of ferric oxide, which should not change moistened red litmus-paper (evidence of the absence of alkaline citrates).

Quinia ferric citrate is slowly but freely soluble in cold, and readily in hot, water, but insoluble in alcohol and ether; its solution is neutral or slightly acid, and has a bitter, mild, chalybeate taste; it gives a white precipitate of quinia with aqua ammoniæ, at common temperature, and the solution assumes a deeper color; but no ferric oxide is thrown down; when the precipitate is collected upon a filter, washed with a few drops of cold water, and then dissolved in a little chlorine-water, the solution will turn emerald-green upon the addition of a few drops of aqua ammoniæ (evidence of the presence of quinia, and distinction from cinchonia). Solution of quinia ferric citrate gives a brown precipitate with liquor potassæ or sodæ, and with aqua ammoniæ, when heated; a blue one with solution of potassium ferrocyanide, when acidulated with a mineral acid, and a grayish-black one with tannic acid.

Examination:

The absence or admixture of cheaper scaled ferric salts may be ascertained:

1. By the bitter taste, while the other scaled ferric salts have a more or less sweetish taste.

2. By the formation of a white precipitate with aqua ammoniæ, in the cold, while the ferric citrates and tartrates yield, with the same reagent, a brown precipitate, and the alkaline ferric salts yield no precipitate at all at common temperature.

3. By giving no odor of ammonia, nor white fumes with a glass rod, moistened with acetic acid, when heated in a test-tube with liquor potassæ. Any admixture of ammonio-ferric salt would be recognized by this test.

In order approximately to determine the purity of quinia ferric citrate, 50 grains of the sample are dissolved in one ounce of water, and treated with a slight excess of aqua ammoniæ, in the cold; the ensuing white precipitate is, after standing about half an hour, collected upon a tared filter, is washed with a few drops of very cold water, containing a slight addition of aqua ammoniæ, and is dried at a temperature not exceeding 80° C.; it should weigh eight grains, and should be soluble in six fluid-drachms of warm ether, and this solution, when set aside for 24 hours, in a corked test-tube, should not separate any crystalline deposit of the less soluble associated alkaloids of quinia (quinidia, cinchonia, and cinchonidia).

FERRI FERROCYANIDUM.

FERRUM FERROCYANATUM.

Ferrocyanide of Iron. Prussian Blue. Ferri-Ferrocyanide.

A deep-blue, tasteless powder, or hard, brittle, blue masses, showing, on the freshly-fractured surfaces, a beautiful bronzed lustre, which disappears when they are powdered. When heated, it emits colorless, inflammable vapors, and the odor of ammonia and hydrocyanic acid; exposed to a high temperature in a closed vessel, it gives off water, ammonium cyanide, and ammonium carbonate, and carbide of iron is left behind.

Ferri-ferrocyanide is insoluble in water, glycerin, and alcohol, and in diluted acids, with the exception of oxalic acid, which dissolves it, with a deep-blue color. Concentrated sulphuric acid dissolves it without decomposition, forming a paste, from which the ferri-ferrocyanide may be re-precipitated by dilution with water. Concentrated hydrochloric acid decomposes it, with the formation of ferric chloride and the evolution of hydro-ferrocyanic acid. It is also decomposed by concentrated nitric acid. Alkaline hydrates and carbonates decompose it, dissolving alkaline ferrocyanide, and leaving rust-brown ferric hydrate behind.

Commercial Prussian blue is not invariably pure ferri-ferrocyanide, but generally contains aluminium and potassium salts, and frequently some uncombined ferric hydrate. These impurities may be detected by boiling the triturated Prussian blue with dilute hydrochloric acid, and precipitating the filtrate with ammonium hydrate, when aluminium hydrate and ferric hydrate are precipitated, while pure ferri-ferrocyanide, treated in this manner, yields no precipitate. If the precipitate has to be examined for aluminium salts, it is collected upon a filter, washed, and treated with warm liquor potassæ; a few drops of this filtrate, allowed to fall into a solution of ammonium chloride, will indicate aluminium salts by a white precipitate.

Examination:

Mineral Admixtures.—About half a drachm of the ferri-ferrocyanide is heated, in a porcelain crucible, to redness; when cool, the residue is dissolved in concentrated hydrochloric acid, and should render a complete and clear solution, with slight effervescence; a residue would indicate fixed mineral admixtures (calcium or barium sulphates).

Metals.—To the solution obtained in the preceding test, a little potassium chlorate is added, and the solution boiled until the odor of chlorine ceases; it is then diluted, filtered, and the filtrate divided into two portions; these are heated, and the one is precipitated with liquor potassæ, the other with aqua ammoniæ; after a while, they are filtered, and each of the alkaline filtrates is tested with ammonium sulphydrate; a black precipitate, in the potassium-solution, would indicate *lead;* a brownish-black precipitate and a blue coloration, in the ammoniacal liquid, *copper;* a white turbidity, in either of the liquids, shows *zinc* to be present.

Earthy Carbonates.—The ammoniacal liquid of the preceding test, regardless of the result of the test, is filtered, and tested with sodium carbonate; an ensuing white precipitate would indicate carbonates of *lime*, *barium*, or *magnesia*.

FERRI HYPOPHOSPHIS.

FERRUM HYPOPHOSPHOROSUM.

Hypophosphite of Iron. Ferric Hypophosphite.

A yellowish-white, odorless powder; when heated, in a dry test-tube, it gives off inflammable phosphorous vapors, with considerable intumescence, leaving behind ferric pyrophosphate. Since hypophosphorous acid is very prone to absorb oxygen, the salt is readily decomposed by all oxidizing agents. Ferric hypophosphite is insoluble in cold water, but soluble in diluted hydrochloric acid, forming a yellow solution, which, when greatly diluted, gives a blue precipitate with potassium ferrocyanide; it is readily soluble in solutions of ferric sulphate and of sodium hypophosphite.

FERRI IODIDUM.

FERRUM IODATUM.

Ferrous Iodide.

Opaque plates or masses, of an iron-gray color, metallic lustre, and crystalline fracture. When heated in a dry test-tube, ferrous iodide fuses, and emits violet iodine vapors, finally leaving behind ferric oxide.

Ferrous iodide is very deliquescent; it is soluble in its own weight of water, and also in alcohol and glycerin, forming yellowish-green solutions, with a styptic taste; its aqueous solution gives a copious blue precipitate with potassium ferricyanide, and acquires a blue color upon addition of a little mucilage of starch, and, afterward, of a minute quantity of chlorine-water. Ferrous iodide and its solutions rapidly oxidize, the latter forming a rust-brown sediment, the former becoming less soluble in water, and yielding a brown solution, one drop of which, when added to a few drops of chloroform, and subsequently shaken with some water, colors the chloroform light carmine-red, which is not the case when the ferrous iodide is fresh, and not yet partly oxidized.

Ferrous iodide is decomposed by, and therefore incompatible with, acids, the alkaline hydrates and carbonates, and those metallic salts which form insoluble iodides.

The oxidation of ferrous iodide is greatly obviated by its mixture with sugar. Upon this fact, the preparation of FERRI IODIDUM SACCHARATUM and of SYRUPUS FERRI IODIDI are based; both share the chemical properties and reactions of the ferrous iodide. The syrup may be preserved without decomposition, when kept in a sunny place, in small, well-corked vials, containing a piece of clean, bright iron wire, or when a few grains of tartaric or citric acid are added for each ounce, and the whole kept in small, well-corked vials.

FERRI LACTAS.

FERRUM LACTICUM.

Lactate of Iron. Ferrous Lactate.

Greenish-white, crystalline crusts or grains, or a greenish-white powder;* when heated, the salt froths up, with the evolution of white, acid, inflammable fumes, and becomes black, and, when completely incinerated, ferric oxide is left behind, which, when cold, should not act upon moist red litmus-paper (evidence of the absence of alkaline salts).

Ferrous lactate is soluble in about 48 parts of cold, and 12 parts of boiling, water, and almost insoluble in alcohol; its aqueous solution is more or less turbid and of a yellowish-green color, and has a mild, sweetish, chalybeate taste; when filtered and diluted, it gives a rust-brown precipitate with the alkaline hydrates, and a blue one with potassium ferrocyanide, and yields a white, granular deposit of mucic acid, on cooling, when boiled for fifteen minutes with nitric acid; plumbic acetate should cause, in a solution of ferrous lactate, only an opalescence (evidence of the absence of sulphuric, hydrochloric, tartaric, citric, and pomic acids).

* The ferrous lactate of the German manufacturers and shops occurs as a yellowish or grayish-green powder, and is obtained by the following process, which is least subject to the formation of peroxide: An alcoholic solution of sodium lactate is exactly decomposed by a concentrated aqueous solution of ferrous chloride. Allowed to stand for 24 hours, in a filled and closely-stoppered bottle, in a cool place, the ferrous lactate separates in a thick, crystalline crust, which, after the menstruum has been removed, is broken by a wooden spatula, and then transferred to a cloth washed with a little alcohol, and afterward subjected to a moderate pressure, under a small screw-press. The resulting salt cake is broken, and dried at a gentle warmth, and finally triturated.

Examination:

Mineral Impurities.—About half a drachm of the ferrous lactate is completely incinerated in a porcelain crucible, or upon an iron spoon; the remainder is extracted with boiling water; the filtrate must neither act upon test-paper, nor leave any residue upon evaporation on platinum-foil. The insoluble residue is then agitated with a little warm diluted acetic acid, the liquid filtered, and a little chlorine-water added; it is then precipitated with aqua ammoniæ, and afterward filtered. The filtrate is examined with a few drops of a solution of sodium phosphate; a white turbidity would indicate *magnesium salts.*

Gummi, Dextrin, Sugar of Milk.—About 20 grains of the ferrous lactate are dissolved in one ounce of warm water, and subsequently precipitated by an excess of liquor sodæ; after warming and shaking, the precipitate is filtered off, and a few drops of a solution of cupric sulphate are added to one portion of the filtrate; the ensuing turbidity must completely redissolve on agitation; a remaining white coagulation would indicate *gummi;* if this be the case, the solution is filtered, and heated to near boiling; an ensuing red precipitate would indicate *sugar of milk.* An excess of concentrated sulphuric acid is added to the rest of the first alkaline filtrate, and the whole evaporated almost to dryness, upon a porcelain capsule, at a temperature not exceeding 100° C. A black, more or less charred, residue would confirm the presence of either gummi or sugar, or both.

FERRI OXIDUM HYDRATUM.

FERRI PEROXIDUM HYDRATUM. FERRUM OXYDATUM RUBRUM. CROCUS MARTIS ADSTRINGENS.

Hydrated Oxide of Iron. Peroxyhydrate of Iron. Ferric Hydrate.

A reddish-brown, tasteless powder, destitute of grittiness; when heated in a dry test-tube, it emits moisture, but no acid vapors; it is slowly but wholly soluble, without effervescence, in warm hydrochloric acid diluted with half its volume of water, forming a yellow solution, which, when diluted with water, gives only a white turbidity with hydrosulphuric acid, and a blue precipitate with potassium ferrocyanide, but no reaction with ferricyanide; when completely precipitated by alkaline hydrates, the filtrate should be colorless, and should render no reaction either with ammonium sulphydrate or with sodium carbonate (evidence of the absence of zinc, magnesium, and calcium).

Examination:

Alkaline sulphates may be detected by agitating the ferric hydrate with a little warm water, and acidulating the filtrate with one drop of hydrochloric acid, and testing it with barium chloride.

Copper may be detected by a blue coloration of solution of ammonium sesqui-carbonate, when agitated with the ferric hydrate, and subsequently filtered; its presence may be confirmed or recognized, when the result of the preceding test is uncertain, by over-saturating the filtrate with acetic acid, and testing it with potassium ferrocyanide; a reddish-brown precipitate would indicate or confirm copper.

FERRI OXIDUM MAGNETICUM.

FERRI OXIDUM NIGRUM. FERRUM OXYDULATO-OXYDATUM. ÆTHIOPS MARTIALIS.

Black Magnetic Oxide of Iron. Ferro-Ferric Hydrate.

A brownish-black or black powder, destitute of taste, and attracted by the magnet; heated in the air, it gives off water, absorbs oxygen, and forms red ferric oxide. It dissolves readily and completely, without effervescence, upon warming, in a mixture of equal parts of water and hydrochloric acid, forming a yellowish solution which gives a blue precipitate with both potassium ferrocyanide and ferricyanide, and a white turbidity with hydrosulphuric acid.

Examination:

Alkaline Sulphates or Chlorides.—About one drachm of the ferro-ferric hydrate is shaken with about four drachms of warm water; the filtrate, when evaporated upon platinum-foil, must leave no fixed residue, nor yield, when acidulated with a few drops of acetic acid, any white precipitate with barium chloride or with argentic nitrate.

Metals.—The remaining ferro-ferric hydrate from the preceding test is boiled in a flask with some dilute liquor potassæ; when cold, the mixture is filtered, and is subsequently saturated with hydrosulphuric-acid gas, or else some sodium-sulphide solution is added: an ensuing black precipitate would indicate *lead*, a white one, *zinc;* if no reaction takes place, the liquid is over-saturated with hydrochloric acid; a slight white turbidity (sulphur) will occur; a yellow precipitate would indicate *arsenic*, an orange one, *antimony*.

The remaining ferro-ferric hydrate, left behind from its treatment with liquor potassæ, is washed, and then digested for some hours with a little strong aqua ammoniæ; when filtered, the filtrate will have a bluish tint, and give a dark turbidity upon the addition of a few drops of ammonium sulphydrate, if *copper* be present.

FERRI PHOSPHAS.

FERRUM PHOSPHORICUM. FERRUM OXYDULATO-OXYDATUM PHOSPHORICUM.

Phosphate of Iron. Ferrous-Ferric Phosphate.

A fine, amorphous powder, of a slate-blue color when cold, and grayish green when hot; heated in a dry test-tube, it gives off water, and leaves a black residue; it is insoluble in water, but soluble in the mineral acids, forming yellow solutions, which, when highly diluted, yield a blue precipitate with both potassium ferricyanide and ferrocyanide, and a white turbidity of sulphur with hydrosulphuric acid. When ferrous-ferric phosphate is boiled in a solution of sodium carbonate, and filtered, a filtrate is obtained which, when neutralized with acetic acid, gives a yellow precipitate with argentic nitrate, and a white one with ammoniated solution of magnesium sulphate.

Examination:

Sodium sulphate, left from insufficient washing, may be detected, when a little of the powder is shaken with some hot water, and the filtrate tested with barium chloride.

Metals.—A strong solution of the powder in hydrochloric acid is mixed with a comparatively large volume of hydrosulphuric acid, and set aside for a few hours, in a closed flask, in a warm place; a slight white turbidity (sulphur) will occur; a dark one would indicate *copper*, a yellow one, *arsenic*, which

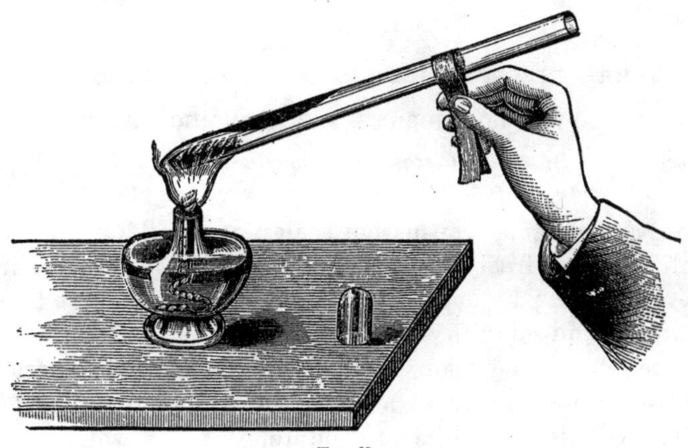

Fig. 63.

latter can be confirmed by the odor when a little of the salt is heated upon charcoal, before the blow-pipe, or by the formation of a metallic mirror, in a narrow tube, upon heating the dried precipitate with potassium cyanide (Fig. 63).

FERRI PYROPHOSPHAS.

FERRUM PYROPHOSPHORICUM.

Bi-basic Phosphate of Iron. Ferric Pyrophosphate.

A white, tasteless powder; when heated in a dry test-tube, it loses moisture, but remains white, and decreases in volume.

It is insoluble in water, but soluble in hydrochloric acid, and in solutions of sodium pyrophosphate and of alkaline citrates. When boiled with a solution of sodium carbonate, ferric pyrophosphate assumes a reddish-brown color, and yields a filtrate of the same tint, but which becomes almost decolorized upon slight over-saturation with acetic acid, and which gives a dense, white precipitate with argentic nitrate (distinction from ferric phosphate, which gives a yellow precipitate, and from ferric metaphosphate, which gives a white gelatinous one).

FERRI PYROPHOSPHAS ET AMMONII CITRAS.

FERRUM PYROPHOSPHORICUM CUM AMMONIO CITRICO.

Pyrophosphate of Iron with Citrate of Ammonium. Ammoniated Pyrophosphate of Iron. Pyrophosphate of Iron in Scales.

Thin, apple-green, transparent scales, of a mild, acidulous, and saline taste: freely soluble in water or glycerin, but insoluble in alcohol; by exposure to the air, the scales lose their transparency and solubility in water; the latter, however, may be restored by the addition of aqua ammoniæ. The solution is not precipitated either by acids or by ammonium hydrate, but gives a yellowish-white precipitate, with the evolution of ammonia, when boiled with liquor potassæ, and a dense white precipitate, when acidulated with acetic acid, and tested with argentic nitrate. Potassium ferrocyanide, when added to the dilute solution of the salt, occasions a pale-blue color, but produces no precipitate.

FERRI SUBCARBONAS.

FERRUM HYDRICUM. FERRUM OXYDATUM FUSCUM. FERRUM SUBCARBONICUM. CROCUS MARTIS APERITIVUS.

Carbonate of Iron.

A fine, amorphous, reddish-brown powder, without odor or taste; when heated in a dry test-tube, it emits watery vapors, which condense in the cooler parts of the tube, and which, when tested with blue litmus-paper, should not alter its color.

Carbonate of iron is insoluble in water, but readily and freely soluble, with slight effervescence, in warm, diluted hydrochloric acid, forming a yellow solution, a few drops of which, when added to water, impart to this the property of yielding a blue precipitate with both potassium ferrocyanide and ferricyanide. The solution therefore affords, with reagents, the reactions of both ferrous and ferric salts.

Examination:

Ferri Subcarbonas is distinguished from Ferri Oxidum Hydratum by its reddish-brown color, by its readier solubility in hydrochloric acid, by effervescence with acids, and by its reaction with potassium ferricyanide.

Sodium or Ammonium Sulphate.—About 10 grains of the powder are agitated, in a test-tube, with about two drachms of warm water; the filtrate should leave no residue upon evaporation on platinum-foil, nor should it render a white turbidity with barium nitrate.

Metals.—The undissolved ferric hydrate remaining upon the filter, in the preceding test, is digested with solution of ammonium sesqui-carbonate for one or two hours; the subsequent filtrate, when tested with a few drops of ammonium sulphydrate, would show *copper* by a bluish tint of the liquid, and *zinc* by a white precipitate.

FERRI SULPHAS.

FERRUM SULPHURICUM.

Sulphate of Iron. Ferrous Sulphate.

Transparent, oblique-rhombic prisms, of a pale bluish-green color and salty, styptic taste, or, when obtained by precipitation, by means of alcohol, a pale-green, crystalline powder (*Ferri Sulphas granulata*). The crystals are efflorescent in the air, and absorb oxygen; when heated, they undergo aqueous fusion; they contain seven molecules (45.56 per cent.) of water of crystallization, six-sevenths of which they lose at a moderate heat, leaving a greenish or grayish white powder (*Ferri Sulphas exsiccata*). At a red heat, the seventh molecule of water, and also the acid, is expelled, leaving behind red, anhydrous, ferric oxide (*Caput Mortuum*).

One hundred parts of water, at 10° C., dissolve 61 parts, at 25° C., 103 parts, and, at 100° C., 280 parts, of crystallized ferrous sulphate; it is insoluble in alcohol; its aqueous solution has a slight acid reaction, readily absorbs oxygen, and becomes turbid by the formation of an insoluble basic ferric sulphate, while a neutral ferric sulphate remains in solution, with the undecomposed ferrous sulphate. The solution of ferrous sulphate, when greatly diluted, gives a white precipitate with barium chloride, a blue one with potassium ferricyanide, and, when not yet oxidized, a white one with the ferrocyanide; it renders no precipitate with hydrosulphuric acid.

Examination:

Metals.—About one drachm of the crystals is dissolved in about two drachms of boiling water, acidulated with a few drops of sulphuric acid; the solution is then added to about 1½ ounce of hydrosulphuric acid, in a stoppered 2-ounce vial, and shaken; after several hours, only a white turbidity (sulphur) should have been formed; a dark turbidity would indicate *copper*, and perhaps other metals. The liquid is then filtered, and evaporated, in a porcelain capsule, until all odor ceases; when this is the case, about half a drachm of concentrated sulphuric acid and 20 drops of concentrated nitric acid are added, and the whole evaporated nearly to dryness. The residue is dissolved in water, and aqua ammoniæ added with constant stirring, until the ensuing precipitate ceases to be redissolved; then from four to five drachms of Liquor Ammonii Acetici are added, the whole heated to boiling, and filtered while hot. An excess of aqua ammoniæ is then added to the colorless filtrate, and, if any more precipitation occurs, it is once more filtered, and subsequently tested with ammonium sulphydrate; an ensuing white turbidity would indicate *zinc*, a reddish-white one, *manganese*. Finally, the liquid, after having been filtered, if such reactions have occurred, is tested with ammonium phosphate; a crystalline, white precipitate, occurring after some time, would indicate *magnesia.*

Crude commercial sulphate of iron is generally considerably contaminated with metallic and earthy salts, and not fit for medicinal use; it contains, besides, the sulphates of zinc, aluminium, and magnesium, and generally so much sulphate of copper as to deposit a metallic cupreous film upon a bright blade of an iron knife or spatula, when immersed for some hours in the aqueous solution, acidulated with a few drops of sulphuric acid.

FERRI VALERIANAS.

FERRUM VALERIANICUM.

Valerianate of Iron. Ferric Valerianate.

A dark, tile-red, amorphous powder, or thin, reddish scales, having the faint odor and taste of valerianic acid. When heated in a porcelain capsule, the salt melts, emits inflammable vapors, and, when incinerated, leaves behind ferric oxide, which should not color moistened turmeric-paper brown, nor dissolve in warm dilute acetic acid.

Ferric valerianate is soluble in alcohol (distinction from citrate and tartrate of iron), but insoluble in water; boiling water decomposes it, extracting the valerianic acid, and leaving the ferric hydrate behind; the filtrate is colorless, reddens litmus-paper, and gives no reaction with ammonium hydrate and subsequent addition of hydrosulphuric acid. Acids decompose ferric valerianate, forming soluble ferric salts, and setting free the valerianic acid.

Examination:

Admixtures of ferric tartrate or citrate, impregnated with oil of valerian, may readily be recognized by their solubility in water and insolubility in strong alcohol; the latter dissolves only the oil of valerian, if such be present, readily recognized by its odor, when a portion of the alcohol is evaporated upon the warm hand.

FERRUM.

Iron.

The source of the medicinal preparations of iron is the refined malleable wrought-iron, of which the piano-forte wire is among the best commercial varieties. When iron filings or turnings are employed instead of wire, care has to be taken that they are not derived from crude cast or pig iron, that they are free from rust, and that they are not contaminated with copper or brass filings from the workshops. Cast or pig iron may be recognized by the evolution of gas of a noxious odor, and by a considerable black residue, when the filings or turnings are dissolved, in a test-tube, in a mixture of equal parts of concentrated hydrochloric acid and water. An admixture of copper or brass filings may be recognized, with approximate certainty, by close inspection, with a magnifying-glass, and by chemical tests, as hereafter described.

In the preparation of solutions of iron, which are subsequently filtered, filings of cast-iron are not exactly objectionable; iron filings or turnings, however, which may contain, or are liable to contain, copper or brass filings, ought not to be employed for medicinal preparations, since copper is dissolved by acids, notwithstanding an excess of iron.

Ferrum Pulveratum.—A fine, gray powder, of a dull, metallic appearance; when heated to redness, it oxidizes, forming a reddish powder, and increasing in weight, if the powder employed was pure and dry. Iron powder dissolves in a mixture of equal parts of hydrochloric acid and water, evolving impure hydrogen gas, of a faint odor, and leaving only a small insoluble black residue; the filtered solution has a light-green color, and affords, when greatly diluted with water, a deep-blue turbidity with potassium ferricyanide, and almost white precipitates with the alkaline hydrates and carbonates, which, however, rapidly oxidize, and become green, and ultimately brown.

Ferrum Reductum.—Iron powder, obtained by reduction of ferric oxide or oxy-hydrate by hydrogen, at a strong heat, forms a very fine, black powder, which reoxidizes when heated to redness, and which is readily and wholly soluble in warm diluted hydrochloric acid, with the evolution of pure hydrogen gas, forming a solution which has the same properties and deportment with reagents as that of powdered iron. If the solution takes place without a copious evolution of gas, and has, when filtered, a yellowish appearance instead of a light-green one, the powder was more or less oxidized, or even so much so as to consist almost wholly of a mixture of ferrous and ferric oxides.

Examination of Iron:

Sulphur may be recognized by the black coloration of a solution of plumbic acetate, when a little of the iron is dissolved in a mixture of equal parts of hydrochloric acid and water, either in a test-tube, loosely closed with a bunch of cotton moistened with solution of plumbic acetate (Fig. 29, page 72), or in a small flask (Fig. 64), whence the evolved gas is passed through a solution of plumbic acetate in a test-tube. If the plumbic solution in either case becomes black, sulphur is indicated.

FIG. 64.

Sulphur, *phosphorus*, and *arsenic*, may be detected, in iron powder, filings, turnings, or wire, by submitting about 100

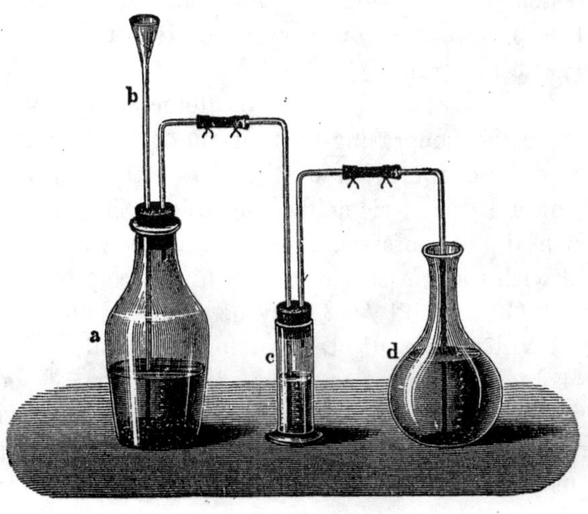

FIG. 65.

grains of the iron, in a generating-flask, *a* (Fig. 65), to the action of about one ounce, or a little more, of a mixture of equal parts of concentrated hydrochloric acid and water, and by passing the evolved gas through a dilute solution of cadmium sulphate in *c*, and through a dilute solution of argentic nitrate, contained in *d*. The acid is, little by little, added to the iron, through the funnel-tube *b*, so that the evolution of the gas and its passage through the fluids are slow. The solution in *c* is prepared by adding to an aqueous solution of cadmium sulphate (1:100) so much of ammonium hydrate as is required to redissolve the precipitate, formed upon the first addition of the alkaline hydrate.

The occurrence of a yellow precipitate in the cadmium solution would indicate *sulphur*, and of a black one in the argentic solution, arsenic or phosphorus. In order to distinguish either arsenic or phosphorus, or both, in the iron, the blackened argentic solution is warmed, and completely precipitated with hydrochloric acid; when cool, it is filtered and the filtrate mixed with an equal volume of hydrosulphuric acid, and allowed to stand, in a corked vial, for several hours; an ensuing yellow precipitate would indicate *arsenic;* the liquid is then filtered, and boiled in a porcelain capsule, until it becomes odorless; it is then over-saturated with aqua ammoniæ, filtered, and tested with solution of ammoniated magnesium sulphate; a white precipitate would be indicative of the presence of *phosphorus* in the iron.

Metallic Impurities.—Zinc may be detected in the solution obtained in the generating-flask of the preceding test, by filtering it, and by evaporation of the filtrate to dryness, after the addition of a little nitric acid; the residue is extracted by boiling it with liquor potassæ, and the filtered liquid, when cool, is tested with sodium sulphide; a white precipitate would indicate zinc. *Copper* and *lead* may be detected in the iron, by treating it with warm diluted nitric acid, and by testing part of this solution with a few drops of a solution of sodium sulphate for lead, and by evaporating another part of the solution to dryness, and subsequently digesting the residue with an excess of aqua ammoniæ; a blue coloration of the filtrate would indicate *copper*.

GLYCERINUM.

GLYCERINA.

Glycerin. Triatomic Propenyl Alcohol.

A colorless and odorless, thick, viscid, neutral liquid, of an intensely sweet taste; when anhydrous, its spec. grav. is 1.267; that of commercial glycerin, containing from 10 to 4 per cent. of water, is from 1.22 to 1.25. Glycerin is not volatile at common temperatures, but distils unchanged, with water-vapor, under pressure; it boils at 290° C., and decomposes at a temperature a little above its boiling-point. Exposed to a strong heat, on platinum-foil or in a porcelain capsule, it emits white, irritating, inflammable vapors, which burn with a blue flame, leaving, finally, a black residue, which, however, is completely dissipated at a red heat.

Glycerin is miscible, in all proportions, with water, aqua ammoniæ, liquor potassæ, alcohol, and ether diluted with alcohol, but not with pure ether, chloroform, carbon bisulphide, or benzol; it mixes with concentrated sulphuric acid without change, but suffers decomposition with concentrated nitric and hydrochloric acids. Glycerin possesses extensive powers as a solvent; it dissolves most substances which are soluble in water, although in a less degree, but is a better solvent for bromine, iodine, and carbolic acid.

It does not mix with fatty oils, and dissolves essential oils only to a limited extent.

Examination:

A *fatty* or *empyreumatic odor* of glycerin is recognized by the smell, especially upon gently warming a little of the sample on a watch-glass or porcelain capsule.

Sugar, *glucose*, and *mucilages*, are indicated by a more or less brown coloration of the glycerin, when mixed with twice its bulk of concentrated sulphuric acid, or when mixed and heated with a strong solution of potassium hydrate.

Glucose may be detected by the occurrence of a brick-red precipitate, when a mixture of equal volumes of the glycerin and of liquor potassæ, with two or three drops of solution of cupric sulphate, is heated in a test-tube.

Sugar is detected by the same reaction, when the glycerin is boiled for a few minutes with an equal volume of very dilute solution of tartaric acid, and the hot mixture tested with Fehling's cupric solution.

Mucilages of gum, dextrin, or *glue*, are indicated by the formation of a white turbidity, gelatinous, or flocculent, when one volume of the glycerin is mixed with four volumes of alcohol.

Metallic salts are detected by agitating one volume of the glycerin with three volumes of hydrosulphuric acid; any impairing of the colorlessness or transparency of the mixture would indicate metallic impurities; they may be distinguished, as to what group of metals they belong to, by dividing the liquid into two portions, and adding to the one a little hydrochloric acid, and to the other aqua ammoniæ. *Copper, lead*, and *tin*, will be indicated by the first test; *iron, zinc*, and *aluminium*, by the second. If a precipitate appears in either case, and the nature of the impurity has to be ascertained, the test must be repeated on a larger scale, and the metallic impurity be determined in the mode described on pages 41 to 44.

Calcium salts may be detected in the diluted glycerin, by a white turbidity when tested with ammonium oxalate.

Acids and their Salts.—When diluted with twice its volume of water, the solution must leave litmus-paper unchanged; it is then examined in four separate portions: for *hydrochloric acid* and *chlorides*, by acidulating with nitric acid, and testing with argentic nitrate; for *sulphuric acid* and *sulphates*, by testing the second portion, also acidulated with nitric acid, with barium nitrate; for *oxalic acid*, by testing the third portion, acidulated with acetic acid, with calcium acetate; and for *nitric acid* and *nitrates*, by adding to the fourth portion a little acetic acid and one drop of neutral indigo-solution, and then warming the mixture by dipping the test-tube into hot water; a decoloration of the bluish or bluish-green tint of the liquid will indicate free nitric acid; when the color remains unaltered, a few drops of concentrated sulphuric acid are added to the mixture while still warm; if decoloration takes place now (and the glycerin is free from chlorides and chlorates), nitrates are indicated.

Another very sensitive test for nitric acid and nitrates, combining the test for *chlorine*, is to mix, in a test-tube, about one fluid-drachm of sulphuric-acid mucilage of starch with a few drops of solution of potassium iodide (free from iodate), and then to add about one fluid-drachm of glycerin; when mixed together with a glass rod, the liquid must remain colorless; a blue color would indicate *chlorine;* when the mixture remains colorless, a thin rod of bright zinc is immersed in the centre of the fluid, with care not to agitate the test-tube; if traces of nitric acid or nitrates be present, a bluish coloration, issuing from the zinc, will appear.

Formic acid may be detected by the formation of a black deposit, when a mixture of the glycerine with an equal volume of diluted aqua ammoniæ and a little solution of argentic nitrate is allowed to stand in a corked test-tube, protected from the light, for 24 hours.

Butyric acid, and analogous fatty acids, may be recognized by the odor of ethyl butyrate (similar to that of artificial essence of pine-apple), when a mixture of two volumes of glycerin with one volume of a mixture of equal parts, by volume, of strong alcohol and concentrated sulphuric acid, is gently warmed by dipping the flask or test-tube into boiling water.

Fig. 66.

Ammonium salts, occasioned by the neutralization of an originally slightly acidulous glycerin with aqua ammoniæ, may be detected by heating the glycerin in a large test-tube, half filled with it, and provided, by means of a rubber cork, with a delivery-tube reaching to the bottom of a receiving test-tube which contains a little water, and is cooled in ice-water (Fig. 66); the generating tube or flask, containing the glycerin, is immersed, for about half an hour, in a boiling concentrated solution of culinary salt; after that time, the contents of the recipient are tested with red litmus-paper, which should not be altered thereby, and are then divided into two separate portions; a few grains of dry potassium hydrate are dissolved in the one portion; upon heating it, ammonia will be recognized by the odor, and by white vapors when a glass rod moistened with acetic acid is held at the orifice of the tube; to the other portion, a few drops of liquor potassæ and, subsequently, one or two drops of solution of potassio-mercuric iodide are added; a yellowish-brown precipitate or coloration would afford an additional indication of ammonium salts.

TABLE

OF THE QUANTITY BY WEIGHT OF WATER CONTAINED IN 100 PARTS BY WEIGHT OF GLYCERIN AT DIFFERENT DENSITIES.

TEMPERATURE 17.5° C.

Specific Gravity.	Percent. of Water.	Specific Gravity.	Percent. of Water.	Specific Gravity.	Percent. of Water.	Specific Gravity.	Percent. of Water.
1.267	0	1.224	13	1.185	26	1.147	39
1.264	1	1.221	14	1.182	27	1.145	40
1.260	2	1.218	15	1.179	28	1.142	41
1.257	3	1.215	16	1.176	29	1.139	42
1.254	4	1.212	17	1.173	30	1.136	43
1.250	5	1.209	18	1.170	31	1.134	44
1.247	6	1.206	19	1.167	32	1.131	45
1.244	7	1.203	20	1.164	33	1.128	46
1.240	8	1.200	21	1.161	34	1.126	47
1.237	9	1.197	22	1.159	35	1.123	48
1.234	10	1.194	23	1.156	36	1.120	49
1.231	11	1.191	24	1.153	37	1.118	50
1.228	12	1.188	25	1.150	38		

HYDRARGYRI CHLORIDUM CORROSIVUM.

HYDRARGYRI PERCHLORIDUM. HYDRARGYRUM BICHLORATUM. HYDRARGYRUM CORROSIVUM SUBLIMATUM.

Corrosive Sublimate. Perchloride or Bichloride of Mercury. Mercuric Chloride.

Colorless, translucent, heavy, crystalline masses, when obtained by sublimation, or small, prismatic crystals, when obtained by crystallization, of a spec. grav. of 5.43; they are permanent in the air, give a dull, white streak when scratched with a knife, are slightly volatile at common temperatures, fuse at 265° C., and volatilize wholly at 295° C., forming dense, white vapors, which, on cooling, solidify in small, shining needles.

Mercuric chloride is soluble in water, requiring, at 0° C., 16 parts, at 20° C., 13½ parts, at 80° C., 4 parts of cold, and 2½ parts of boiling, water for solution; it is less soluble in glycerin, 100 parts of which dissolve about 7 parts of the salt; it is freely soluble in alcohol and ether, requiring 2⅓ parts of cold, and 1⅙ part of boiling alcohol, and 3 parts of ether. The aqueous solution reddens blue litmus-paper, and has an acrid, styptic taste.

In the aqueous solution of mercuric chloride, the fixed alkaline and earthy hydrates and alkaline carbonates produce, when added in small quantity, a reddish-brown precipitate; when added in excess, a yellow one; ammonium hydrate gives a white one; argentic nitrate, a curdy white one; iodides, when added in small quantity, a yellowish, and in larger quantity, a vermilion-red one, soluble in an excess of the precipitant; stannous chloride, when added in small quantity, a white, and when added in excess, a gray precipitate. When hydrosulphuric acid is gradually added to a solution of mercuric chloride, the precipitation takes place according to the proportions of the reagent and the chloride, in progressive variation of color from white to yellow, orange, reddish brown, and black; an excess of the reagent produces at once a complete black precipitation. When the aqueous solution of mercuric chloride is rubbed upon bright copper, it coats the latter with a brilliant metallic film. It forms white, insoluble or sparingly soluble compounds with many organic substances, as albumen, fibrin, gluten, etc.

Mercuric chloride is soluble, without decomposition, in nitric, hydrochloric, and sulphuric acids, and crystallizes from the solutions on cooling, if they were saturated while hot.

HYDRARGYRI CHLORIDUM MITE.

HYDRARGYRI SUBCHLORIDUM. HYDRARGYRUM CHLORATUM MITE. MERCURIUS DULCIS.

Calomel. Sub- or Proto-Chloride of Mercury. Mercurous Chloride.

Mercurous chloride varies in the minuteness of its particles, and accordingly in its appearance and in the energy of its physiological action.

When obtained by *sublimation*, it forms ponderous, yellowish-white masses or cakes, of a fibrous, crystalline fracture, yielding a lemon-yellow streak when scratched with a knife, and having a specific gravity of 7.176. When reduced to a

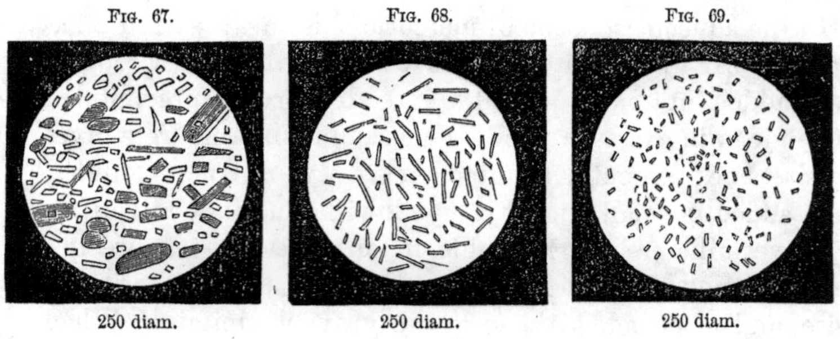

Fig. 67. 250 diam. Fig. 68. 250 diam. Fig. 69. 250 diam.

MERCUROUS CHLORIDE:

By Trituration. *By Condensation.* *By Precipitation.*

fine powder by trituration and levigation, it has a dull-white appearance with a yellowish tint; it becomes slightly yellowish when triturated with strong pressure in a porcelain mortar, and consists, when seen under the microscope, of comparatively large, transparent, crystalline fragments (Fig. 67).

Prepared by *sublimation* and by *condensation* of the vapor by a current of air or steam, mercurous chloride (Hydrargyri Chloridum Mite vapore paratum) forms a perfectly white and less ponderous powder, consisting of smaller laminar particles, when seen under the microscope (Fig. 68).

Prepared by *precipitation*, mercurous chloride forms a fine, snow-white powder, consisting of minute, amorphous * particles (Fig. 69), which are not transparent.

With regard to therapeutical action, mercurous chloride, obtained by sublimation and subsequent trituration and levigation, and consisting of the largest particles, has the mildest effect; next to this comes the calomel obtained by sublimation and condensation by air or steam; † that obtained by precipitation, and having the minutest division of its particles, has the most powerful physiological action.

Mercurous chloride, when heated in a dry test-tube, is slowly but completely volatilized with a faint noise and without fusion. It is insoluble in the common solvents, but soluble to some extent in saliva, in the pancreatic juice, in albumen, and animal secretions. When agitated with hot water, with alcohol, or with dilute acetic, hydrochloric, or nitric acids, it is not acted upon by any of them. When boiled for some time with water, it suffers slow decomposition into metallic mercury and mercuric chloride. The alkaline hydrates and carbonates, and the hydrates of the alkaline earths, reduce mercurous chloride to black oxide. Concentrated boiling hydrochloric and sulphuric acid decompose the salt; the former producing metallic mercury and mercuric chloride, the latter mercuric sulphate and chloride. Warm concentrated nitric acid also dissolves it gradually, with the evolution of nitric-oxide vapors, forming a solution of mercuric chloride and nitrate, which solutions blacken bright copper when dropped upon it, and coat it brilliantly when rubbed upon it. Mercurous chloride is also soluble in chlorine-water without acquiring a transient or permanent yellow color (distinction from mercurous bromide).

* When obtained by precipitating a solution of mercuric chloride with sulphurous-acid gas, the resulting mercurous chloride is of a crystalline structure.

† No other kinds of mercurous chloride can as yet be considered officinal, and no others should be dispensed for internal use, unless ordered or prescribed as "*Calomel via humida paratum*," or "*Calomel precipitatione paratum.*"

Examination:

When heated in a narrow test-tube, mercurous chloride must completely sublime, without previous fusion and without emitting ammoniacal odors or yellow nitrous vapors.

Mercuric chloride may be detected by triturating some of the calomel with diluted alcohol, and shaking the mixture in a test-tube, and by subsequent filtration through a moist double filter; the filtrate must impart no stain to bright copper, nor yield any reaction with hydrosulphuric acid or with argentic nitrate.

Ammonio-mercuric Chloride.—The mercurous chloride of the preceding test, remaining upon the filter, is rinsed with diluted acetic acid through the broken filter into a test-tube, and the mixture is agitated for a few minutes and filtered. The filtrate is then tested in separate portions with hydrosulphuric acid and argentic nitrate; a black turbidity in the first instance, and a white one in the second, would indicate ammonio-mercuric chloride.

HYDRARGYRI CYANIDUM.

HYDRARGYRUM CYANATUM.

Cyanide of Mercury. Mercuric Cyanide.

Small, colorless, anhydrous, prismatic crystals, transparent when freshly prepared, but soon assuming a white and opaque appearance; when perfectly dry, they become black when exposed to heat in a dry tube, and emit vapors of mercury and a colorless inflammable gas (cyanogen), which burns, when ignited, with a purple flame; a black residue of paracyanogen, intermingled with globules of mercury, is left behind; when the salt is humid, traces of hydrocyanic acid, of carbonic acid, and of ammonia, are also formed and evolved.

Mercuric cyanide is soluble in eight parts of water and about twenty parts of alcohol; its aqueous solution evolves hydrocyanic acid upon the addition of hydrochloric acid, and gives a black precipitate with hydrosulphuric acid, but is not precipitated by the alkaline hydrates and carbonates, nor by argentic nitrate, nor by albumen; stannous chloride, containing free hydrochloric acid, precipitates metallic mercury with the evolution of hydrocyanic acid. The solution of mercuric cyanide affords no mercuric stain upon bright copper. unless acidulated with hydrochloric acid.

Examination:

Basic oxy-mercuric cyanide is indicated by an alkaline reaction of the solution upon turmeric-paper.

Mercuric chloride and other *soluble mercuric salts* may be detected in the solution, by the occurrence of a transient turbidity upon the gradual addition of single drops of solution of potassium iodide.

HYDRARGYRI IODIDUM RUBRUM.

HYDRARGYRUM BIIODATUM RUBRUM.

Biniodide of Mercury. Mercuric Iodide.

A heavy, crystalline powder, or small, brilliant, octahedral crystals, of a vermilion color, becoming yellow when gently heated; when heated in a dry tube (Fig. 70), mercuric iodide fuses to a yellowish-brown liquid, and sublimes into yellow rhombic scales, which pass into the red modification of octahedral crystals, slowly on cooling, and at once by concussion.

Mercuric iodide is nearly insoluble in cold, and only sparingly soluble in boiling, water; it is soluble in 130 parts of cold, and 15 parts of boiling, alcohol, less soluble in ether, and very little in glycerin and in oils. Concentrated acids, and the solutions of the alkaline hydrates, decompose it; it is freely soluble in aqueous solutions of potassium and sodium hydrates, of potassium iodide and cyanide, of mercuric chloride, and of sodium chloride and sodium hyposulphite; the latter solution deposits upon heating, if the solvent is not in excess, red mercuric sulphide. All its solutions form a black precipitate with an excess of hydrosulphuric acid, either at once, or upon the addition of an acid.

Mercuric iodide is partly decomposed when shaken with chlorine-water; the obtained filtrate, when rubbed upon bright

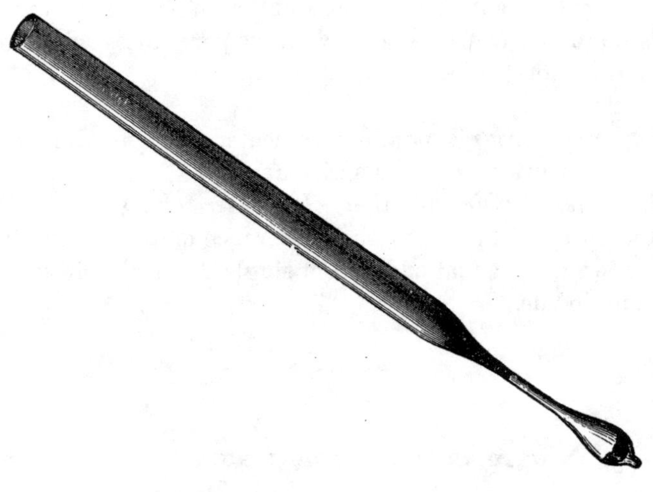

Fig. 70.

copper, coats it with a brilliant metallic film, and, when shaken with a little chloroform, imparts to it a purple color.

Examination:

Mercuric sulphide, red oxide of lead, or other fraudulent admixtures, will remain undissolved upon solution of the powder, either in solution of potassium iodide, or in 20 to 25 parts of boiling alcohol. If a residue is left, it is washed with water and subsequently treated with warm nitric acid, and filtered; the filtrate is slightly diluted, and tested with a few drops of diluted sulphuric acid; an ensuing white precipitate indicates *lead*. If a red residue remains, insoluble in nitric acid, it is tested by heating it upon platinum-foil to redness; if it is wholly volatile, *mercuric sulphide* is recognized, and, if a residue remains, *fixed admixtures* are indicated.

HYDRARGYRI IODIDUM VIRIDE.

HYDRARGYRUM IODATUM FLAVUM.

Green Iodide of Mercury. Protoiodide of Mercury. Mercurous Iodide.

A heavy, dirty-greenish yellow powder, which suffers gradual decomposition, and becomes brownish on exposure to light, heat, and air. When heated in a dry tube (Fig. 70, page 230), it becomes brownish red, fuses, and is completely volatilized, yielding a sublimate of minute yellow and scarlet crystals of mercuric iodide, intermixed with metallic mercury.

Mercurous iodide is not quite insoluble in water, but insoluble in alcohol and in ether; it is decomposed by concentrated acids, by the alkaline hydrates, and also by boiling solutions of chlorides, bromides, and iodides, which convert it into mercuric iodide and metallic mercury. When mercurous iodide is agitated in a little water to which a few drops of ammonium sulphydrate have been added, and the subsequent filtrate is mixed with one drop of liquor ferri perchloridi and then agitated with a little chloroform, the latter will acquire a yellow or reddish color, which will appear still more distinct upon the subsequent addition of a little water.

Examination:

Mercuric iodide may be detected when about 10 grains of the green powder are rubbed with about 2 drachms of alcohol; the filtrate should yield no black precipitate, when added to about one ounce of hydrosulphuric acid; such a reaction would indicate biniodide.

Fixed impurities will remain behind upon complete volatilization of the mercurous iodide in a dry tube; such would be very likely to originate from either the mercury or the iodine, and, if required, their nature may be determined by the methods as described on pages 238–240 and 243, 244.

HYDRARGYRI OXIDUM FLAVUM.

HYDRARGYRUM OXYDATUM VIA HUMIDA PARATUM.

Yellow Oxide of Mercury. Precipitated Mercuric Oxide.

A heavy, orange-yellow powder, without crystalline structure when seen under the microscope; it assumes a red color on being heated; it is more readily acted upon by reagents than the coarser red oxide; the latter remains unchanged when agitated with a warm solution of oxalic acid, while the yellow oxide combines with the oxalic acid, forming white mercuric oxalate; when agitated with a hot alcoholic solution of mercuric chloride, the yellow oxide becomes at once black (oxychloride), while the red oxide remains unchanged for some time.

The chemical reactions of the precipitated yellow mercuric oxide, and its deportment with reagents, correspond with those of the red oxide.

HYDRARGYRI OXIDUM RUBRUM.

HYDRARGYRUM OXYDATUM RUBRUM.

Red Oxide of Mercury. Mercuric Oxide.

Heavy, coherent masses, consisting of bright, brick-red, crystalline scales, which, when finely pulverized, form a dull orange-red powder, of a specific gravity of 11.30; when heated in a dry tube, red mercuric oxide assumes a dark-brown appearance, but regains its original color on cooling; at a heat a little above the boiling point of mercury (360° C.), it is resolved into its constituents, and is entirely volatilized below red heat.

Mercuric oxide is slightly soluble in water, so that, when agitated with boiling water, the filtrate yields a faint reaction with hydrosulphuric acid; it is insoluble in pure glycerin, in alcohol, ether, and chloroform, somewhat soluble in saliva and in albuminous animal secretions, and entirely soluble in strong and in somewhat diluted acids. The fixed alkaline and earthy hydrates and alkaline carbonates produce in solutions of mercuric oxide and its salts, when added in small quantity, a reddish brown, when added in excess, a yellow precipitate; ammonium hydrate, a white one; iodides (provided that the solution does not contain a large excess of acid), when added in a small quantity, a yellowish, and in a larger quantity, a vermilion-red one, soluble in an excess of the precipitant; stannous chloride, when added in a small quantity, gives a white, and, in an excess, a gray precipitate. When hydrosulphuric acid is gradually added to the solution, a precipitate is formed which appears, according to the proportion of the reagent, successively white, yellow, orange, reddish-brown, and finally, with an excess of the precipitant, black.

Examination:

Mercuric nitrate is indicated by disengagement of red nitrous vapors, when the oxide is heated in a dry test-tube. As

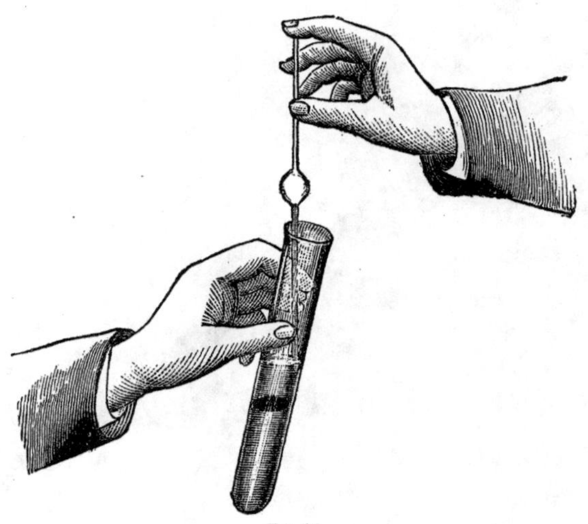

Fig. 71.

a confirmatory test, one drachm of the oxide may be agitated for a few minutes with about 2 drachms of boiling water wherein 10 grains of sodium carbonate have been dissolved; the mixture is then filtered, the filtrate nearly neutralized by a few drops of sulphuric acid, and evaporated to about half its bulk; this is tested in two portions for nitric acid; the one by tingeing it faintly blue with sulphuric acid solution of indigo, and subsequent gentle heating; the other by mixing it with a little concentrated solution of ferrous sulphate, and by placing the liquid carefully upon concentrated sulphuric acid (Fig. 71); a decoloration in the first test, and the occurrence of a dark-brown color upon the line of junction between the two liquids in the second, would confirm the presence of oxyacids of nitrogen.

Admixtures.—About 20 grains of the oxide are dissolved in about one drachm of strong nitric acid diluted with an equal volume of water; with the aid of heat, a complete solution must take place; if the oxide be very old, a slight residue of reduced mercury might remain, which, when separated and heated in a porcelain capsule, should wholly volatilize. If a red or brown residue is left from the solution, an admixture of mineral substances (brick-powder, mercuric sulphide, or red

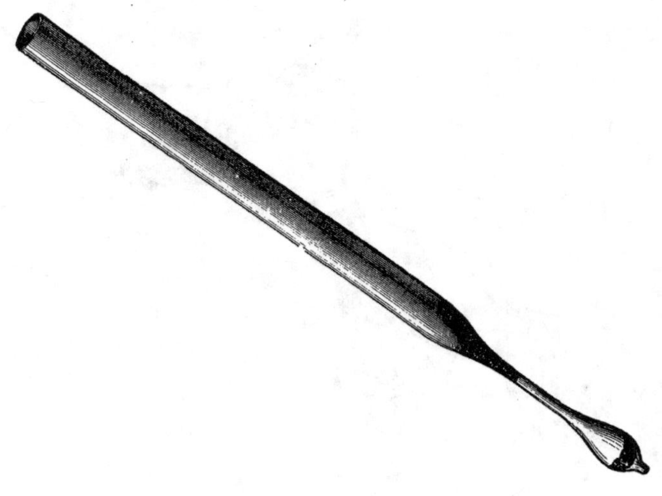

Fig. 72.

oxide of lead) would be indicated. If the nature of such a residue has to be ascertained, a somewhat larger quantity may be obtained, which, when washed and dried, may be heated in a reducing-tube (Fig. 72); vermilion volatilizes, forming a fine red sublimate; red oxide of lead fuses, and exhibits, when cooled, a yellow vitrified appearance, and dissolves, when boiled in concentrated nitric acid diluted with an equal bulk of water, leaving behind silicious mineral substances, if such be present.

HYDRARGYRI SUBNITRAS.

HYDRARGYRUM NITRICUM OXYDULATUM.

Subnitrate of Mercury. Mercurous Nitrate.

Colorless, transparent rhombic or prismatic crystals,* which, when gradually heated in a dry tube, emit yellow nitrous vapors, become yellow, then red, and are finally resolved into metallic mercury; the crystals become grayish black when moistened with lime-water.

Mercurous nitrate is soluble in water with partial decomposition, and with the separation of a yellowish basic salt; it is, however, entirely soluble in water acidulated with nitric acid, forming a colorless solution, which, when rubbed on bright copper, coats it with a white, metallic film, and, when greatly diluted, yields a white precipitate with hydrochloric acid, and a black one with aqua ammoniæ or lime-water.

Liquor Hydrargyri Nitrici Oxydulati of the Pharmacopœa Germanica is a solution of this salt, containing 10 per cent. of mercurous nitrate.

Examination:

Mercuric nitrate may be detected by complete precipitation of the solution with diluted hydrochloric acid, and by testing the filtrate in separate portions with hydrosulphuric acid and with stannous chloride, and warming; a black precipitate with the first reagent, and a gray one with the second, would indicate mercuric nitrate.

* According to the proportion between the mercury and the nitric acid employed in the preparation, there are a normal and a basic mercurous nitrate, which correspond in their chemical and therapeutical properties, except that, when rubbed with a little sodium chloride, the normal salt remains white, while the basic salt gives a grayish-green mixture.

HYDRARGYRI SULPHAS FLAVA.

HYDRARGYRUM SULPHURICUM FLAVUM. TURPETHUM MINERALE.

Yellow Subsulphate of Mercury. Basic Mercuric Sulphate.

A heavy, lemon-yellow powder, of a crystalline structure when seen under the microscope; when heated in a dry tube, it assumes a reddish-brown hue, but regains its original color on cooling; at a higher temperature it volatilizes without fusion, yielding a white sublimate (mercuric sulphate) intermingled with gray metallic mercury; it is decomposed and entirely volatilized at a red heat, evolving vapors of mercury and of sulphurous acid.

Basic mercuric sulphate is almost insoluble in cold, and sparingly soluble in hot, water, but soluble in diluted hydrochloric and nitric acids, forming colorless solutions, which, when diluted, give a white precipitate with barium salts, and which otherwise, in their deportment with reagents, resemble the solutions of mercuric chloride and oxide (pages 225 and 233).

HYDRARGYRI SULPHURETUM RUBRUM.

HYDRARGYRUM SULPHURATUM RUBRUM. CINNABARIS.

Cinnabar. Vermilion. Red Sulphide of Mercury. Mercuric Sulphide.

Heavy masses, or cakes, of a specific gravity of 8.1, and of a dull blackish-red color and a brilliant crystalline texture, yielding a red streak when scratched with a knife, and a magnificent scarlet powder, which becomes black when moistened with an ammoniacal solution of argentic nitrate. When heated in a porcelain capsule, or upon charcoal, cinnabar assumes an almost black color, but turns red again after cooling; at a strong heat, it is wholly dissipated, burning with a bluish flame, and emitting the odor of sulphurous acid; heated in a reducing-tube or in close vessels, it sublimes below red heat without decomposition.

Mercuric sulphide is insoluble in the common solvents, nor is it acted upon by concentrated acids or by alkaline hydrates, at common temperatures; a mixture of concentrated nitric and hydrochloric acid dissolves it gradually, forming a colorless solution which, when diluted with water, gives a white precipitate with barium chloride, coats metallic copper with a film of mercury, and corresponds in its deportment with reagents to solutions of mercuric salts (pages 225 and 233).

Examination:

Oxides of Lead and Iron.—About 20 grains of the cinnabar are agitated with about two fluid-drachms of concentrated nitric acid; the scarlet color must remain unaltered, as change to a darker tint would indicate red oxide of lead; the mixture is then gently heated by immersing the test-tube in hot water, and is subsequently diluted with twice its volume of water, and filtered; the filtrate should be colorless; a yellowish appearance would indicate red *basic plumbic chromate*, or mercuric chromate (chromic cinnabar); it is then tested in separate portions with hydrosulphuric acid, with sulphuric acid, and with potassium iodide, for lead; another portion is tested with potassium ferrocyanide for ferric oxide; if this be present, the yellowish color of the nitric acid, agitated with the cinnabar, may be due only to iron.

Chromates may be detected or confirmed by the occurrence of red irritating fumes of chloro-chromic acid, when about ten grains of the cinnabar are carefully mixed and heated in a test-tube with a few small fragments of dry sodium chloride and a few drops of concentrated sulphuric acid.

Mercuric Iodide, Realgar, and Antimonic Cinnabar.—About 30 grains of the cinnabar are agitated with two fluid-drachms of warm liquor potassæ; the filtrate should be colorless, should cause neither a coloration nor a turbidity when dropped into chlorine-water, and should yield a white precipitate when dropped into a dilute solution of plumbic acetate. A yellow or reddish coloration of the chlorine-water would indicate mercuric iodide, and a yellow precipitate with plumbic acetate, red arsenic sulphide (Realgar), or antimonic oxy-sulphide (Antimonic Cinnabar). If either of the latter two be indicated, the alkaline filtrate will give, upon supersaturation with hydrochloric acid, a yellow precipitate when the first compound is present, and an orange-red one with the second.

HYDRARGYRUM.

Mercury. Quicksilver.

A silver-white and brilliantly-lustrous metal, of a specific gravity of 13.596 at 15.5° C.; liquid at common temperatures, and easily divisible into spherical globules; it solidifies when cooled to —39.5° C., forming at and below that temperature a soft and malleable metal of a specific gravity of 14.4 at —40° C.; it boils at 350° C., forming a transparent, colorless vapor; it is, however, volatile to a perceptible extent at all temperatures above 20° C.; when pure, it is unalterable at common temperatures, and remains bright and brilliant.

Mercury is insoluble in the common solvents, in concentrated hydrochloric acid, and at common temperatures also in sulphuric acid; but it is dissolved by the latter when boiled with it, and is readily dissolved without residue by nitric acid, forming a solution, which contains mercuric nitrate when heat is applied and an excess of concentrated acid, and mercurous nitrate when the metal is in excess or is acted upon by cold and diluted nitric acid.

Examination:

Mercury amalgamates with many metals, and, to a certain extent, without change of its appearance and properties; the most common of such metallic impurities are lead and tin, and occasionally zinc and bismuth; their presence in the commercial metal is indicated by a dull, tarnished appearance, and a black, powdery coating of the surfaces of the metal, and of the inside of the vessels containing it, and by lead-gray streaks upon white paper when a few globules of the metal are allowed to roll over it.

Such contamination may be ascertained by agitating for a few minutes a little of the mercury, in a strong one-ounce bottle, with a mixture of one drachm of Liquor Ferri Persulphatis (free from ferrous salt) and one drachm of water; after subsiding, the aqueous liquid is poured into a test-tube, diluted with an equal volume of water, and tested with a few drops of potassium ferri cyanide; a blue turbidity will indicate the above-mentioned metallic impurities.

When their nature has to be determined, the following method is practicable and simple: about one ounce of the metal, including as much of the powdery coating on the surfaces of the metal and the bottle as can be collected, is heated and volatilized in a small porcelain crucible, in a place where the vapors are readily removed by draft; if a non-volatile residue remains, it is heated to redness. A small part of the residue is then heated in a test-tube with a few drops of concentrated hydrochloric acid; the solution is decanted from the insoluble residue, and one drop of solution of auric chloride is added; an ensuing gray or grayish-purple turbidity would indicate *tin.*

The rest of the residue in the crucible is treated with warm concentrated nitric acid; if only a partial solution takes place,

FIG. 73.

and at the same time a white precipitate is formed, this may be oxide of tin or antimony; in order to distinguish them, the precipitate is separated from the acid solution, washed with a little water, and subsequently heated upon charcoal before the blow-pipe; *stannous oxide* remains unchanged, while *tetroxide of antimony* volatilizes in white fumes, forming a white concentric incrustation on the coal (Fig. 73).

The nitric-acid solution is diluted with an equal bulk of water, and part of it is tested with solution of sodium sulphate;

a white precipitate would indicate *lead;* another part is poured into a large beaker full of water; a white opalescence or turbidity of the water indicates *bismuth.*

If lead be present, the rest of the nitric-acid solution is saturated and completely precipitated with hydrosulphuric-acid gas, and allowed to stand in a corked test-tube for some hours; it is then filtered and over-saturated with aqua ammoniæ; a white precipitate would indicate *zinc.*

If the precipitate is not quite white, and the lead has been completely removed, it might be due to traces of *iron*, of which metal, however, mercury can only contain traces, since it does not amalgamate with it.

HYDRARGYRUM AMMONIATUM.

HYDRARGYRUM AMIDATO-BICHLORATUM. HYDRARGYRUM AMMONIATUM BICHLORATUM. HYDRARGYRUM PRÆCIPITATUM ALBUM.

White Precipitate. Ammoniated Mercury. Ammonio-mercuric Chloride.

White, pulverulent, friable masses, or an opaque white powder, which, when heated, is entirely volatile without fusion; it becomes black when moistened with hydrosulphuric acid, gray when boiled with solution of stannous chloride, and pale yellow, with the evolution of ammonia, when heated with liquor potassæ.

Ammonio-mercuric chloride is insoluble in the common solvents, but is gradually decomposed by boiling with water, forming a yellow precipitate; it is readily and wholly soluble without effervescence in warm hydrochloric, nitric, and acetic acids, forming colorless solutions, which yield a white precipitate with liquor potassæ and with argentic nitrate, a black one with an excess of hydrosulphuric acid, and a red one with potassium iodide, and which produce a black stain upon bright, metallic copper, coating it, when rubbed thereon, with a brilliant metallic film.

Examination:

Mercuric chloride is detected by agitating about 10 grains of the powder with about two drachms of diluted alcohol; the filtrate is mixed with about half an ounce of hydrosulphuric acid; a black precipitate indicates mercuric chloride.

Mercurous chloride may be detected by a black coloration of the powder, when it is triturated with lime-water, or by dissolving about 10 grains of the powder in warm diluted nitric acid; if an insoluble residue remains, it is washed by decantation, and, when the water ceases to act on blue litmus-paper, the residue is agitated with limewater; if mercurous chloride, it will become black.

Plumbic Carbonate and Chloride, and Calcium Carbonate.—Carbonates are indicated by effervescence of the powder with acids, and plumbic chloride by its insolubility in diluted nitric acid; if a residue remains, it is washed, and boiled in strong acetic acid, and the liquid tested with one or two drops of sulphuric acid for lead; which may further be confirmed by agitating the powder with warm acetic acid, and testing the filtrate in two separate portions, with sodium sulphate and with potassium iodide, drop by drop, which both form precipitates with compounds of lead, the former a white, insoluble one, the latter a yellow one, soluble in an excess of the reagent.

These and all other non-volatile admixtures are also indicated by remaining behind when a few grains of the ammoniated mercury are heated and volatilized, in a narrow, dry test-tube. *Diammonio-mercuric chloride*, or fusible white precipitate, will be indicated in this test by a partial or complete fusion of the powder, previous to its volatilization, provided that the ammonio-mercuric chloride be free from any fixed fusible admixture.

Zinc and *magnesium oxides* may be detected, in the solution of the powder in nitric or acetic acid, by complete precipitation with hydrosulphuric acid, and by subsequent over-saturation of the filtrate with ammonium hydrate; a white precipitate will indicate either of these oxides; if the nature of the precipitate has to be determined, it is collected and washed upon a filter, and examined by the method described on page 43.

Starch.—An admixture of starch is detected by the microscope, and also by a blue coloration, when about five grains of the powder are triturated, and subsequently heated to boiling, with about two drachms of water, and then tested with one drop of iodinized potassium iodide.

IODOFORMUM.

IODOFORMIUM.

Iodoform. Teriodide of Formyl. Methenyl Iodide.

Small, lemon-yellow, friable, six-sided scales, of a pearly lustre, a peculiar, penetrating, and persistent odor, and a sweetish taste, and with a somewhat unctuous feel to the touch. Iodoform has a spec. grav. of 2.0, is volatile at common temperatures, and when heated in a dry tube, by immersing it in boiling water, sublimes at about 95° C., solidifying in small scales; it fuses at about 115° C., and is decomposed at 120° C., forming violet vapors, and being resolved into iodine and hydroiodic acid, with a residue of carbon, which burns away at a stronger heat.

Iodoform is almost insoluble in water, glycerin, diluted acids, and aqueous solutions of the alkaline and earthy hydrates, but is soluble in 80 parts of cold, and 12 parts of boiling, alcohol, in 20 parts of ether, and readily in chloroform, in carbon bisulphide, and in the fixed and volatile oils. The concentrated mineral acids, when cold, have no action on iodoform; when heated, it remains unchanged with hydrochloric acid, gives a reddish-brown solution with nitric acid, remaining limpid and brown on dilution with water; it is freely dissolved, with a violet color, by hot sulphuric acid; upon dilution, however, the color disappears, and the iodoform is separated again in small yellow scales. It is not acted upon by the aqueous solutions of the alkaline hydrates, but their alcoholic solutions dissolve and decompose it, forming alkaline iodide and formiate.

IODUM.

IODINUM. IODINIUM.

Iodine.

Heavy, brilliant, crystalline plates or scales, of an opaque bluish-black appearance and imperfect metallic lustre, and of a peculiar odor, resembling faintly that of chlorine. Its specific gravity is 4.948; it melts at 107° C., and boils at 175° C.; it is, however, slowly volatile at common temperatures. When heated in a dry tube (Fig. 74), iodine melts and rises in deep violet vapors, which condense in the cooler parts of the tube to small, brilliant crystals.

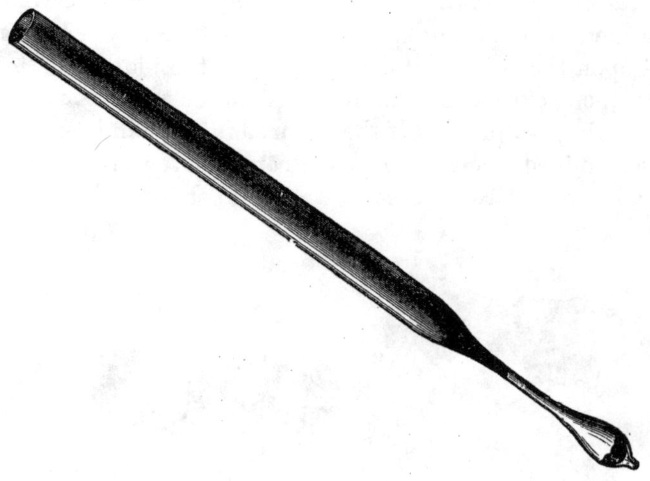

Fig. 74.

Iodine is but sparingly soluble in water, requiring 5,524 parts of it at 12° C., and imparting to it a faint yellowish tinge. It is more soluble in glycerin, 100 parts of which dissolve a little more than 1½ part of iodine. It is also soluble to some extent in the aqueous solutions of certain salts, as for instance of ammonium chloride and nitrate. Aqueous solutions of hydriodic acid and of the alkaline iodides and bromides, dissolve iodine freely, as do also alcohol and ether, with a reddish-brown color, benzol and chloroform with a violet-red, and carbon bisulphide with a rich purple. An aqueous solution of sodium hyposulphite dissolves iodine at first without color, and afterward with a brownish-red tint.

Chloroform and carbon bisulphide, when shaken with an aqueous solution of iodine, deprive it of most of the iodine, and assume, when the fluids have separated, a more or less red color, while the aqueous solution appears almost colorless.

Iodine forms with starch a deep-blue compound, which offers a very delicate test for iodine in all solutions and in bodies which contain it in the free state.

Examination:

Moisture is indicated in iodine by its adhering to the surface of the bottles, and by a sticky coherence of the scales.

Fixed and *insoluble admixtures* (graphite, coal, carburet of iron, metallic oxides or sulphides) are detected by remaining behind upon the volatilization of a little of the iodine in a test-tube, or upon solution of it in alcohol or in an aqueous solution of sodium hyposulphite. If the nature of such admixtures has to be determined, the residue is collected and washed upon a filter, and afterward treated with warm hydrochloric acid

FIG. 75.

diluted with an equal bulk of water, which dissolves metallic oxides, and to some extent the sulphides, with the evolution of hydrosulphuric acid. The obtained solution may further be examined for metals, as described in the systematic course of analysis (pages 41–43). The insoluble residue left from the solution in hydrochloric acid is levigated (Fig. 75), whereby graphite and carburet of iron may be separated and distinguished from heavier mineral substances.

Cyanogen iodide may be detected by triturating about 30 grains of the iodine with about 2 drachms of tepid water, and by subsequent agitation of the mixture for a few minutes; it is then filtered and washed with a few drops of water, and to the filtrate is added drop by drop so much of an aqueous solution of sulphurous acid as to decolorize it; then a few drops of solution of ferrous sulphate, and subsequently of liquor potassæ, are added, and the turbid liquid is then slightly oversaturated with diluted hydrochloric acid; if a blue precipitate takes place now, either at once or gradually, and, upon warming, cyanogen iodide is indicated.

The volumetric estimation of iodine has already been described on page 64.

LIQUOR AMMONII ACETATIS.

LIQUOR AMMONII ACETICI.

Solution of Ammonium Acetate.

A clear, colorless liquid, without empyreumatic odor, and of a mild, saline taste; it contains about six per cent. of neutral ammonium acetate, and has a spec. grav. of from 1.012 to 1.015 (1.028–1.032, Pharm. German.); it is wholly volatile upon evaporation, and emits the odor of ammonia when heated with potassium hydrate, and that of acetic acid when heated with sulphuric acid; it assumes a red color upon the addition of a trace of ferric chloride.

Examination:

Metallic impurities may be detected by mixing the solution with an equal bulk of hydrosulphuric acid, and then over-saturating it, first with acetic acid, and subsequently with aqua ammoniæ.

Sulphates and *chlorides* may be detected by a white turbidity, when the liquid is acidulated with acetic acid and tested, in separate portions, with barium nitrate for the former salts, and with argentic nitrate for the latter.

LIQUOR ANTIMONII CHLORIDI.

LIQUOR STIBII CHLORATI. BUTYRUM ANTIMONII CHLORIDI.

Solution of Trichloride of Antimony or of Antimonious Chloride.

A dense, transparent, colorless or pale-yellow liquid, of 1.470 spec. grav. (1.36, Pharmacopœa Germanica). Dropped into water, it gives a white, bulky precipitate (antimonious chloride with antimonious oxide—*Algaroth's Powder*), which is redissolved upon addition of potassium hydrate or tartaric acid. The solution with potassium hydrate remains unchanged, or gives only a slight turbidity, with hydrosulphuric acid, and yields a black precipitate with argentic nitrate, while the solution with tartaric acid gives a copious orange-red precipitate with hydrosulphuric acid, and a white one with argentic nitrate.

Examination:

About 40 drops of the liquor antimonii chloridi are added to a solution of half a drachm of tartaric acid in half an ounce of water; this solution may serve for the following tests:

Antimonic chloride is detected by a yellow coloration of the solution upon the addition of a few drops of potassium iodide.

Lead and *copper* are detected in the solution, the former by a white precipitate with dilute solution of sodium sulphate, the latter by a bluish coloration with aqua ammoniæ in excess.

Arsenic may be detected by heating, in a test-tube, about two drachms of a mixture consisting of equal volumes of the liquor antimonii chloridi and concentrated hydrochloric acid, with a strip of tin-foil (real tin) or with about 20 drops of concentrated solution of stannous chloride; a brown precipitate would indicate arsenic, which may further be identified by collecting the precipitate by decantation, and by subsequent washing with a little solution of tartaric acid; it is then dried, and may be examined by heating part of it with exsiccated sodium carbonate, upon charcoal, before the blow-pipe (Fig. 73, p. 239), as well as by heating another portion of it, with a little potassium cyanide, in a reduction-tube (Fig. 63, p. 213); arsenic will be recognized by its smell in the first test, and by a metallic mirror in the second.

LIQUOR CALCIS.

AQUA CALCIS. AQUA CALCARIÆ.

Lime-Water. Solution of Calcium Hydrate.

A saturated solution of calcium hydrate, containing nearly one grain of the hydrate, or nearly three-quarters of a grain of calcium oxide, in each ounce of water. Lime-water has an alkaline reaction upon test-paper, and absorbs carbonic acid from the air, forming on its surface a pellicle consisting of minute plates of calcium hydrocarbonate; its alkaline reaction disappears when an excess of carbonic-acid gas has been passed through it, and the excess has been expelled afterward by boiling.

Lime-water affords no precipitate with sulphuric acid (distinction from solutions of barium or strontium hydrate), but it forms white precipitates with carbonic, boracic, phosphoric, arsenious and arsenic, oxalic, and tartaric acids and their salts, and precipitates the solutions of those salts whose metallic oxides or hydrates are insoluble in water.

The quality of lime-water is best ascertained by its property, when warmed in a test-tube, of separating nearly half the quantity of calcium hydrate in minute hexagonal prisms; upon cooling, the crystals redissolve, and the water becomes perfectly clear again.

LIQUOR FERRI ACETATIS.

LIQUOR FERRI ACETICI.*

Solution of Ferric Acetate.

A transparent, dark, reddish-brown liquid, of from 1.134 to 1.138 spec. grav.; it has a faint odor of acetic acid, which ap-

* *Liquor Ferri acetici*, of the Pharmacopœa Germanica, is prepared by complete precipitation of 10 parts of solution of ferric sulphate, of 1.31 spec. grav., with eight parts of aqua ammoniæ, of 0.960 spec. grav., both greatly diluted with water; the precipitate is collected, and washed upon a flannel or felt filter, and, when the water has dropped off, is subjected to a gentle pressure; the soft, humid mass of ferric hydrate is then transferred into a flask, and dissolved in six parts of diluted acetic acid, of 1.040 spec. grav.

pears, however, strongly upon warming; this may also be recognized by the formation of white vapors, when a glass rod, moistened with aqua ammoniæ, is brought near the liquid; it produces a deep-blue precipitate, when a few drops are allowed to fall into a mixture of several ounces of water with a few drops of hydrochloric acid and of solution of potassium ferrocyanide.

Examination:

About one fluid-drachm of the liquid is diluted with two drachms of water, and completely precipitated with aqua ammoniæ; the filtrate must be wholly volatile when a few drops of it are evaporated in a porcelain capsule; a viscid residue, which becomes charred, at a stronger heat, with the evolution of vapors having the odor of caramel, would indicate *sugar* or *fruit-acids*, which, when present in considerable quantities, prevent the complete precipitation of the ferric solution by ammonium hydrate. A bluish tint of the filtrate would indicate *copper*, which, with other metallic impurities, may further be ascertained by mixing it with hydrosulphuric acid, and subsequently over-saturating with acetic acid.

LIQUOR FERRI NITRATIS.

LIQUOR FERRI NITRICI. LIQUOR FERRI PERNITRATIS.

Solution of Pernitrate of Iron. Solution of Ferric Nitrate.

A pale-yellow liquid, of a spec. grav. between 1.060 and 1.070, and of a chalybeate, astringent, acid taste. Added to water, it gives a deep-blue precipitate with potassium ferrocyanide, but none with potassium ferricyanide, and yields a reddish-brown precipitate with aqua ammoniæ; when a few drops of a concentrated solution of ferrous sulphate are added to a little of the solution of ferric nitrate, and the mixture is carefully transferred upon concentrated sulphuric acid (Fig. 71, p. 233), a dark zone, indicating nitric acid, will ensue upon the line of contact between the two liquids.

One fluidounce of the liquid, completely precipitated by aqua ammoniæ, yields a reddish-brown precipitate of ferric hydrate, which, when washed, dried, and ignited, weighs between 8 and 10 grains.

Examination:

Metals.—In the ammoniated filtrate of the preceding quantitative test, or in the liquor ferri nitrici, when completely precipitated at common temperature, by aqua ammoniæ, and filtered, *copper* will be indicated by a blue color of the liquid, and *zinc* by a white turbidity upon the addition of a few drops of ammonium sulphide.

Chloride and *sulphate* may be detected, in the diluted solution, by white precipitates when tested with argentic and with barium nitrates.

LIQUOR FERRI SULPHATIS.

LIQUOR FERRI SULPHURICI OXYDATI.

Solution of Ferric Sulphate.

The U. S. Pharmacopœia has two solutions of ferric sulphate, *Liquor Ferri subsulphatis*, having a spec. grav. of 1.552, and *Liquor Ferri tersulphatis*, having a spec. grav. of 1.320. Both are solutions of ferric sulphate, the former one being less acid, and containing some basic ferric sulphate, and usually some nitrate. The *Liquor Ferri persulphatis* of the British Pharmacopœia has the spec. grav. of 1.441, and that of the Pharmacopœa Germanica a spec. grav. of 1.319.

They all are transparent, red or reddish-brown liquids, without odor, of an astringent, metallic taste, and miscible in all proportions, with water, alcohol, and glycerin. A few drops of either of them, added to water, form a mixture in which potassium ferricyanide produces no reaction, but ferrocyanide gives a dark-blue precipitate, aqua ammoniæ a bulky, reddish-brown one, and barium chloride a white one.

Examination:

Copper and Zinc.—About two drachms of the liquor ferri are diluted with about two ounces of water, and completely precipitated with aqua ammoniæ; the whole is heated, and subsequently filtered; the filtrate will appear bluish, if *copper* be present, and should be entirely volatile upon platinum-foil; a fixed residue would indicate *alkaline*, *earthy*, or *metallic impurities*. Part of the filtrate is mixed with an equal volume of hydrosulphuric acid; an ensuing white precipitate would indicate *zinc*, and a dark one, insoluble upon over-saturation with hydrochloric acid, *copper*.

Nitric acid and *nitrates* may be detected in a portion of the filtrate of the preceding test, by over-saturating it with concentrated sulphuric acid, and by subsequently adding one drop of a solution of potassium permanganate, or indigo solution, and gently warming. A decoloration will indicate *nitric acid* and *nitrates*.

LIQUOR HYDRARGYRI NITRATIS.

LIQUOR HYDRARGYRI NITRICI OXYDATI.

Solution of Pernitrate of Mercury. Solution of Mercuric Nitrate.

A dense, transparent, nearly colorless, acid liquid, of a spec. grav. of 2.165, when prepared according to the U. S. Pharmacopœia, and of 2.246, when prepared according to the British Pharmacopœia. When a few drops of it are evaporated at a gentle heat, upon platinum-foil, they leave a white residue, which, upon increased heat, becomes successively yellow, red, brown, and is finally wholly dissipated. The solution remains limpid on the addition of water or of diluted hydrochloric acid (evidence of the absence of subnitrate); it gives a dull yellow precipitate with an excess of the fixed alkaline and earthy hydrates, a white one with ammonium hydrate, and a black one with an excess of hydrosulphuric acid; it deposits a brilliant metallic coating on bright copper, and shares, in its deportment with reagents, the general characteristics of mercuric salts, as described under mercuric chloride and oxide (pages 225 and 233). It causes a crystal of ferrous sulphate, dropped into it, as well as the liquid around the salt, to assume a deep-brown color.

When diluted with about ten times its bulk of water, it should not give a turbidity when tested, in separate portions, with a few drops of solutions of argentic and of barium nitrates.

LIQUOR PLUMBI SUBACETATIS.

LIQUOR PLUMBI SUBACETICI. ACETUM PLUMBICUM.

Solution of Subacetate of Lead. Solution of Triplumbic Acetate.

A dense, clear, colorless liquid, of 1.267 spec. grav. (1.235–1.240 Pharmacopœa Germanica), having an alkaline reaction and a sweet, astringent taste, and becoming turbid by absorption of atmospheric carbonic acid, and by dilution with water containing carbonates, sulphates, or carbonic acid. It is precipitated, whether diluted with water or not, by the alkaline and alkaline-earthy hydrates and carbonates, by sulphuric, hydrochloric, oxalic, tannic, and other acids and their salts, and by almost all neutral salts; it forms white, opaque, insoluble compounds with vegetable gums, mucilages, and extracts, and with vegetable and albuminous substances.

Liquor plumbi subacetatis gives a yellow precipitate with potassium iodide, and a black one with hydrosulphuric acid; it forms an opaque, white jelly when mixed with mucilage of gum; it may be recognized as containing an acetate, by evolving the odor of acetic acid, when heated with a few drops of sulphuric acid.

Traces of copper are indicated by a faint greenish color of the liquid, and may be further recognized by a bluish coloration of the filtrate, when a little of the liquor plumbi subacetatis is mixed with an excess of aqua ammoniæ.

LIQUOR POTASSÆ.

LIQUOR POTASSII HYDRICI. LIQUOR KALI CAUSTICI.

Solution of Potassa. Solution of Potassium Hydrate.

A colorless, limpid liquid, without odor, and of an acrid, caustic taste and a soapy feel when rubbed between the fingers; it has a destructive action on vegetable and animal substances, and is a powerful solvent for many organic and mineral compounds; it absorbs carbonic acid from the air; its spec. grav. is 1.065, and it contains 5.80 per cent. of potassium hydrate.*

Examination:

Carbonate is indicated by effervescence or by the formation of gas-bubbles, when an equal volume of acetic acid is added to the liquor potassæ, or it may also be detected by the formation of a white precipitate when a little of the liquor potassæ is mixed with an equal bulk of water, and is then added to lime-water.

Potassium chloride, sulphide, and *hyposulphite,* may be detected by dropping a little of the liquor potassæ into diluted solution of argentic nitrate; a grayish-brown precipitate will take place, completely soluble upon addition of nitric acid in excess; if the precipitate does not wholly dissolve, and leaves behind a white residue, *chloride* is indicated; when the residue is black, *sulphide* or *hyposulphite.*

Sulphate, Silicate, and Alumina.—A little of the liquor potassæ is slightly over-saturated with diluted nitric acid; part of the solution is tested with barium nitrate for *sulphate;* another part may also be tested with argentic nitrate for *chloride;* the rest of the solution is evaporated, in a porcelain capsule, to dryness; the remaining salt must yield a limpid solution with water; a white turbidity would indicate silicate; the solution, when necessary, is filtered, and then tested with a few drops of ammonium chloride and aqua ammoniæ for *alumina,* which, when present, will cause a white precipitate.

Calcium salts may be detected, in the diluted liquor potas-

* Liquor Kali caustici, of the Pharmacopœa Germanica, has a spec. grav. of from 1.330 to 1.334, and contains 33.3 per cent. of potassium hydrate, or from 28 to 29 per cent. of potassium oxide.

sæ, by a white precipitate with ammonium oxalate, or with sodium carbonate.

Metallic impurities are indicated by a dark coloration or turbidity when the liquor potassæ is mixed with twice its volume of hydrosulphuric acid, and subsequently over-saturated with acetic acid.

For *Volumetric Estimation*, *see* page 58.

TABLE

OF THE QUANTITY BY WEIGHT OF POTASSIUM OXIDE CONTAINED IN 100 PARTS BY WEIGHT OF SOLUTION (LIQUOR POTASSÆ) AT DIFFERENT DENSITIES.

Temperature, 17.5° C.

Specific Gravity.	Percent. of Pot. Oxide.	Specific Gravity.	Percent. of Pot. Oxide.	Specific Gravity.	Percent. of Pot. Oxide.	Specific Gravity.	Percent. of Pot. Oxide.
1.576	45	1.414	34	1.269	23	1.135	12
1.568	44.5	1.407	33.5	1.263	22.5	1.129	11.5
1.560	44	1.400	33	1.257	22	1.123	11
1.553	43.5	1.393	32.5	1.250	21.5	1.117	10.5
1.545	43	1.386	32	1.244	21	1.111	10
1.537	42.5	1.379	31.5	1.238	20.5	1.105	9.5
1.530	42	1.372	31	1.231	20	1.099	9
1.522	41.5	1.365	30.5	1.225	19.5	1.094	8.5
1.514	41	1.358	30	1.219	19	1.088	8
1.507	40.5	1.352	29.5	1.213	18.5	1.082	7.5
1.500	40	1.345	29	1.207	18	1.076	7
1.492	39.5	1.339	28.5	1.201	17.5	1.070	6.5
1.484	39	1.332	28	1.195	17	1.065	6
1.477	38.5	1.326	27.5	1.189	16.5	1.059	5.5
1.470	38	1.320	27	1.183	16	1.054	5
1.463	37.5	1.313	26.5	1.177	15.5	1.048	4.5
1.456	37	1.307	26	1.171	15	1.042	4
1.449	36.5	1.301	25.5	1.165	14.5	1.037	3.5
1.442	36	1.294	25	1.159	14	1.031	3
1.435	35.5	1.288	24.5	1.153	13.5	1.026	2.5
1.428	35	1.282	24	1.147	13	1.021	2
1.421	34.5	1.275	23.5	1.141	12.5	1.015	1.5

With the decrease and increase of temperature, the density of the solution suffers a corresponding increase or decrease, amounting, for each degree of the centigrade thermometer, in either direction—

For solution of a specific gravity of 1.576 to that of 1.500, to about 0.00055.
" " " " 1.484 " 1.358 " 0.0005.
" " " " 1.345 " 1.231 " 0.0004.
" " " " 1.219 " 1.111 " 0.00033.

LIQUOR POTASSII ARSENITIS.

LIQUOR POTASSII ARSENICOSI. LIQUOR KALI ARSENICOSI.

Solution of Potassium Arsenite. Fowler's Solution.

The solution of potassium arsenite has a slight alkaline reaction; it gives, with nitrate of silver, a bright-yellow precipitate, soluble in aqua ammoniæ; this solution, when gently warmed for some time, by immersing the test-tube in hot water, suffers a reduction of the silver salt, and deposits the metal, as a brilliant coating, upon the walls of the test-tube. Hydrosulphuric acid produces no immediate precipitate in the solution of potassium arsenite, but, upon the addition of hydrochloric acid, there at once appears a lemon-yellow precipitate, soluble in ammonium hydrate or carbonate.

A quantitative estimation of the arsenious acid contained in liquor potassii arsenitis may be made, by completely precipitating, with hydrosulphuric acid, 10 drachms of the solution, diluted with an equal volume of water, and acidulated with hydrochloric acid; the precipitate is collected and washed upon a tared filter, and, when completely dried, is weighed. The weight of the arsenious sulphide, divided by 1.242, gives the quantity of arsenious acid contained in 10 drachms of the solution, which should be five grains.

The quantitative estimation may also be made by the volumetric test: 68.30 grammes of the solution of potassium arsenite, neutralized with about 25 grains of sodium bicarbonate, and diluted with an equal bulk of water to which a little mucilage of starch has been added, will require 100 cubic centimetres of the test-solution of iodine (page 63).

LIQUOR SODÆ.

LIQUOR SODII HYDRICI. LIQUOR NATRI CAUSTICI.

Solution of Soda. Solution of Sodium Hydrate.

A colorless, limpid liquid, of an acrid, caustic taste, a soapy feel, and a strong alkaline reaction; it has a destructive and solvent action on vegetable and animal matters, absorbs carbonic acid from the atmosphere, has a spec. grav. of 1.071, and contains 5.7 per cent. of sodium hydrate.*

Examination:

Sodium carbonate is indicated by effervescence, or by the formation of gas-bubbles, when the liquid is mixed with concentrated hydrochloric acid; it may also be detected by the formation of a white precipitate upon mixing a little of the liquid with twice its volume of lime-water.

Sodium sulphate and *chloride* are indicated by white precipitates, when the diluted liquid is slightly over-saturated with diluted nitric acid, and tested with barium nitrate for sulphate, and with argentic nitrate for chloride.

Calcium salts may be detected by a white precipitate, when the diluted liquid is tested with solution of sodium carbonate.

Potassium hydrate may be recognized by a white, granular precipitate, on dropping the liquid into a strong solution of tartaric acid, allowing the latter to remain in excess.

TABLE

OF THE QUANTITY BY WEIGHT OF SODIUM OXIDE CONTAINED IN 100 PARTS BY WEIGHT OF SOLUTION (LIQUOR SODÆ) AT DIFFERENT DENSITIES.

TEMPERATURE 17.5° C.

Specific Gravity.	Percent. of Sod. Oxide.	Specific Gravity.	Percent. of Sod. Oxide.	Specific Gravity.	Percent. of Sod. Oxide.	Specific Gravity.	Percent. of Sod. Oxide.
1.500	35	1.389	27.5	1.281	20	1.174	12.5
1.492	34.5	1.382	27	1.274	19.5	1.167	12
1.485	34	1.375	26.5	1.266	19	1.160	11.5
1.477	33.5	1.367	26	1.259	18.5	1.153	11
1.470	33	1.360	25.5	1.252	18	1.146	10.5
1.463	32.5	1.353	25	1.245	17.5	1.139	10
1.455	32	1.345	24.5	1.238	17	1.132	9.5
1.448	31.5	1.338	24	1.231	16.5	1.125	9
1.440	31	1.331	23.5	1.224	16	1.118	8.5
1.433	30.5	1.324	23	1.217	15.5	1.111	8
1.426	30	1.317	22.5	1.210	15	1.104	7.5
1.418	29.5	1.309	22	1.203	14.5	1.097	7
1.411	29	1.302	21.5	1.195	14	1.090	6.5
1.404	28.5	1.295	21	1.188	13.5	1.083	6
1.396	28	1.288	20.5	1.181	13	1.076	5.5

* Liquor Natri caustici of the Pharmacopœa Germanica has a spec. grav. of from 1.330 to 1.334, and contains from 30 to 31 per cent. of sodium hydrate, or about 23.5 per cent. of sodium oxide.

With the decrease and increase of temperature, the density of the solution suffers a corresponding increase or decrease, amounting, for each degree of the centigrade thermometer, in either direction—

For solution of a specific gravity of 1.500 to that of 1.353, to about 0.00045.
" " " " 1.345 " 1.210 " 0.0004.
" " " " 1.203 " 1.076 " 0.00033.

LITHII CARBONAS.

LITHIUM CARBONICUM.

Carbonate of Lithium. Lithium Carbonate.

An odorless, white, granular powder, fusible at a high temperature; when heated in an alcohol-flame, upon the looped end of platinum wire, previously moistened with hydrochloric acid, it imparts a crimson color to the flame.

Lithium carbonate is but sparingly soluble in water or alcohol, requiring about 108 parts of the former, at 16° C., for solution; it is readily dissolved, with effervescence, by diluted acids; its solution in diluted hydrochloric acid, when evaporated to dryness, leaves a residue which is readily and completely soluble in a few drops of a mixture of equal parts of alcohol and ether (distinction from potassium and sodium chlorides). When this latter solution is poured into a small porcelain capsule, and ignited, it burns with a red flame; the residue left in the capsule after the ignition is then dissolved in a few drops of water, and added to a very dilute solution of sodium phosphate, to which one drop of solution of ammonium chloride has been added; a white, crystalline precipitate of lithium phosphate, readily soluble in hydrochloric acid, will appear.

Examination:

One grain of the lithium carbonate is placed in a small test-tube, and two drachms of cold water, exactly weighed, are gradually added to the carbonate, with frequent agitation, closely observing the point when complete solution of the salt takes place; it must not occur until nearly the whole of the water is added; otherwise an admixture of alkaline or other salts is indicated; in this case, the presence of the former may be ascertained by the above-described method, depending upon the solubility of lithium chloride in a mixture of alcohol and ether.

Calcium salts may be detected in the aqueous solution of the lithium carbonate, previously neutralized with hydrochloric acid, by a white turbidity with ammonium oxalate.

Magnesium and *aluminium* may be detected in the solution, neutralized with hydrochloric acid, by testing it, in two separate portions, with lime-water and with sodium carbonate; a white turbidity would indicate the presence of compounds of either of these elements.

Metallic impurities are detected, in the aqueous solution, by hydrosulphuric acid, and subsequent acidulation with hydrochloric acid.

LITHII CITRAS.

LITHIUM CITRICUM.

Citrate of Lithium. Lithium Citrate.

A white, amorphous, deliquescent powder, soluble in 25 parts of water, and also soluble in alcohol. Heated in a porcelain capsule, it blackens, evolves inflammable vapors, and leaves a white residue, which, when dissolved in a little alcohol, with one or two drops of hydrochloric acid, and ignited, imparts a crimson color to the flame.

When the aqueous solution of lithium citrate is completely precipitated with calcium chloride, the filtrate, when heated, will become turbid, and when filtered after cooling, and the filtrate reheated to boiling, it becomes turbid again (evidence of the presence of citric acid).

Ten grains of lithium citrate, heated for about 15 minutes, in a tared porcelain crucible, at a low red heat, with free access of air, leave 5.3 grains of a white residue of lithium carbonate.

Examination:

Potassium salts are detected in the concentrated solution of the citrate, by a white, crystalline precipitate, upon the addition of a few drops of concentrated solution of sodium bitartrate.

Sodium salts are detected in the solution, by a white precipitate when tested with potassium antimoniate.

The presence of potassium and sodium salts may also be ascertained by dissolving, in one or two drops of diluted hydrochloric acid, the residue of lithium carbonate obtained by incineration of the citrate; this solution is evaporated to dryness, and is subsequently dissolved in a few drops of a mixture of equal parts of alcohol and ether; a complete solution should take place, as an insoluble residue would indicate potassium or sodium chlorides.

Metallic impurities may be detected, in the solution, by hydrosulphuric acid.

MAGNESIA.

MAGNESIA USTA. MAGNESII OXIDUM. MAGNESIUM OXYDATUM.

Magnesia. Calcined Magnesia. Magnesium Oxide.

A white, inodorous, bulky, more or less light powder, of a slightly alkaline taste, and an alkaline reaction upon moist blue litmus-paper; when exposed to heat, it suffers no change whatever.

Magnesia is almost insoluble in water, but unites readily with one equivalent of water, at once when mixed with it, or slowly on exposure to the atmosphere, forming a hydrate which is soluble in about 5,000 parts of cold, and 36,000 parts of boiling, water, and which attracts carbonic acid from the air.

When triturated with water, magnesia must dissolve without effervescence, upon the addition of sulphuric acid (evidence of the absence of carbonate), and must form a clear solution (evidence of the absence of calcium, barium, and strontium oxides); this solution may be divided into two portions, one of which is mixed with an equal volume of hydrosulphuric acid, and is subsequently slightly over-saturated with aqua ammoniæ; a dark coloration or turbidity, either before or after the addition of the ammonium hydrate, would indicate metallic impurities, and a white precipitate, on the addition of the aqua ammoniæ, *zinc oxide;* the filtrate is tested with ammonium oxalate for *calcium*. The second portion of the solution in sulphuric acid is over-saturated with aqua ammoniæ, and tested with sodium phosphate, which will produce a copious, white, crystalline precipitate of ammonio-magnesium phosphate.

Magnesia is liable to contain the impurities of the magnesium carbonate from which it has been obtained, and may be examined for them, if they have not been ascertained by the preceding tests for identity and purity, by the methods described on page 260.

MAGNESII CARBONAS.

MAGNESIA CARBONICA. MAGNESIA ALBA.

Carbonate of Magnesium. Magnesium Carbonate.

White, bulky, pulverulent masses, commonly in square cakes, or a light, white powder, smooth to the touch, and nearly insoluble in water, but soluble with effervescence in dilute acids, yielding limpid, colorless solutions; these, after the addition of a little solution of ammonium chloride, are not precipitated upon slight over-saturation with ammonium hydrate, and render a copious white precipitate of ammonio-magnesium phosphate, upon the addition of sodium phosphate.

Magnesium carbonate is decomposed at a red heat, and also by all acids, and by the fixed alkaline hydrates.

One drachm of magnesium carbonate requires for saturation 7.93 grains of citric, and 94.23 grains of tartaric, acid; 100 parts of it, when calcined at a red heat, leave 40 to 43 parts of magnesium oxide.

Examination:

About 20 grains of the powdered magnesium carbonate are mixed and agitated with about one ounce of warm water; the filtrate is tested with turmeric-paper, and, if this becomes brown, *alkaline carbonates* are indicated; when a few drops of the filtrate are evaporated upon platinum-foil, only a very slight residue should remain. The magnesium carbonate left on the filter is rinsed into a flask, by means of a washing-bottle; the mixture is warmed, and sulphuric acid added, drop by drop, until solution is effected; a remaining slight turbidity would indicate traces of *silicic acid*. The solution is filtered, if necessary, and saturated with hydrosulphuric-acid gas, and is then rendered alkaline by the addition of a strong solution of ammonium carbonate; an ensuing greenish turbidity would indicate salts of *iron;* a light reddish one, salts of *manganese;* a white one, not disappearing upon the addition of aqua ammoniæ, salts of *aluminium* or *zinc* (the incidental presence of phosphates would also give a white precipitate). In order to distinguish them, the precipitate is washed, and subsequently dissolved in a few drops of liquor potassæ, and the solution diluted, and tested with ammonium chloride, which precipitates *aluminium hydrate*, while *zinc* remains in solution, and may be recognized by reprecipitation with hydrosulphuric acid.

The ammoniacal filtrate is then tested with a few drops of ammonium oxalate; a white precipitate, insoluble upon the addition of ammonium chloride, would indicate salts of *calcium.*

Chlorides and *sulphates* may be detected, in the diluted solution of the magnesium carbonate in diluted sulphuric acid, by testing the same in separate portions, with barium nitrate for sulphates, and with argentic nitrate for chloride.

MAGNESII SULPHAS.

MAGNESIA SULPHURICA.

Epsom Salt. Sulphate of Magnesium. Magnesium Sulphate.

Colorless, transparent, rhombic prisms, but usually met with in commerce as small, acicular needles; they contain seven molecules (51.22 per cent.) of water of crystallization, six of which pass off at 120° C.; the last molecule is not expelled below 220° C.; the crystals do not effloresce at common temperatures and in ordinary atmospheric humidity, but they do so slowly in warm, dry air. When heated, they undergo aqueous fusion, give out their water of crystallization, and at a red heat undergo igneous fusion, with partial decomposition.

Magnesium sulphate dissolves in three parts of cold, and in one part of boiling, water, but is insoluble in alcohol; its aqueous solution has a nauseous, bitter taste, and a neutral reaction on test-paper; it is decomposed, and gives white precipitates, with the fixed alkaline hydrates and carbonates, and also with the earthy hydrates and their soluble salts; ammonium hydrate and carbonate do not at once cause a precipitate in dilute solutions of magnesium sulphate, or, if so, only an incomplete one, since ammonium salts, when present or formed in acidulous solutions, hinder or retain this reaction; but, on addition of phosphoric acid or solutions of tri-basic phosphates, a complete precipitation takes place, which precipitate, however, is soluble in dilute acids.

The crystals of magnesium sulphate are isomorphous with those of zinc sulphate, and cannot be distinguished from them by the eye; it is easy, however, to discriminate between them, not only by the difference in taste, but also by the action of a few drops of ammonium sulphydrate on their aqueous solutions; that of magnesium sulphate, in this case, remains unchanged, while solution of zinc sulphate yields a white precipitate.

Examination:

Metals may be detected by the occurrence of a turbidity, when the concentrated solution of magnesium sulphate is mixed with twice its bulk of hydrosulphuric acid, and when subsequently a few drops of ammonium sulphydrate are added; a white precipitate with the latter reagent would indicate *zinc;* when a dark precipitate is formed, *copper* and *iron* are indicated, and may be confirmed in the slightly-acidulated solution of the salt, the former by a reddish-brown precipitate, the latter by a blue one, with potassium ferrocyanide.

Alkaline sulphates may be detected by boiling, in a porcelain capsule, to about half its original bulk, a solution of about 20 grains of the magnesium sulphate in three ounces of water, with one drachm of barium carbonate; *ammonium sulphate* will be recognized during the ebullition by the odor of ammonia, and by white fumes when a glass rod, moistened with acetic acid, is held over the hot liquid; *sodium sulphate* will be indicated by the alkaline reaction of the filtrate with litmus and turmeric papers. *Calcium* salts may be detected, in the diluted solution of the salt, by a white turbidity with ammonium oxalate.

One hundred grains of magnesium sulphate, dissolved in boiling water, and completely precipitated by a boiling solution of sodium carbonate, yield a precipitate which, when washed and dried, weighs 34 grains, and when calcined at a red heat, 16.26 grains.

A *quantitative estimation of magnesium sulphate* may be afforded, by making a solution of 100 grains of the salt in water to which subsequently have been added a little solution of ammonium chloride and some aqua ammoniæ, and completely precipitating it with sodium phosphate; the mixture is allowed to stand for 10 or 12 hours, when the precipitate is collected upon a filter, washed with very dilute aqua ammoniæ, and, when dry, completely incinerated in a porcelain crucible. The weight of the residue, divided by 2.775, shows the percentage of magnesium oxide; and when the number thus obtained is multiplied by 2.216, the product represents the percentage of crystallized magnesium sulphate in the salt under examination.

MANGANESII OXIDUM NIGRUM.

MANGANUM HYPEROXYDATUM. MANGANESIUM OXYDATUM NATIVUM.

Black Oxide of Manganese. Pyrolusite. Manganese Dioxide.

Heavy, compact masses, of a dull-black or brownish-black, earthy appearance, or masses of acicular or rhombic crystals of a black, metallic lustre, and, if pure pyrolusite, of a spec. grav. of 4.9. In commerce, it occurs usually ground, as a coarse, dull, black powder, consisting of manganese dioxide, sesqui-oxide, and monoxide, and is contaminated with the gangue (quartz, felspar, barytes, limestone, etc.), which frequently amount to 40 or 50 per cent.

Manganese dioxide is infusible and unalterable by heat, except that it loses oxygen ; it does not combine with acids, but is decomposed by them ; it is insoluble in water. When a particle of it is heated to redness upon platinum-foil, with a few grains of potassium hydrate and nitrate or chlorate, it yields a dark-green fuse, which dissolves in water, with a green color, changing to purple when the solution is boiled. When heated in a test-tube, with hydrochloric acid, chlorine-gas is evolved, and a brown solution obtained, which, when filtered, and saturated with aqua ammoniæ, gives a flesh-colored precipitate with hydrosulphuric acid ; the color of this precipitate is, however, frequently rendered darker, or even brownish black, by the presence of oxides of iron and other metals.

Since the value of pyrolusite, for its application in the arts and trades, depends less upon the nature of its impurities than upon the percentage of real manganese dioxide, an examination of the mineral is invariably required before its application, and is mainly directed to the determination of the amount of dioxide.

Among the several methods of conducting the assay, the two following are simple and accurate, the one being an approximate, the other a quantitative one:

1. Ten grains of the finely-powdered black manganese dioxide are added, in a small flask, to a solution of 40 grains of granular ferrous sulphate in two drachms of water, and, when mixed by gentle agitation, one fluid-drachm of concentrated hydrochloric acid is added, and the mixture allowed to stand in a warm place, with occasional gentle agitation, for several hours ; a few drops of diluted hydrochloric acid are then added, and the mixture heated to boiling, and, after a while, filtered ; the filtrate is diluted, and tested with potassium ferricyanide ; if it gives no blue precipitate, the test bears evidence that the pyrolusite contains at least 60 per cent. of real manganese dioxide ; if a blue precipitate takes place, the peroxide is wanting in that strength in proportion to the amount of the precipitate.

2. Fifty grains of the black oxide of manganese, in a fine powder, are carefully introduced into the flask A (Fig. 76) of the little apparatus described on page 60, into which previously have been poured 100 grains of concentrated hydrochloric acid

and about half an ounce of water; 50 grains of pure, crystallized oxalic acid are then added, the cork carrying the tubes is fitted, and the whole apparatus quickly weighed or counterpoised; the flask B is charged with a little concentrated sulphuric acid, through which the evolved carbonic-acid gas has to pass, and which absorbs and retains the moisture; gentle heat is applied to the flask A, as long as a brisk evolution of gas takes place; the process is completed when this action and the passage of gas-bubbles through the sulphuric acid both cease, and the black color of the mixture has changed to a more or less brown one; the residual gas is then driven off, by momentary ebullition, and the apparatus weighed. Every two molecules of carbonic acid evolved correspond to one molecule of manganese dioxide decomposed; the molecular weight of the latter (87) being so nearly equal to twice that of carbonic acid (44), that the loss of weight suffered by the apparatus may be taken to represent the quantity of real manganese dioxide in 50 grains of the sample; and it has only to be doubled in order to express the percentage.

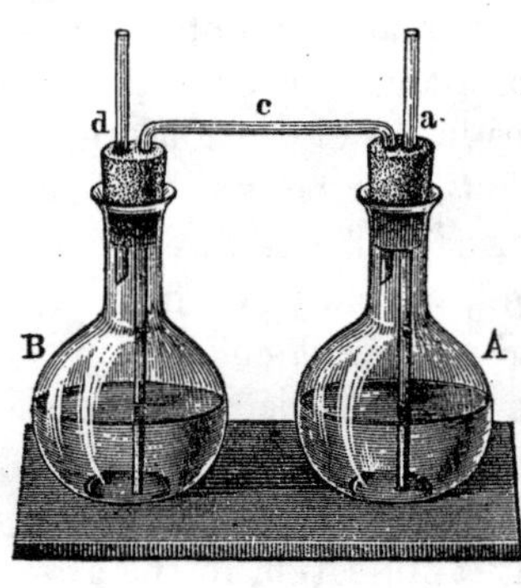

FIG. 76.

MANGANESII SULPHAS.

MANGANUM SULPHURICUM.

Sulphate of Manganese. Manganous Sulphate.

Colorless or pale rose-colored prismatic crystals, occurring in three different forms, with different quantities of water of crystallization: (1.) Oblique-rhombic prisms (isomorphous with ferrous sulphate), containing seven molecules of water of crystallization, and obtained when crystallized at a temperature below 6° C.; (2.) Rhomboidal prisms (isomorphous with cupric sulphate), containing five molecules of water of crystallization, obtained when crystallized at a temperature between 7° and 20° C.; and (3.) Right-rhombic prisms (isomorphous with magnesium sulphate), containing four molecules of water of crystallization, and obtained when crystallized between 20° and 30° C.

The latter salt is the one commonly met with. The crystals are permanent in the air, though slightly efflorescent in air that is dry and warm; they are soluble in nearly their own weight of cold water, but insoluble in alcohol; the aqueous solution is neutral and colorless, or has, when concentrated, a faint rose-color; its taste is astringent, and it affords, with the alkaline hydrates and carbonates, white precipitates, of which those with the hydrates gradually become yellow, and finally dark brown, by oxidation; ammonium sulphydrate produces a flesh-colored precipitate soluble in acids; hydrosulphuric acid throws down the same precipitate, but not before the addition of an alkaline hydrate or carbonate; tannic acid or tincture of nutgall will not act upon the solution; potassium ferrocyanide and barium nitrate produce white precipitates, and potassium ferricyanide a brown one.

When a fragment of a crystal of manganous sulphate is heated with one or two drops of liquor potassæ, upon platinum-foil, it yields a bluish-green fuse.

Examination:

Ferrous and *cupric sulphates* are detected, in the diluted solution, acidulated with hydrochloric acid, the former by a blue precipitate with potassium ferrocyanide, the latter by a black one with hydrosulphuric acid.

Magnesium and *alkaline sulphates* may be detected by completely precipitating the dilute solution of the salt with ammonium sulphydrate, and by testing part of the filtrate with sodium phosphate; a white, crystalline precipitate will indicate *magnesium sulphate;* if no reaction has taken place, another portion of the filtrate is evaporated in a porcelain capsule, and the residue heated to redness upon platinum-foil; a fixed remainder would indicate *potassium* and *sodium* salts.

MORPHIA.

MORPHIUM. MORPHINUM.

Morphia. Morphine.

Small, brilliant, prismatic crystals, transparent and colorless, or a white, crystalline powder. Heated in a dry test-tube, morphia loses its transparency and its water, and fuses to a yellow mass, which becomes white and crystalline on cooling; heated on platinum-foil, it burns away, leaving a carbonaceous residue which is wholly dissipated at a red heat.

Strong sulphuric acid dissolves morphia without coloration; the solution becomes green on the addition of one drop of solution of potassium bichromate, or purple with one drop of nitric acid. Concentrated nitric acid, diluted with an equal volume of water, dissolves morphia, with a yellow color, which, after a while, or at once upon heating, becomes purple; this yellow solution in dilute nitric acid remains unchanged upon the addition of a few drops of stannous chloride (distinction from brucia, which yields a violet coloration). When a few particles of morphia are added to a little diluted neutral liquor ferri chloridi or sulphatis, a deep-blue color is produced. This reaction takes place also in solutions of morphia, if they are not too dilute.

Morphia is but sparingly soluble in cold, but a little more in boiling, water, forming a solution of a bitter taste and a faint alkaline reaction; it is almost insoluble in ether (distinction from narcotia and codeia), in benzol, and in amylic alcohol, somewhat soluble in chloroform, and quite so in about 90 parts of cold, or 30 parts of boiling, alcohol; it dissolves freely in dilute acids, in the fixed alkaline hydrates, and in lime-water, but is almost insoluble in ammonium hydrate.

Dilute solutions of morphia in acidulated water are not precipitated by liquor potassæ or sodæ, when added in excess (distinction from narcotia), nor by potassium bicarbonate (distinction from the cinchona alkaloids), nor by tannic acid; they discharge the color of solution of potassium permanganate quickly, and decompose potassium iodate at once, liberating iodine, which may be extracted by agitating the solution with a little chloroform or carbon bisulphide, which, on subsiding, acquires a scarlet color, while the aqueous solution remains brown. The acidulated solution of morphia affords a white precipitate with potassio-mercuric iodide, and a brown one with iodinized potassium iodide.

The ready solubility of morphia in liquor potassæ and sodæ, and its reducing action upon iodic acid and potassium permanganate, distinguish it from almost all other vegetable alkaloids.

Examination:

Narcotia is indicated by a white, crystalline residue, left upon evaporating, on a watch-glass, a little pure ether agitated with a few grains of the morphia.

Mineral impurities or admixtures may be detected by a fixed residue, upon complete incineration of a little of the morphia on platinum-foil, as well as by their insolubility, when about three grains of the morphia are dissolved in two fluid-drachms of boiling alcohol.

Estimation of the Morphia Strength of Opium:

Since the therapeutical and commercial value of opium mainly depends upon the quantity of morphia, an examination of opium is invariably required before its introduction into the market or its application for the manufacture of the opium alkaloids, or for medication. Among the various methods for the estimation of the morphia strength of opium, the following are simple in execution, require comparatively little time, and render approximately correct results.

Staples's Process.—One hundred grains of the powdered and dried opium are exhausted upon a filter with some warm, pure benzol, until the drops of the benzol pass through colorless. The opium is subsequently dried upon the filter until the odor of benzol has entirely disappeared, and the powder has acquired a dry appearance; it is then triturated, and rinsed into a flask with so much water as to measure 10 fluid-drachms; the mixture is macerated, with occasional agitation, for 12 hours, and is then poured upon a moistened filter, and tepid water allowed to percolate through the opium, until the washings are quite or nearly colorless. The aqueous solution is then evaporated in a beaker, upon the water-bath, at a moderate heat, to about half a fluidounce, and this is mixed with an equal bulk of alcohol of a spec. grav. of 0.835, filtered through a small filter, and the latter washed with a little dilute alcohol. Then, one-half of a mixture, consisting of 60 drops of aqua ammoniæ and two fluid-drachms of alcohol, is added, with agitation, and the mixture allowed to stand, in a closed flask, for six hours, when the remainder of the ammonia is added, and the mixture permitted to rest again for 24 hours. The crystalline deposit in the flask being detached from its sides, the entire contents are gradually poured upon a small tared filter, and the crude, crystalline morphia washed with a few drops of cold water, and dried at a temperature not exceeding 80° C., when the morphia is exhausted on the same filter, with a little warm pure ether, and dried again at the same temperature, until, upon repeated weighings, the weight remains unaltered. The weight indicates the percentage of morphia in the opium.

Hager-Jacobsen's Process.—Six and a half grammes (100.321 grains) of the powdered and dried opium are triturated with three grammes (46.302 grains) of dry calcium hydrate, and so much water as to form a soft mass; this is rinsed into a tared flask of about 100 cubic centimetres (3½ ounces) capacity, with so much water that the whole weighs 74½ grammes (2 ounces and 3 drachms). The flask is then loosely corked, and digested, on a water-bath, with occasional agitation, for about one hour. After cooling, the flask is replaced upon the balance, and the amount of the evaporated water is exactly restored; the liquid is then passed through a small moist filter. The filtrate is collected in a test-tube, of about one inch in width and from six to seven inches in length, upon which there has been previously made a mark indicating the volume of 50 cubic centimetres (the bulk of 13 drachms of distilled water, at about 16° C.). This volume is generally obtained from the filter; in case it be a little more, the funnel is withdrawn when the filtrate reaches the mark. A mixture of eight drops of benzol and three cubic centimetres (46⅓ fluid-grains) of ether is then added, and the test-tube corked and agitated; which process is continued after the subsequent addition of 4½ grammes (70 grains) of powdered ammonium chloride, until this is dissolved. The mixture is allowed to stand for three or four hours; the crystalline deposit in the test-tube is then detached, if necessary, and the whole gradually transferred to a small tared and moistened filter; the crystalline mass is washed with a few drops of water, and then dried, at a temperature not exceeding 80° C., and is subsequently washed upon the same filter with a little chloroform. Finally the filter is completely dried, at a temperature not exceeding 80° C., and weighed. The weight of the precipitate indicates the percentage of morphia in the opium.

Schneider's Process.—Ten grammes (154.340 grains) of the powdered and dried opium are exhausted with a mixture of 150 grammes (4 ounces, 6½ drachms) of water, and 20 grammes (5 drachms, 9 grains) of pure hydrochloric acid; the residue, after extraction, washing, and drying, should not exceed 4.5 grammes (1 drachm, 9½ grains) in weight; to the acid fluid, 20 grammes (5 drachms, 9 grains) of common salt are added, and the liquid, after standing for 24 hours, is passed through a filter, and this and the deposit of narcotia washed with a little dilute solution of common salt; aqua ammoniæ is then added to the filtrate in a slight excess, and the whole allowed to stand for 24 hours more; the crystalline deposit is then collected, redissolved in diluted acetic acid, and precipitated with diluted aqua ammoniæ; the precipitate is collected upon a moist tared filter, washed with a little cold water, dried at a heat not exceeding 80° C., and weighed; its weight should be not less than one gramme (15.434 grains), corresponding to 10 per cent. of morphia.

Estimation of the Morphia Strength of Tincture of Opium:

When *tincture of opium* has to be examined for the quantity of morphia it contains, this may be ascertained by either of the two following methods:

I. Twenty-one and one-third fluid-drachms (representing 100 grains of opium) of the Tinctura Opii (each fluidounce of which, when prepared according to the U. S. Pharmacopœia, represents 37.5 grains of opium) are evaporated in a porcelain capsule or a beaker, on a water-bath, at a moderate temperature, to about one-third of the original volume; then, after standing in a cool place for twenty-four hours, the liquid is decanted from the resinous deposit, and the latter washed with altogether about 4 fluid-drachms of cold water. The entire liquid is then reduced to about one fluidounce, by evaporation at a moderate temperature, and, when cold, is passed through a small moist filter, the capsule or beaker being washed, by means of a washing-bottle, with about one drachm of water. To the filtrate is then added an equal volume of alcohol of a spec. grav. of 0.835, and, subsequently, one-half of a mixture consisting of 60 drops of aqua ammoniæ and two fluid-drachms of alcohol.

The operation is then continued in the same mode as described under *Staples's* morphiometric process on page 267.

II. Twenty-one and one-third fluid-drachms (representing 100 grains of opium) of the Tinctura Opii are diluted, in a beaker, with about nine ounces of tepid water, and then completely precipitated with a strong solution of plumbic acetate; the mixture is allowed to stand in a warm place for two hours; it is then filtered, and the precipitate washed upon the filter with tepid water, until this ceases to redden blue litmus-paper. The filtrate is next saturated with hydrosulphuric-acid gas, and allowed to stand for one hour, when it is filtered and washed again. Then so much of a solution of potassium bicarbonate (free from carbonate) is added to the filtrate as to render it slightly alkaline; it is then filtered, and the precipitate washed with a little carbonic-acid water; the filtrate is then over-saturated with acetic acid, and evaporated in a tared porcelain capsule, on a water-bath, to about three ounces by weight; when nearly cold, it is completely precipitated with a solution of one part of potassium carbonate in two parts of water, and is then allowed to stand for 24 hours, when the crystalline deposit is collected upon a small tared filter, and washed with a few drops of cold water, or until this ceases to change the color of red litmus-paper at once. The precipitate is dried on the filter at a temperature not exceeding 80° C., and is then exhausted on the filter with a little warm, pure ether, and subsequently dried again at the same temperature.

In both tests the precipitate, when completely dry, should weigh at least 10 grains, indicating the employment of an opium containing 10 per cent. of morphia, and a morphia strength of the tincture amounting to 3.75 grains in each fluidounce.

MORPHIÆ ACETAS.

MORPHIUM SEU MORPHINUM ACETICUM.

Acetate of Morphia or Morphine. Morphia Acetate.

A white or nearly white powder, with a feeble odor of acetic acid, and permanent in the air; when heated upon platinum-foil, it fuses, and is wholly dissipated at a red heat. It remains colorless when moistened with concentrated sulphuric acid, and dissolves in concentrated nitric acid, with a scarlet color.

Morphia acetate is soluble in about 24 parts of cold water, acidulated with a few drops of acetic acid, and freely soluble in boiling water, and in diluted acids; it is little soluble in cold, but more so in warm, alcohol, but almost insoluble in ether. Its aqueous solution has a very bitter taste and a slightly acid reaction; it is rendered turbid by tannic acid, but becomes transparent again upon the addition of diluted sulphuric acid (evidence of the absence of narcotia); it is not permanently precipitated by potassium hydrate, when added in a slight excess (distinction from most alkaloids), nor by potassium bicarbonate (further distinction from narcotia and from the cinchona alkaloids), and assumes a greenish-blue coloration with dilute solution of ferric chloride.

MORPHIÆ HYDROCHLORAS.

MORPHIUM SEU MORPHINUM HYDROCHLORICUM. MORPHIÆ MURIAS.

Hydrochlorate of Morphia or Morphine. Morphia Hydrochloride.

Colorless, transparent, flexible, acicular crystals, of a silky lustre, or a crystalline powder, containing 14 per cent. of water of crystallization, which evaporates at a moderate temperature. Heated upon platinum-foil, morphia hydrochloride fuses, and, at a higher temperature, burns away without residue.

Morphia hydrochloride is soluble in 20 parts of cold, and in nearly its own weight of boiling, water, and in 60 parts of cold, and 10 parts of boiling, alcohol. Its aqueous solution is neutral, has a bitter taste, and assumes a yellowish-red coloration upon the addition of strong nitric acid, with gentle heating, and a bluish one with dilute solution of ferric chloride; it renders no permanent turbidity with dilute liquor potassæ in a slight excess, but is precipitated by ammonium hydrate. When acidulated with hydrochloric acid, the solution yields no precipitate with tannic acid, nor with potassium bicarbonate (distinction from narcotia and from the cinchona alkaloids). With argentic nitrate, the aqueous solution of morphia hydrochloride gives a white, curdy precipitate, insoluble in nitric acid, but soluble in aqua ammoniæ, which solution, when heated in a test-tube, separates metallic silver.

When a little dry morphia hydrochloride is added to a mixture of two parts of concentrated sulphuric acid and one part of water, in a small test-tube, no change of color of the liquid takes place, either at common temperatures, or when gently warmed by immersing the test-tube in hot water (evidence of the absence of salicin and other bitter substances); when this liquid is divided into two portions, and one drop of strong nitric acid is added to the one part, a red coloration occurs, and on adding a trace of potassium bichromate to the other part, only a slight yellowish-green coloration takes place.

Morphia hydrochloride dissolves in chlorine-water, with a yellowish color, which becomes brown upon addition of aqua ammoniæ (distinction from quinia, which yields an emerald-green coloration).

MORPHIÆ SULPHAS.

MORPHIUM SEU MORPHINUM SULPHURICUM.

Sulphate of Morphia or Morphine. Morphia Sulphate.

Colorless, transparent, fasciculate, feathery crystals, permanent in the air; they contain about 14 per cent. of water, of which 12 per cent., water of crystallization, are given off at

100° C. When heated upon platinum-foil, morphia sulphate fuses, and burns away without residue.

Morphia sulphate is soluble in twice its weight of cold, and less than its own weight of boiling, water (distinction from quinia sulphate); it is less soluble in alcohol, and almost insoluble in ether and in chloroform. Its aqueous solution is neutral and very bitter; it gives no permanent precipitate with potassium hydrate when added in a slight excess, nor with potassium bicarbonate (distinction from the cinchona alkaloids), but a white one with ammonium hydrate or carbonate. It gives a bluish reaction with ferric chloride, and a white precipitate, insoluble in acids, with barium chloride.

Morphia sulphate dissolves in strong sulphuric acid without coloration, even when gently warmed by dipping the test-tube in warm water (evidence of the absence of salicin and other bitter glucosides); it dissolves in concentrated nitric acid with a yellowish-red coloration (distinction from quinia). When dissolved in a little chlorine-water, morphia sulphate yields a greenish-yellow solution, which becomes dark-brown upon addition of aqua ammoniæ (further distinction from quinia, which yields an emerald-green reaction).

NICOTIA.

NICOTINUM.

Nicotine. Nicotia.

A colorless or nearly colorless, oily, and volatile liquid, of 1.027 spec. grav., with a pungent odor, resembling that of tobacco, and an acrid, burning taste. By exposure to the air, it becomes gradually brown and thick; when heated, it volatilizes, forming irritating vapors, which, when ignited, burn with a bright flame. When dropped into concentrated sulphuric acid, it dissolves, with a red color, and, when one drop of solution of potassium bichromate is added, the solution becomes brown, and subsequently green. Nicotia produces white fumes with hydrochloric and acetic acids, precisely like ammonium hydrate; when dropped into concentrated hydrochloric acid, and heated, it dissolves with a deep-violet color, and likewise in nitric acid, with an orange-yellow color.

Nicotia sinks when dropped into water (distinction from conia, which floats); it is miscible with water, alcohol, ether, carbon bisulphide, and chloroform, and with most fixed and essential oils; its solutions have an alkaline reaction, and an acrid, burning taste; they are precipitated by solutions of tannic acid and of potassio-mercuric iodide; the alcoholic solution should yield no turbidity with diluted sulphuric acid (evidence of the absence of ammonium hydrate). The aqueous solution of nicotia, when applied to the eye, causes the pupil alternately to dilate (mydriasis) and to contract (stenocoriasis).

OLEUM AMYGDALARUM AMARARUM.

Oil of Bitter Almonds.

A thin, colorless, or golden-yellow liquid, of great refractive power, and of the odor of bitter almonds, when triturated with water. Exposed to the air, it greedily absorbs oxygen, with the formation of crystals of benzoic acid; its spec. grav. is from 1.04 to 1.06; its boiling-point, 180° C.

When dropped into water, oil of bitter almonds sinks, but dissolves upon shaking, unless too much oil has been used. When a few drops of liquor potassæ are added to its aqueous solution, and afterward one or two drops of ferrous chloride or sulphate, and finally, after agitation, a slight excess of hydrochloric acid, there will appear a blue coloration, and, after a while, a blue precipitate.

Oil of bitter almonds is miscible, in all proportions, with alcohol, ether, chloroform, carbon bisulphide, and essential and fatty oils; it is also soluble in concentrated nitric acid, without color, and without the evolution of nitrous fumes; it forms a thick, crimson liquid with concentrated sulphuric acid.

Among the several compounds of which oil of bitter almonds consists, it contains from 3 to 14 per cent. of hydrocyanic acid, partly free and partly combined; the crude oil derived by distillation, without further rectification, generally is the richest in acid.

Estimation of the Available Quantity of Hydrocyanic Acid in Oil of Bitter Almonds:

1. Fifty grains of the oil are mixed with half an ounce of strong alcohol in a small flask; then four ounces of water are added, and 50 drops of liquor potassæ, or so much as to render the liquid strongly alkaline. Volumetric test-solution of argentic nitrate is then delivered into this liquid (Fig. 77), with constant stirring, until the ensuing precipitate ceases to be redissolved, and therefore a slight permanent turbidity occurs. The quantity of argentic nitrate employed represents exactly half the amount of hydrocyanic acid, and has, therefore, to be multiplied by four, to obtain the percentage.

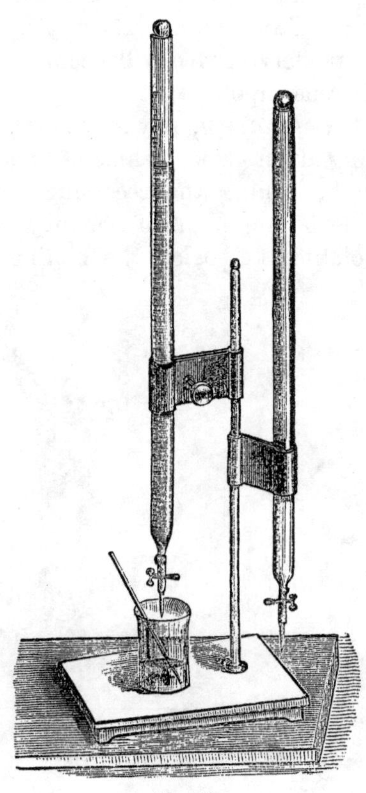

Fig. 77.

2. Another method consists in dissolving 50 grains of the oil in half a fluidounce of alcohol, in a small flask, and adding about four ounces of water; to this solution is then added a solution of 30 grains of argentic nitrate in four fluid-drachms of aqua ammoniæ; the mixture is then gently warmed to about 50° C., and, after repeated agitation, an excess of nitric acid is added, and the whole gently warmed by immersing the flask in hot water. The precipitate of argentic cyanide is collected upon a tared and moist filter, washed until the washings cease to redden blue litmus-paper, and then dried and weighed. The number of grains of the weight of the

dry argentic cyanide being divided by five, the quotient, multiplied by two, gives the percentage of anhydrous hydrocyanic acid contained in the oil.

Examination:

Alcohol may be detected in oil of bitter almonds by agitating it with three times its volume of concentrated nitric acid, and subsequently warming the mixture by dipping the test-tube into hot water. No reaction takes place with pure oil; but, if it has an admixture of more than three per cent. of alcohol, effervescence will occur, with disengagement of yellowish nitrous vapors.

Chloroform, as well as *alcohol*, can be detected by submitting about two drachms of the oil to distillation from a water-bath, cooling the receiving test-tube in ice-water (Fig. 78). The boiling-point of the oil being at 180° C., only admixtures volatile at or below the boiling-point of water will distil, with

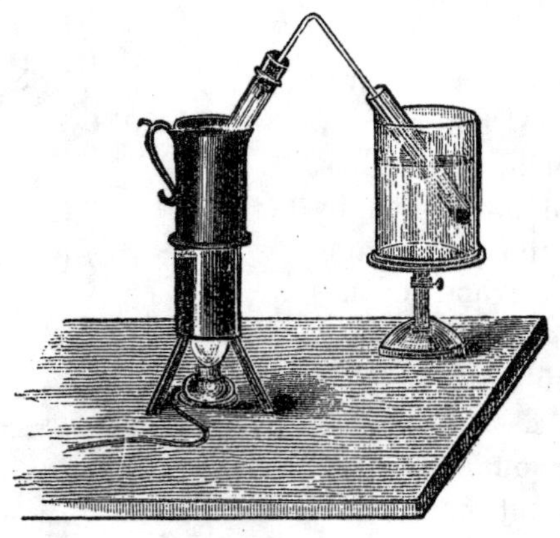

Fig. 78.

but small traces of the constituents of the oil. The obtained distillate is mixed with a little iodine-water; if chloroform be present, and no alcohol, it will absorb the iodine, and separate, with a rose-color. The colorless, aqueous liquid is decanted, and then warmed by dipping the test-tube in warm water; one drop of solution of iodinized potassium iodide is added, and then one drop of liquor potassæ, or so many as nearly to decolorize the liquid. If alcohol be present, minute yellow crystals of iodoform will be produced (Fig. 79), which, after subsiding, in a conical glass, may be recognized by the examination of the sediment under the microscope.

Nitrobenzol may be detected by dissolving 15 grains of fused potassium hydrate in about two fluid-drachms of alcohol; the solution may be aided by dipping the test-tube into hot water; when it is complete, from 15 to 20 drops of the oil are added, and the heating continued for a minute; if the oil be pure, the mixture will remain colorless or nearly so, and will form, on cooling, a white, crystalline deposit of potassium benzoate; if it is contaminated with nitrobenzol, the liquid will assume a dark-brown color.

FIG. 79.

Or the oil may be tested by agitating about one drachm of it with half a drachm of fused potassium hydrate. If nitrobenzol is present, the yellow color of the oil will give place to a red tint, which passes into green, if much nitrobenzol is present; the green color disappears after about twelve hours and returns to the red tint.

Essential Oils.—Adulteration with cheaper essential oils, as well as with nitrobenzol, may be detected by the property of oil of bitter almonds to dissolve in a concentrated aqueous solution of sodium bisulphite when added drop by drop and agitated; whereas such admixtures remain undissolved, floating upon the aqueous solution after dilution with a little tepid water.

OLEUM SINAPIS.

Volatile Oil of Mustard.

A colorless or pale-yellow liquid, of a most penetrating, pungent odor, and of 1.01 to 1.015 spec. grav.; when dropped into water, it sinks slowly, and dissolves in about 50 parts of it, which solution forms a black precipitate, when heated with a few drops of solution of argentic nitrate.

Mustard oil is miscible with alcohol, ether, chloroform, carbon bisulphide, and benzol, and with fatty and essential oils; it suffers decomposition with concentrated nitric acid, with the evolution of nitrous vapors, and the formation of a resinous residue. When dropped into concentrated sulphuric acid, mustard oil dissolves, without color, and without the evolution of heat, and when mixed with concentrated sulphuric acid, in the proportion of one part of oil to three parts of acid, taking care that the mixture is kept cool, sulphurous acid is evolved, and, after 12 hours, a colorless or but slightly brown, thick liquid or crystalline mass is formed, devoid of the odor of mustard oil.

Examination:

Admixtures of *essential oils*, *carbon bisulphide*, *nitrobenzol*, and *alcohol*, are indicated by becoming warm and dark-colored when about five or six drops of the oil are added to about 50 or 60 drops of concentrated sulphuric acid, in a dry test-tube.

Admixtures of *alcohol*, *benzol*, and other *hydrocarbons*, are also indicated when two or three drops of the oil are allowed to fall into a test-tube, filled about one-third with cold distilled water; they should sink slowly to the bottom, remaining clear and transparent, until, after gently inclining the tube two or three times, they become opalescent. When contaminated with only a few per cent. of the above adulterations, the drops lose their transparency, and become opalescent, as soon as they fall into the water.

PLUMBI ACETAS.

PLUMBUM ACETICUM.

Acetate of Lead. Sugar of Lead. Plumbic Acetate.

Colorless, transparent, brilliant prisms or plates, or, as generally met with, heavy, compact crystalline masses, somewhat resembling loaf-sugar, having an acetous odor and a sweet, astringent taste; they contain three molecules (14.21 per cent.) of water of crystallization, and effloresce slowly and absorb carbonic acid when exposed to the air; they become black when in contact with gaseous or dissolved hydrosulphuric acid. When gently heated, plumbic acetate fuses at 75.5° C., loses water and acetic acid, and is decomposed at a higher temperature, leaving a black residue, which is reduced, at a red heat, to plumbic oxide or to metallic lead.

Plumbic acetate is soluble in about two parts of cold, and in half its weight of boiling, water, and in eight parts of alcohol, but insoluble in ether and in chloroform; its solution in water has generally a slightly turbid appearance from traces of plumbic carbonate, which, however, disappears upon the addition of acetic acid; the aqueous solution has a feeble acid reaction, forms white precipitates with the alkaline hydrates (soluble in excess of potassium and sodium hydrates), with the alkaline carbonates, and with sulphates and chlorides, a yellow one with iodides, and a black one with hydrosulphuric acid and with sulphides. When completely precipitated by sodium chloride, the colorless filtrate will assume a deep-red tint with a few drops of solution of ferric chloride.

Examination:

Calcium and *barium acetates* may be detected, in a solution of 20 grains of the plumbic acetate in one ounce of warm water, by precipitating it with 20 drops of concentrated hydrochloric acid; the filtrate is completely precipitated with hydrosulphuric acid, is filtered, and then over-saturated with sodium carbonate, and gently warmed; a white turbidity would indicate calcium or barium salts.

Copper is indicated by a blue coloration of the solution of plumbic acetate, when precipitated with aqua ammoniæ.

PLUMBI CARBONAS.

PLUMBI SUBCARBONAS. PLUMBUM CARBONICUM. CERUSSA.

Carbonate or Subcarbonate of Lead. White Lead. Basic Plumbic Carbonate.

A heavy, white, opaque powder, or friable lumps, which are blackened by hydrosulphuric acid. Heated upon charcoal, before the blow-pipe, plumbic carbonate turns yellow, fuses, and is finally reduced to soft, malleable metallic globules. It is insoluble in pure water, but somewhat soluble in water containing much carbonic acid or alkaline bicarbonates; it is wholly dissolved, with effervescence, in diluted acetic and nitric acids, rendering colorless solutions, of a sweet, astringent taste; these solutions yield white precipitates with dilute sulphuric and hydrochloric acids, and with soluble sulphates and chlorides; they also give a white precipitate with liquor potassæ or sodæ, soluble in an excess of the precipitant, a yellow one with potassium iodide, and a black one with hydrosulphuric acid.

Examination:

Admixtures of *barium*, *calcium*, and *plumbic sulphates*, remain behind, upon solution of the white lead in dilute nitric acid. Their quantity may be ascertained by dissolving 100 grains of the sample in a sufficient quantity of warm diluted nitric acid, and by collecting and washing the insoluble residue upon a tared filter; when completely dry, the weight indicates the percentage of such admixtures.

If the nature of the admixture has to be ascertained, the residue is transferred to a flask, and is digested, for some days, with occasional agitation, with a solution of ammonium sesquicarbonate; if plumbic sulphate be present, it is decomposed, and plumbic carbonate formed; as an evidence of the presence of the former, part of the supernatant liquid may be acidulated with nitric acid, and tested with barium nitrate for sulphuric acid.

Admixtures of *calcium carbonate* or *phosphate*, *barium carbonate*, and *oxide of zinc*, are more or less soluble in nitric acid. In order to detect and to distinguish them, the nitric-acid solution of the sample is freely diluted with water, and is subsequently saturated and completely precipitated with hydrosulphuric-acid gas; it is then filtered, and warmed, to expel the excess of gas, and a small portion of the solution is oversaturated with sodium carbonate; an ensuing white precipitate will confirm the above admixtures; in order to ascertain their nature, the rest of the solution is nearly neutralized with a few drops of liquor sodæ and then tested, in separate portions, with solution of calcium sulphate for *barium*, with oxalic acid, after previous addition of a little sodium acetate, for *calcium*, and by addition first of sodium acetate and subsequently of ammonium sulphydrate, for *zinc*.

PLUMBI IODIDUM.

PLUMBUM IODATUM.

Iodide of Lead. Plumbic Iodide.

A bright-yellow, heavy, inodorous powder, when obtained by precipitation, or shining, golden-yellow, flexible scales, when crystallized; when heated in a dry test-tube, plumbic iodide turns red, and fuses to a thick reddish-brown liquid, which congeals, on cooling, to a yellow mass; at a stronger heat, it is decomposed, with the evolution of violet vapors; and when heated with exsiccated sodium carbonate, on charcoal, before the blow-pipe, it is entirely reduced to metallic globules.

Plumbic iodide is soluble in about 1,250 parts of cold, and 194 parts of boiling, water, and also, to a slight extent, in alcohol; a hot saturated aqueous solution, on cooling, deposits the salt in brilliant yellow scales; it is readily soluble in acetic acid, in liquor potassæ or sodæ, in concentrated solutions of the alkaline or earthy iodides, in warm solution of ammonium chloride, and in solution of sodium hyposulphite, from all which solutions hydrosulphuric acid precipitates black plumbic sulphide.

When shaken in chlorine-water, plumbic iodide suffers partial decomposition, and yields a filtrate from which chloroform or carbon bisulphide will extract iodine, with a red color. When boiled with solutions of alkaline or earthy carbonates, it is decomposed, forming iodides and plumbic carbonate. When boiled with granular or powdered iron or zinc, plumbic iodide is likewise decomposed, forming soluble iron or zinc iodide and metallic lead.

PLUMBI NITRAS.

PLUMBUM NITRICUM.

Nitrate of Lead. Plumbic Nitrate.

Colorless, transparent or opaque, anhydrous, octahedral crystals, permanent in the air. Heated in a dry test-tube, the crystals deflagrate, emit yellow nitrous vapors, and leave a residue of plumbic monoxide.

Plumbic nitrate is soluble in 7½ parts of cold water, but less soluble in alcohol. Its aqueous solution is neutral, has a sweet, astringent taste, and gives a white precipitate with sulphuric or hydrochloric acid, and with solutions of sulphates or chlorides, a yellow one with potassium iodide, and a black one with hydrosulphuric acid. When triturated with concentrated sulphuric acid, and heated, the salt evolves red nitrous fumes.

Examination:

Barium nitrate may be detected in the filtrate from the aqueous solution of the salt, after complete precipitation with hydrosulphuric acid, by over-saturating it with sodium carbonate; a white precipitate would indicate barium.

Copper may be detected, in the aqueous solution of the salt, by completely precipitating it with sodium sulphate, and by testing the filtrate with an excess of ammonium hydrate, when copper will be indicated by a bluish color of the liquid.

PLUMBI OXIDUM.

PLUMBUM OXYDATUM FUSCUM. LITHARGYRUM.

Oxide of Lead. Litharge. Plumbic Monoxide.

Small scales or heavy masses, of a brick-red color and a foliaceous fracture, or a heavy powder of the same color, devoid of taste and odor. It fuses at a red heat, and, when heated upon charcoal, is reduced to the metallic state.

Plumbic monoxide is but sparingly soluble in water, imparting thereto a feeble alkaline reaction; it is soluble in warm solutions of the fixed alkaline hydrates, and in diluted nitric and acetic acids, without effervescence or residue; it slowly absorbs carbonic acid from the atmosphere, and contains the more carbonate the longer it has been exposed to the air; from this cause, when very old, it becomes more or less effervescent with acids. The nitric-acid solution of plumbic monoxide yields white precipitates with dilute sulphuric and hydrochloric acids, with solutions of sulphates and chlorides, and with the alkaline hydrates, which latter, ammonium hydrate excepted, redissolve the precipitate, when added in excess; it gives a black one with hydrosulphuric acid, and, when neutral, a yellow one with potassium iodide.

Examination:

Plumbic carbonate and *red oxide* are detected, when a small quantity of the litharge is triturated with a little water, and the mixture is added, drop by drop, to concentrated nitric acid, in a test-tube; carbonate is recognized by effervescence; red oxide by a brown residue, insoluble in an excess of acid, with gentle warming, but soluble upon the addition of a little oxalic acid or sugar; if this residue, however, does not dissolve, an adulteration with powdered silicates, crude ferric oxide, etc., is indicated.

Silicates are also indicated by a white turbidity or a flocculent precipitate, occurring in the solution after the addition of the oxalic acid, in the preceding test.

Oxides of zinc and *alkaline earths* may be detected by mixing the nitric-acid solution with three times its volume of hydrosulphuric acid, or so much as completely to precipitate the lead; the filtrate is evaporated to half its volume, and one portion of it is over-saturated with sodium carbonate; an ensuing white precipitate would indicate oxides of zinc, calcium, or barium; the other part is neutralized with liquor sodæ, and tested, in separate portions, with sodium acetate, and subsequent addition of hydrosulphuric acid, for zinc; with ammonium oxalate for calcium, and with calcium sulphate for barium.

Copper and Sulphates.—About 30 grains of the powdered litharge are digested, for one hour, with occasional agitation, in about half an ounce of a mixture consisting of equal parts of aqua ammoniæ and dilute solution of ammonium sesqui-carbonate; a bluish tint, in the obtained filtrate, would indicate copper, and a white precipitate, when slightly over-saturated with diluted nitric acid, and subsequently tested with barium nitrate, would show sulphate.

Metallic Lead.—About 20 grains of the litharge are digested in a test-tube, with occasional agitation, for half an hour, with a solution of plumbic nitrate; a few drops of the decanted liquid are then introduced into a little water, to which previously have been added a few drops of sulphuric-acid mucilage of starch, and a few drops of solution of potassium iodide. If the sample contains even traces of metallic lead, this gives rise to the formation of plumbic nitrite, which will decompose the potassium iodide, with the liberation of iodine, which at once produces a blue color with the starch.

PLUMBI OXIDUM RUBRUM.

PLUMBUM OXYDATUM RUBRUM. MINIUM.

Red Oxide of Lead. Minium. Triplumbic Tetroxide. Plumbic Plumbate.

A heavy, orange-red powder, which becomes dark when heated, but regains its original color on cooling. Heated upon charcoal, before the blow-pipe, it fuses, and is reduced to metallic globules. Warm diluted nitric or acetic acid will dissolve red oxide of lead only partially, leaving a brown residue, which is soluble, however, upon the addition of a little oxalic acid or sugar. A slight remaining turbidity, of a whitish aspect, is due to silicic acid, with which red oxide of lead is generally more or less contaminated; any insoluble red or brown residue, however, would indicate impurities.

The impurities and admixtures which red oxide of lead is liable to contain, and the methods of detecting them, are the same as mentioned and described under litharge, on the preceding page.

POTASSII ACETAS.

POTASSIUM SEU KALIUM ACETICUM.

Acetate of Potassium. Potassium Acetate.

A snow-white, very deliquescent salt, of a foliaceous or fibrous satiny appearance, unctuous to the touch, and of a warm, pungent, saline taste; it fuses at about 280° C., and is decomposed at a higher temperature, leaving behind a mixture of carbon and potassium carbonate.

Potassium acetate is soluble in its own weight of cold water, in 4 parts of alcohol, and in 3 parts of glycerin; its dilute aqueous solution assumes a deep red color with one or two drops of solution of ferric chloride, and renders a white granular precipitate with solution of sodium bitartrate or of tartaric acid. Potassium acetate disengages the vapor of acetic acid with concentrated sulphuric acid, and the vapor of acetic ether when heated with a mixture consisting of equal parts of alcohol and sulphuric acid.

Examination:

Metals are detected in the aqueous solution by a dark coloration or precipitate with hydrosulphuric acid; if such reaction takes place, the solution of the salt may be acidulated with a few drops of hydrochloric acid, and tested with potassium ferrocyanide; *copper* will be indicated by a reddish coloration, *iron* by a blue one.

Tartrates, sulphates, and chlorides, are indicated by the occurrence of a turbidity when a concentrated aqueous solution of the salt is dropped into strong or absolute alcohol; the latter two are also recognized in the diluted solution, acidulated with nitric acid, by white precipitates when tested in separate portions with argentic nitrate and barium nitrate respectively.

For *Volumetric Estimation*, *see* page 58.

POTASSII BICARBONAS.

POTASSIUM SEU KALIUM BICARBONICUM.

Bicarbonate of Potassium. Potassium Bicarbonate.

Transparent, colorless crystals, having the form of irregular eight-sided prisms with two-sided summits; they are permanent in the air, but, when exposed to a red heat, they lose 30.7 per cent. of their weight, comprising half their quantity of carbonic acid and the whole (6 molecules, or 9 per cent.) of their water, and return to the state of carbonate.

Potassium bicarbonate dissolves in 4 to 5 parts of cold water, forming a slightly alkaline solution which effervesces with acids and evolves carbonic-acid gas when heated to boiling; it gives a white granular precipitate with excess of tartaric acid, but no precipitate with magnesium sulphate unless when heated. It is almost insoluble in alcohol.

One drachm of potassium bicarbonate requires for exact saturation 42 grains of citric, or 45 grains of tartaric, acid.

For *Volumetric Estimation*, *see* page 58.

Examination:

Carbonate may be detected in the cold aqueous solution by means of magnesium sulphate, and in another portion by mercuric chloride; a white precipitate with the first reagent, soluble upon addition of ammonium chloride, and a brick-red one with the second, would indicate carbonate.

Other Impurities.—The aqueous solution is slightly over-saturated with diluted nitric acid; a white turbidity, occurring immediately or after a while, would indicate *silicates;* the solution, if necessary, is filtered, and tested in separate portions with argentic nitrate for *chloride*, and with barium nitrate for *sulphate;* a white turbidity with argentic nitrate, gradually turning dark, would indicate *potassium hyposulphite;* in this case, as a confirmatory test, 10 grains of the potassium bicarbonate may be dissolved in one drachm of water, and the solution slightly over-saturated with acetic acid; then a few drops of mucilage of starch, and subsequently of solution of iodinized potassium iodide, are added drop by drop; the first drops of the latter reagent should at once produce the blue reaction, which will not take place immediately if potassium hyposulphite be present in the salt.

Metals are detected by testing two portions of the solution of potassium bicarbonate, one of which is over-saturated by means of acetic acid, with hydrosulphuric acid : a dark reaction in either instance would indicate the presence of metallic impurities; if hyposulphite be present, a white turbidity will take place in the acid solution.

POTASSII BICHROMAS.

POTASSIUM SEU KALIUM BICHROMICUM. KALIUM CHROMICUM RUBRUM.

Bichromate of Potassium. Potassium Bichromate.

Large, transparent, orange-red, four-sided, tabular crystals, anhydrous and permanent in the air; exposed to heat, they fuse below redness, and are decomposed at a red heat with the evolution of oxygen, leaving a residue consisting of green chromic oxide and yellow potassium chromate; these may be separated by dissolving the latter in water.

Potassium bichromate is soluble in 10 parts of cold, and in 2½ parts of boiling, water, yielding an intensely orange-yellow solution, with a cooling, bitter taste, and an acid reaction; the solution becomes lemon-yellow with the alkaline hydrates and carbonates, and green or almost colorless, with the formation of a brown precipitate, when heated with reducing agents; it forms insoluble colored bichromates and chromates with the solutions of most metallic salts. When heated with concentrated hydrochloric or sulphuric acid and a little alcohol, a vehement reduction takes place, and the liquid acquires a deep green color. A concentrated solution of the salt gives a white, granular precipitate with a concentrated solution of sodium bitartrate.

Examination:

Sulphate may be detected by heating to boiling a mixture

of the aqueous solution with an equal volume of concentrated hydrochloric acid and a few drops of alcohol; when subsequently diluted with water and tested with barium chloride, a white precipitate will ensue if sulphate be present.

Chloride may be detected when the aqueous solution of the salt is mixed with about one-third of its volume of concentrated sulphuric acid, and when afterward a little alcohol is added; the mixture will become green, with spontaneous ebullition; it is then heated, and subsequently diluted with water, and tested with argentic nitrate for chloride.

POTASSII BITARTRAS.

POTASSIUM SEU KALIUM BITARTARICUM. TARTARUS DEPURATUS. CREMOR TARTARI.

Bitartrate of Potassium. Cream of Tartar. Potassium Bitartrate.

White, semitransparent, hard, prismatic crystals, or aggregated groups of crystals, or a white, gritty powder, permanent in the air, and having a spec. grav. of 1.97, a sour taste and an acid reaction. When exposed to heat, in a porcelain crucible, potassium bitartrate is decomposed, with the evolution of empyreumatic, inflammable vapors, leaving a black residue of carbon and pure potassium carbonate; this residue, when dissolved in a little water, gives a filtrate which effervesces with acids, and forms a white, granular precipitate with an excess of tartaric acid.

Potassium bitartrate is soluble in from 18 to 20 parts of boiling water, in about 200 parts of water of 15° C., and in 250 parts of water of 10° C.; it is insoluble in alcohol, but dissolves wholly in hydrochloric acid, in dilutē solutions of the alkaline hydrates and carbonates, of boracic acid, and of sodium biborate.

Examination:

Insoluble admixtures (like *terra alba*, and similar crude adulterations) are indicated by a residue left when small samples of the powder are dissolved separately in warm, diluted liquor potassæ, and in dilute hydrochloric acid.

Sulphates and *chlorides* are detected by agitating about one drachm of the salt in one ounce of warm water, and by testing portions of the clear liquid, when cool, and after the addition of a few drops of nitric acid, with barium nitrate for sulphates, and with argentic nitrate for chlorides.

Alum.—An adulteration of powdered cream of tartar with alum is at once indicated by a greater solubility of the salt in water, by its intumescence upon incineration, and by the incomplete solubility of the fused residue in water, as also by the odor of ammonia, and the production of white fumes from a glass rod, moistened with acetic acid, when the powder is heated in liquor potassæ, and by the formation of a white precipitate, when a few drops of this alkaline solution are allowed to fall into a dilute solution of ammonium chloride.

Metals may be detected in the rest of the solution of the preceding test, by hydrosulphuric acid.

Calcium and *iron* salts are detected when about 10 grains of the salt are dissolved in one ounce of aqua ammoniæ, by testing the solution, in two portions, with oxalic acid and with ammonium sulphydrate; a white precipitate with the former will indicate salts of calcium, and a dark coloration or turbidity with the latter reagent, iron.

The presence of calcium salts in potassium bitartrate, if it be free from aluminium salts, may be confirmed by dissolving five grains of the potassium bitartrate, with five grains of crystallized sodium carbonate, in about four drachms of boiling water; a complete and clear solution must take place, as a remaining white turbidity would prove the presence of calcium salts.

Iron salts may be confirmed, in a hydrochloric-acid solution of the salt, by a blue reaction with potassium ferrocyanide.

For *Volumetric Estimation*, *see* page 58.

POTASSII BROMIDUM.

POTASSIUM SEU KALIUM BROMATUM.

Bromide of Potassium. Potassium Bromide.

Anhydrous, colorless, semitransparent, cubical crystals, sometimes elongated into prisms, or flattened to plates, permanent in the air, and of a spec. grav. of 2.415; when exposed to heat, they decrepitate, and fuse at a little below a red heat, without decomposition. When a few grains are triturated and subsequently heated, in a dry tube, with a little potassium bisulphate, yellowish-red vapors of bromine are evolved.

Potassium bromide is soluble in about three parts of cold, and in its own weight of boiling, water, but only sparingly in alcohol; its aqueous solution has a strong saline taste, is neutral, and, when dropped into a very dilute solution of argentic nitrate, causes a yellowish-white curdy precipitate, which is insoluble upon addition of nitric acid, but soluble in an excess of ammonium hydrate (distinction from argentic iodide); when dropped into a very dilute solution of mercuric chloride, no reaction takes place (additional distinction from potassium iodide); it gives a white, granular precipitate with a saturated solution of sodium bitartrate.

Potassium bromide and its solution may also be distinguished from the iodide by adding to the solution a little mucilage of starch, and subsequently a few drops of chlorine-water; the solution of bromide becomes light yellow; that of iodide will at once assume a deep-blue color.

Examination:

Moisture which may be contained in the crystals, as well as in the granular form of the salt, is recognized, and may be determined, by the loss of weight when the salt is dried at about 80° C.

Potassium carbonate is detected by a white turbidity occurring upon the addition of a little of the concentrated solution of the salt to lime-water.

Sulphates may be detected, in the aqueous solution, acidulated with a few drops of diluted nitric acid, by a white precipitate with barium nitrate.

Potassium bromate is detected by a red color ensuing upon the addition of concentrated hydrochloric acid to the aqueous solution of the salt.

Potassium and *sodium chlorides* are distinguished from potassium bromide, and may be recognized by adding some chlorine-water to the aqueous solution of the salt; if this is bromide, the mixture assumes at once a yellow color, which, however, will be caused to disappear from the aqueous solution, and be completely taken up, by chloroform, ether, or carbon bisulphide, when agitated with the solution. None of these reactions will take place with potassium or sodium chloride.

If an admixture of potassium chloride, or other salts, be suspected, the purity of the sample may be ascertained by preparing a solution of 10 grains of the dry, crystallized salt in one ounce of water, acidulated with a few drops of diluted nitric acid, and completely precipitating it with a solution of argentic nitrate; the precipitate is collected upon a moist double filter, both being of the same paper and the same size, is washed, dried, and, when completely dry, weighed, the one filter serving to counterpoise the other one, which contains the precipitate. If the salt was pure potassium bromide, the obtained argentic bromide should weigh 15.77 grains; if it contained potassium or sodium chloride, the weight will be greater in proportion to the amount of those admixtures, since their molecular weights are higher; 10 grains of potassium chloride, for instance, would give 19.21 grains of argentic chloride.

The same test may also be used to indicate the purity of the bromide, by ascertaining the quantity of argentic nitrate required to precipitate completely a certain weight of potassium bromide, 10 grains of which require 14.3 grains of argentic nitrate for precipitation.

An additional and sensible mode of detecting an admixture of chlorides is the following: One drachm of the potassium bromide is triturated with one drachm of potassium bichromate, the powder is introduced into a small flask, and six fluiddrachms of concentrated sulphuric acid are added. The mixture is then submitted to distillation at a gentle heat; the delivery-tube leads into about one fluidounce of aqua ammoniæ fortior, contained in a test-tube which is kept cool in ice-water

(Fig. 80). Bromine distils over, and is dissolved by the ammonium hydrate without color; but, if chlorides are present, chloro-bromic acid is produced, distils over, and forms yellow

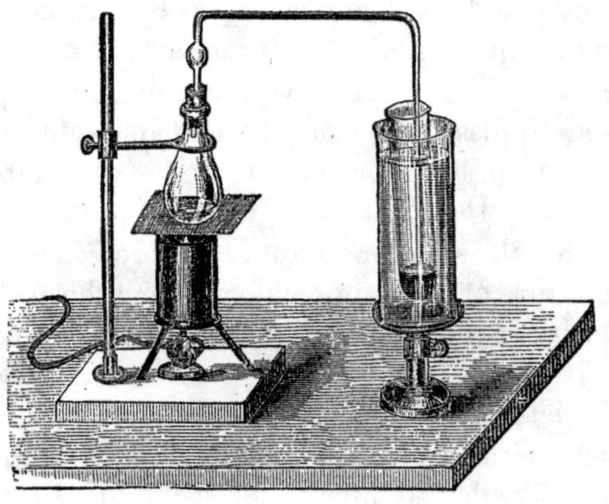

Fig. 80.

ammonium chromate, which imparts a yellowish color to the aqua ammoniæ, and indicates the presence of chlorides.

For *Volumetric Estimation*, *see* page 62.

POTASSII CARBONAS CRUDA.

POTASSIUM SEU KALIUM CARBONICUM CRUDUM.

Crude Carbonate of Potassium. Pearlash.

White masses, or a coarse granular powder intermingled with smaller lumps, somewhat deliquescent, and of a burning, alkaline taste. Water extracts from pearlash the potassium carbonate and hydrate, and the soluble impurities, the greater part of the impurities remaining behind (sulphates, chlorides, silicates, phosphates, and carbonates of potassium, sodium, calcium, and aluminium); the filtered solution effervesces with acids, and renders a white, granular precipitate with an excess of tartaric acid.

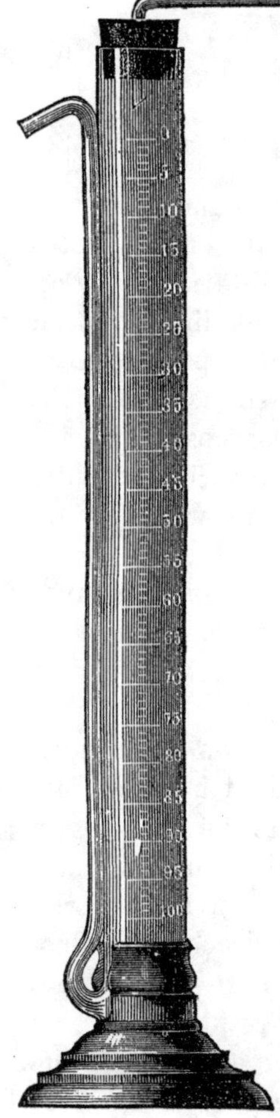

Fig. 81.

The examination of pearlash consists in the determination of the quantity of soluble alkaline carbonate and hydrate, or the available potassium oxide.

Approximate Estimation:

Three and a half drachms of commercial pearlash, when successively exhausted with from four to six ounces of warm water, render a solution which must neutralize at least two drachms of sulphuric acid of 1.843 spec. grav.

Volumetric Estimation:

Five grammes (77.170 grains) of the pearlash are completely exhausted with two or three fluidounces of tepid water; the solution is collected in a beaker-glass, is colored blue with a few drops of neutral litmus-solution, and heated to near boiling; then oxalic-acid test-solution (page 56) is carefully added from a burette (Fig. 81), with constant stirring, until the blue tint begins to appear violet; the liquid is heated once more to boiling, and, if necessary, test-solution is once more added, drop by drop, until the violet tint appears again distinctly. The number of cubic centimetres of the test-solution used, when multiplied by 0.472, gives at once the percentage of potassium oxide contained in the pearlash.

When the quantity of carbonate has to be determined, this may be done by testing 50 or 100 grains of the pearlash by the method described on page 60.

POTASSII CARBONAS DEPURATA.

POTASSIUM SEU KALIUM CARBONICUM DEPURATUM.

Purified Carbonate of Potassium. Purified Pearlash.

A coarse, granular, white, salt-powder, permanent in dry, deliquescent in moist air; heated in a flame, it communicates to it a violet color. Purified pearlash is soluble in nearly an equal weight of water, forming a strong alkaline solution which frequently appears slightly turbid, and deposits gradually a flocculent or gelatinous sediment of silicic acid; it is soluble in alcohol; its aqueous solution is decomposed by acids with effervescence; it gives a white precipitate with argentic nitrate, and also with magnesium sulphate, and a white granular one with an excess of tartaric acid.

Purified pearlash contains about 80 per cent. of potassium carbonate, and not more than 15 to 18 per cent. of water.

For *Volumetric Estimation*, *see* page 58.

Examination:

Potassium hydrate is indicated in the solution of the salt, by a grayish or brownish precipitate, instead of a white one, with argentic nitrate; it may be approximately estimated by agitating 100 grains of the dried salt with about half a fluid-ounce of absolute alcohol; the mixture is then filtered, the residue upon the filter washed with one fluid-drachm of alcohol, and the entire filtrate evaporated to complete dryness in a tared porcelain capsule; the weight of the residue indicates the percentage of potassium hydrate. As an evidence of the absence of carbonate in the residue, one drop of acetic acid may be allowed to fall upon it; effervescence would indicate carbonate.

Foreign Salts.—100 grains of the purified potassium carbonate are dissolved in two drachms of water in a test-tube; the solution should be complete and limpid, or nearly so; it is diluted with an equal volume of water, filtered, and over-saturated with hydrochloric acid; a gelatinous precipitate after a time would indicate *silicic acid;* the liquid is then filtered, and part of the filtrate over-saturated with aqua ammoniæ;

a white turbidity would indicate *aluminium salts;* the other part of the filtrate is tested with barium chloride for *sulphate.*

Chloride and *phosphate* may be detected in the diluted solution of the salt, over-saturated with nitric acid, by testing it in two portions, with argentic nitrate for chloride, and by over-saturation with aqua ammoniæ, and subsequent addition of ammoniated magnesium sulphate, for phosphate.

Sulphite and *hyposulphite* are detected in the filtered solution of the salt, slightly over-saturated with acetic acid, by adding a few drops of mucilage of starch, and subsequently two or three drops of diluted solution of iodinized potassium iodide; the first drop of the latter solution should produce a blue reaction at once, which will not occur before the addition of several drops if the above impurities are contained in the salt.

Sodium carbonate may be detected by a white, crystalline precipitate, occurring at once or after some time, when a hot diluted solution of the potassium carbonate is nearly neutralized with acetic acid, and subsequently tested with potassium antimoniate.

Metallic impurities are detected in the filtered solution of the salt, by dividing it into two parts, one of which is oversaturated with hydrochloric acid; both are then mixed with twice their volume of hydrosulphuric acid; any immediate reaction in either of the fluids would indicate metals.

POTASSII CARBONAS PURA.

POTASSIUM SEU KALIUM CARBONICUM PURUM.

Pure Carbonate of Potassium. Salt of Tartar. Potassium Carbonate.

A white, deliquescent, granular powder, wholly soluble in an equal weight of water, forming a limpid alkaline liquid, which effervesces with acids, and gives a white, crystalline precipitate with an excess of tartaric acid. When exposed to a red heat, dry potassium carbonate loses about 16 per cent. of its weight. One drachm of the dry carbonate requires for saturation 55.88 grains of citric, and 54.78 grains of tartaric acid.

For *Volumetric Estimation*, *see* page 58.

Examination:

Bicarbonate.—$3\frac{1}{3}$ drachms of the carbonate are dissolved in 4 fluid-drachms of water, aided by dipping the test-tube in hot water; the solution should be clear and complete, and remain so after cooling; the separation of a crystalline deposit would indicate potassium bicarbonate.

Purified pearlash.—Part of the obtained solution is slightly over-saturated with diluted nitric acid, and allowed to stand in a corked test-tube for several hours; an ensuing gelatinous precipitate would indicate *silicic acid;* the solution, after filtering, if necessary, is then tested in separate portions, with argentic nitrate for *chloride*, and with barium nitrate for *sulphate*, which impurities would indicate the admixture or substitution of purified pearlash.

Metals.—Part of the solution is tested with hydrosulphuric

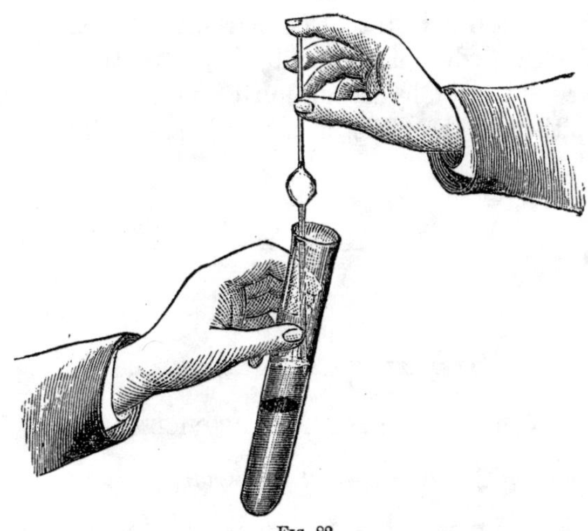

Fig. 82.

acid in two test-tubes, the one after over-saturation with diluted nitric acid. An ensuing dark coloration or precipitate in either of the fluids would indicate *metals*.

Potassium nitrate.—The remainder of the concentrated solution of the salt is over-saturated with diluted sulphuric acid; the clear solution is decanted after a while, and divided into two parts, one of which is mixed with a little ferrous sulphate and then transferred, by means of a pipette, upon concentrated sulphuric acid in a test-tube (Fig. 82); an ensuing purple or brown coloration, at the junction of the two strata of the liquids, would indicate nitrate; the other part is tinged a faint blue with sulphuric-acid solution of indigo, and heated; decoloration would confirm the presence of nitrate.

POTASSII CHLORAS.

POTASSIUM SEU KALIUM CHLORICUM.

Chlorate of Potassium. Potassium Chlorate.

Colorless, transparent, rhomboidal plates, of a pearly lustre, anhydrous, and permanent in the air, and of a spec. grav. of 2.35; when thrown upon burning charcoal they deflagrate, as they also do more or less violently when triturated or heated with readily-combustible substances, as sulphur, carbon, phosphorus, etc. Potassium chlorate fuses at 334° C., and afterward gives off oxygen, until at 352° C. it has parted with the whole of it (39.2 per cent. by weight), leaving behind potassium chloride (60.8 per cent.). When a little sulphuric acid is dropped on the crystals of the chlorate, they become first yellow and then orange-red.

Potassium chlorate dissolves in 16.7 parts of water at 15° C., in about 2½ parts of boiling water, and in 120 parts of alcohol of 0.835 spec. grav. Its saturated aqueous solution has a cooling, saline, slightly acerb taste, and, when mixed with concentrated hydrochloric acid, produces a deep greenish-yellow coloration, with the evolution of chlorine gas. When a few drops of a concentrated solution of potassium chlorate, and subsequently a little concentrated sulphuric acid, are added to a little of a dilute solution of anilin sulphate, upon a watch-glass, the mixture assumes a brilliant deep-violet color. With solution of tartaric acid, the concentrated solution of potassium chlorate gives a white granular precipitate.

Examination:

Potassium Nitrate.—A little of the powdered salt is heated in a porcelain crucible to a full red heat; the residue, when cool, is dissolved in a few drops of water, and the solution tested with turmeric-paper; a brown discoloration of the paper would indicate an admixture of potassium nitrate. As a confirmatory test, a few drops of the solution of the residue may be added to a solution of mercuric chloride; an ensuing red precipitate will confirm the presence of nitrate.

Potassium chloride and *sulphate* are detected in the aqueous solution, acidulated with a few drops of diluted nitric acid, by the occurrence of a white precipitate, in the case of the former with argentic nitrate, of the latter with barium nitrate.

Most commercial potassium chlorate occasions a slight turbidity with argentic nitrate.

POTASSII HYDRAS.

POTASSA. POTASSIUM SEU KALIUM HYDRICUM PURUM. KALIUM CAUSTICUM.

Pure Caustic Potash. Potassa. Potassium Hydrate.

A white, opaque, granular powder, containing, when completely dry, 23 per cent. of water, or, when fused, white, semi-transparent plates or cylindrical sticks, of a fibrous fracture, containing 16 per cent. of water of hydration; exposed to the air, the salt absorbs water and carbonic acid, and gradually deliquesces. It imparts a red color to the flame.

Potassium hydrate dissolves in less than its own weight of water, and also in alcohol, with the evolution of heat, and is slightly soluble in ether; its solution has a soapy feel, a burning, corrosive taste, and a strong alkaline reaction; it gives a grayish-brown precipitate with argentic nitrate, soluble in ammonium hydrate, and precipitates from their solutions most metallic oxides, several of which are redissolved in an excess of the potassium hydrate; it decomposes ammonium salts with the evolution of ammonia.

For *Volumetric Estimation*, *see* page 58.

Examination:

Potassium hydrate must render a clear and complete solution when about 10 grains of it are dissolved in one ounce of alcohol; a residue would indicate foreign salts.

Carbonate may be detected when portions of a concentrated aqueous solution of the hydrate are dropped severally into acetic acid and into lime-water; effervescence with the acid, and a white turbidity with the lime-water, would indicate carbonate.

Nitrate is indicated by the disappearance of the blue color, on heating a little of the aqueous solution which has been mixed with an excess of diluted sulphuric acid, and tinted blue with one drop of sulphuric-acid indigo-solution.

Chloride and *sulphate* are detected in the diluted solution, over-saturated with dilute nitric acid, by testing it in separate portions, with argentic nitrate for chloride, and with barium nitrate for sulphate.

Aluminium salts and *phosphoric acid* may be detected in the diluted solution, by means of aqua ammoniæ, and after filtering, if a precipitate be formed, by subsequent addition of ammoniated magnesium sulphate; a white, gelatinous precipitate with the ammonium hydrate would indicate aluminium salts, and a white, crystalline one with the latter reagent, occurring at once or after several hours, *phosphoric acid*.

Metals are detected by a dark coloration or turbidity of the solution, when mixed with an excess of hydrosulphuric acid, and subsequently over-saturated with an acid.

POTASSII HYDRAS CRUDA.

POTASSIUM SEU KALIUM HYDRICUM CRUDUM.

Crude Potash.

Fused, heavy, compact masses, of a stony appearance, fracture, and hardness, of a soapy feel, burning, corrosive taste, and a destructive action on vegetable and animal matters; its color is mostly greenish or brownish gray; it is deliquescent, and rapidly absorbs water and carbonic acid. Heated to redness, it fuses, but remains unchanged; at a very high heat it is volatile.

Crude potash dissolves, for the most part, in water and in alcohol, with evolution of heat, leaving a more or less considerable residue of impurities; the decanted solution gives a grayish-brown precipitate with argentic nitrate, soluble upon addition of aqua ammoniæ.

The insoluble impurities of crude potash consist chiefly of carbonates, sulphates, silicates, chlorides, and ferric and manganic oxides.

Examination:

In order to ascertain the nature of the impurities, one ounce of the crude potash is triturated and dissolved in two ounces of tepid water, and the whole is allowed to subside in a conical glass measure; two ounces of the clear solution are then mixed with four ounces of strong alcohol, and the mixture allowed to stand for several hours; the solution is then decanted from the precipitate, as far as practicable, and the latter dissolved in hot water; when cool, this solution is filtered, and the insoluble residue washed with a little water, and preserved upon the filter for further examination. The obtained aqueous solution may be examined as follows:

Carbonate and *silicate* are recognized, on dropping a little of the solution into a test-tube containing a mixture of equal parts of water and concentrated nitric acid; the former will be indicated by effervescence, the latter by a white, gelatinous turbidity, ensuing at once or after some hours.

Sulphate and *chloride* may be detected, in separate portions of the solution, by over-saturating it with nitric acid, and subsequently testing with barium nitrate for sulphate, and with argentic nitrate for chloride.

Sulphite and *hyposulphite* are indicated by the occurrence of an insoluble residue, when a portion of the solution is precipitated with argentic nitrate, and the precipitate is redissolved by ammonium hydrate.

Nitrate is detected, in a portion of the solution, after the addition of an excess of sulphuric acid and one drop of sulphuric-acid solution of indigo; the blue tint will disappear upon warming, if nitrate be present.

Phosphate may be detected by a white, crystalline precipitate, when the solution is mixed with an equal volume of aqua ammoniæ, and subsequently tested with magnesium sulphate.

Metallic impurities are recognized by a dark coloration or precipitate, when both the alcoholic solution and the aqueous solution of the residue from the alcoholic one are tested separately with hydrosulphuric acid, and subsequently by over-saturation with hydrochloric acid.

The residue left behind from the solution in alcohol, and preserved upon the filter, is washed with a few drops of alcohol, and then dissolved upon the filter in so much diluted hydrochloric acid as to render a neutral solution; this is tested, in separate portions, with ammonium oxalate for *calcium*, and with barium chloride and a few drops of hydrochloric acid for *sulphate*.

POTASSII FERROCYANIDUM.

POTASSIUM SEU KALIUM FERROCYANATUM.

Ferrocyanide of Potassium. Yellow Prussiate of Potassium. Potassium Ferrocyanide.

Large, transparent, yellow, tabular crystals, derived from an octahedron with a square base; they cleave with facility in a direction parallel to the base of the octahedron, have a peculiar toughness and flexibility, contain three molecules (12.76 per cent.) of water of crystallization, and are permanent in the air. When moderately heated, they give off their water of crystallization, and are converted into a white powder, which is decomposed at a red heat, leaving a residue consisting of potassium cyanide, ferric oxide, and carbon.

Potassium ferrocyanide is soluble in four parts of cold, and two parts of boiling, water, and insoluble in alcohol; its solution has a mild saline taste, gives a white, granular precipitate with a saturated solution of sodium bitartrate, and, when diluted, a blue one with ferric, a brick-red one with cupric, and a white one with ferrous and with plumbic salts; it is not acted upon by hydrosulphuric acid, by tannic acid, nor by the alkaline hydrates and carbonates.

Examination:

Foreign salts are indicated when the potassium ferrocyanide does not yield a complete and clear solution with four parts of water.

Carbonate is indicated by effervescence of the concentrated solution upon the admixture of acetic acid, or upon placing fragments of a crystal in diluted sulphuric acid.

Sulphate is detected, in the diluted solution, acidulated with nitric acid, by a white turbidity with barium nitrate.

Chloride may be detected, when a mixture of 10 grains of the exsiccated salt with 30 grains of fused and powdered ammonium nitrate (free of chloride) is gradually introduced into a red-hot iron spoon, and heated to near redness; when cool, the whole is dissolved in water, and the filtered solution over-saturated with nitric acid, and tested with argentic nitrate, which will indicate chloride by a white precipitate.

POTASSII IODIDUM.

POTASSIUM SEU KALIUM IODATUM.

Iodide of Potassium. Potassium Iodide.

Colorless, anhydrous, semitransparent or opaque crystals, cubical or sometimes elongated in form, permanent in dry, but slightly deliquescent in moist, air, and having a spec. grav. of 2.9. When exposed to heat, potassium iodide decrepitates, and fuses below a red heat; on cooling, it solidifies into a crystalline, pearly mass, without loss of weight, except humidity; at a full red heat, it is slowly volatilized, without decomposition. When a few grains of the salt are heated in concentrated sulphuric acid, or, in a dry test-tube, with a little potassium bisulphate, violet-colored vapors of iodine are evolved; and when a few grains are dissolved in a little chlorine-water, and this shaken with half its volume of chloroform or carbon bisulphide, these will acquire a purple or violet color.

Potassium iodide dissolves in two-thirds of its weight of cold water, in 6½ parts of alcohol, of 0.835 spec. grav., and in 40 parts of absolute alcohol, forming solutions of an acrid, saline taste, and feeble alkaline reaction; the aqueous solution gives, with an excess of tartaric acid, a white, granular precipitate; with argentic nitrate, a yellowish one, which is insoluble in diluted nitric acid, and almost insoluble in ammonium hydrate, but becomes white with the latter; and a vermilion-red precipitate with mercuric chloride, soluble in an excess of either the solution or the reagent; it gives a violet-blue color with a little mucilage of starch, upon subsequent addition of a few drops of chlorine-water (distinction from potassium bromide and chloride), and a white, crystalline precipitate with a saturated solution of sodium bitartrate.

Examination:

Water, which may be contained as interstitial moisture in the crystals, is recognized, and may be quantitatively determined, by the loss of weight upon drying a known weight of the triturated salt at about 80° C.

Impurities and Admixtures.—In order to obtain for examination an average representation of the iodide, several ounces of smaller and larger crystals are selected from the bulk of the salt, and triturated to a granular powder, part of which may serve for the following tests: One drachm of it is dissolved in one fluid-drachm of water; the solution formed must be clear and complete, and remain so after the addition of 1½ fluidounce of strong or absolute alcohol; an ensuing turbidity or crystalline deposit would indicate foreign salts (carbonate, sulphate, iodate, nitrate); if this precipitate is considerable, it may be collected upon a filter, washed with a few drops of alcohol, and then dissolved in a few drops of warm water; the obtained solution may be tested for *carbonate* with turmeric-paper, or by allowing one or two drops of it to fall into concentrated hydrochloric acid; a brown coloration of the paper, and effervescence with the acid, will indicate the presence of carbonate; the rest of the solution is acidulated with a few drops of hydrochloric acid, and tested for *sulphate* with one drop of barium chloride, and subsequently, for *nitrate*, by tingeing it faintly with sulphuric-acid indigo-solution, and heating.

Another method of ascertaining the presence of foreign salts in potassium iodide, without determining their nature, consists in adding a solution of mercuric chloride to the solution of the iodide, the salts being taken in the proportion of four equivalents of potassium iodide to one equivalent of mercuric chloride; five grains of the former are dissolved in three ounces of water in a beaker-glass; a solution of two grains of the latter salt in two ounces of water is then gradually added, with constant stirring, to the solution of the iodide; if this be pure, its excess is just sufficient to redissolve all the mercuric iodide formed; if impure, a deficiency of iodide will give rise to the appearance of a permanent precipitate of mercuric iodide before the whole of the mercuric solution is added, and this result will take place the earlier, the less iodide and the more impurities are contained in the salt.

Potassium iodate may be detected in the aqueous solution of the salt, by adding a few drops of mucilage of starch, and then a few drops of a concentrated solution of tartaric acid, insufficient to cause a precipitate; if iodate be contained in the salt, a violet coloration of the liquid will occur at once. Or the aqueous solution, mixed with a few drops of concentrated solution of tartaric acid, may be shaken with some chloroform, which will assume a red color when iodate is present.

Iodate may also be recognized in potassium iodide by dropping a crystal of tartaric acid into a strong solution of the iodide in previously boiled, distilled water, and allowing it to remain at rest for several minutes; if iodate be present, the crystal will be enveloped after that time in a yellowish-white zone.

Carbonate may be detected by a white turbidity when the aqueous solution of the potassium iodide is mixed with twice its volume of lime-water.

Sulphate may be detected in the diluted solution of the iodide, previously acidulated with hydrochloric acid, by a white precipitate with barium chloride.

Nitrate may be detected in the aqueous solution, if the salt be free from iodate, by adding a little sulphuric-acid mucilage of starch, and by immersing in the liquid a thin piece or rod of bright zinc; if any nitrate be present, the liquid will gradually become blue, the coloration issuing from the zinc.

Chloride and *bromide* are detected by carefully precipitating, with argentic nitrate, a warm, diluted, aqueous solution of the salt; a slight excess of aqua ammoniæ is then added, and, after agitation, the mixture is filtered, and the filtrate over-saturated with nitric acid; since ammonium hydrate dissolves only traces of argentic iodide, the transparency of the liquid must be not at all, or only slightly, impaired; a white turbidity, subsiding to a precipitate, would indicate bromide or chloride.* In order to distinguish these, the precipitate is collected upon a filter and washed with a little water, until this ceases to redden blue litmus-paper; the filter is then pierced by a glass rod, and the precipitate rinsed into a test-tube; after subsidence, the water is, as far as possible, decanted, and chlorine water added and agitated with the precipitate; since chlorine decomposes argentic bromide, dissolving the disengaged bromine with a yellow color, bromide will be recognized by a more or less deep yellow coloring of the fluid, while argentic chloride remains unchanged and without any action on chlorine-water. When chloroform or ether is then added to the fluid and agitated, it will absorb the bromine and the yellow color from the water.

A confirmatory test for the recognition of bromide is, to add to a solution of the salt an excess of solution of cupric sulphate, and subsequently so much of a saturated solution of sulphurous acid as to impart its strong odor to the mixture; the whole is gently warmed, filtered, and heated to boiling, in order to expel the sulphurous acid; when cold, it is re-filtered if necessary, and the filtrate shaken with half its volume of chloroform, which, after subsidence has taken place, will appear yellow if bromide is contained in the salt.

* When a comparatively large quantity of potassium iodate is contained in the iodide, argentic iodate is also formed in this test, and gives rise, upon the addition of the nitric acid, to the formation of a white, *granular* precipitate, which, however, is quite different in appearance from the *curdy* one of the argentic chloride or bromide.

A quantitative estimation of the purity of potassium iodide may be made by completely precipitating with argentic nitrate a warm solution of 10 grains of the salt in about one ounce of water; after cooling and subsidence, the supernatant menstruum is carefully decanted, so as to avoid any loss of the precipitate; one fluid-drachm of aqua ammoniæ, diluted with two drachms of water, is then added to the precipitate, and, after agitation for a few minutes, this latter is collected upon a tared, moist filter, is washed with a little water, and dried at a temperature not exceeding 80° C. If the potassium iodide was pure, 14.1 grains of argentic iodide should be obtained.

For *Volumetric Estimation*, *see* page 62.

POTASSII NITRAS.

POTASSIUM SEU KALIUM NITRICUM.

Nitrate of Potassium. Salpetre. Nitre. Potassium Nitrate.

Long, striated, six-sided, prismatic crystals, with dihedral summits, colorless and transparent, and of a spec. grav. of 2.0; or a white, granular powder, permanent in the air. When thrown upon burning coals, it deflagrates with bright scintillations, leaving an alkaline residue, which, when heated upon the looped end of a platinum wire, imparts a red color to the flame; exposed to heat, it fuses below redness.

Potassium nitrate is soluble in water, 100 parts of which dissolve, at 0° C., 13.3 parts, at 18° C., 29 parts, and at 97° C., 236 parts of the salt; it is far less soluble in glycerin, and almost insoluble in alcohol; its aqueous solution is neutral, has a cooling, saline taste, and forms a white, granular precipitate with concentrated solution of sodium bitartrate; a few drops of it mixed with a solution of ferrous sulphate, and carefully placed upon concentrated sulphuric acid (Fig. 82, page 296), give rise to the formation of a dark coloration upon the line of contact between the two fluids.

Examination:

Chloride and *sulphate* are detected in the diluted solution of the salt by white precipitates when acidulated with nitric acid, and tested in two separate portions, with argentic nitrate for chloride, and with barium nitrate for sulphate.

Chlorate is indicated by a yellow coloration, and the evolution of chlorine, when a concentrated solution of the potassium nitrate is mixed and gently warmed with an equal volume of concentrated hydrochloric acid.

Calcium and *magnesium salts* are detected by a white turbidity when the diluted solution is warmed with dilute solution of sodium carbonate; they may be distinguished by adding a little aqua ammoniæ to the dilute solution of the salt, and testing it, in separate portions, with ammonium oxalate for calcium, and with sodium phosphate for magnesium.

Sodium nitrate may be detected by a white, crystalline precipitate, occurring either at once or after several hours, when a concentrated cold solution of the salt is tested with a few drops of solution of potassium antimoniate, and then allowed to stand for six hours.

If potassium nitrate contains even a few per cent. of sodium nitrate, it will have a moist appearance, arising from the deliquescent character of the latter salt.

POTASSII PERMANGANAS.

POTASSIUM SEU KALIUM PERMANGANICUM SEU HYPERMANGANICUM.

Permanganate of Potassium. Potassium Permanganate.

Slender, prismatic, dark purple crystals, of a metallic lustre, and permanent in the air; they decrepitate when thrown upon burning coals, or when suddenly heated, and detonate with a flame when triturated with equal parts of sulphur, and heated to 177° C.

Potassium permanganate is soluble in 16 parts of cold, and 2 parts of boiling, water, forming a deep purple solution, which, when highly diluted, is lilac-colored; it is insoluble in alcohol and in chloroform; its solution has a sweet, astringent taste, is neutral, and becomes yellowish brown when mixed and heated with alcohol.

Since permanganic acid is readily reduced, the solution of the salt is decomposed and decolorized by most organic substances, and by inorganic reducing agents—e. g., sulphurous, hydrosulphuric, and oxalic acids, and all metallic subsalts. Potassium permanganate is, therefore, a powerful oxidizer, causing more or less violent reactions with many substances, and the combustion of inflammable bodies.

POTASSII ET SODII TARTRAS.

POTASSIUM ET SODIUM TARTARICUM. NATRIO-KALIUM TARTARICUM. SODIUM TARTARATUM. TARTARUS NATRONATUS.

Rochelle Salt. Seignette Salt. Potassium and Sodium Tartrate.

Large, colorless, transparent, prismatic crystals, the faces of which are unequally developed, containing eight molecules (29.86 per cent.) of water of crystallization, and being slightly efflorescent. The salt occurs in commerce generally ground, as a snow-white powder. When heated in a porcelain capsule, it undergoes aqueous fusion, the water evaporates, and the salt is decomposed, with the evolution of inflammable vapors and the odor of caramel, while a black residue is left, which consists of a mixture of potassium and sodium carbonates with carbon, and which colors turmeric-paper brown, and imparts a yellow color to the flame when heated upon the looped end of a platinum wire.

Potassium and sodium tartrate is insoluble in alcohol, but soluble in 2 to 2½ parts of cold, and in less than its own weight of boiling, water; the solution is neutral, has a mild, cooling, saline taste, and forms a white, granular precipitate with acids and with solutions of acidulous salts.

For *Volumetric Estimation, see* page 58.

Examination:

Two drachms of the salt are dissolved in six drachms of warm water; the solution should be clear and complete, and remain so after cooling; it should not act upon litmus-paper, nor effervesce upon addition of hydrochloric acid (evidence of the absence of sodium carbonate or bicarbonate).

Chlorides and *sulphates* may be detected by a white precipitate when the diluted solution of the salt, acidulated with nitric acid, is tested in two separate portions, with argentic nitrate for the former, and with barium nitrate for the latter. In case the solution separates granular potassium bitartrate upon the addition of the acid, sufficient hot water is added to redissolve the precipitate before adding the reagent.

Calcium salts are detected in the diluted solution, by means of ammonium oxalate.

Metallic impurities are detected in the concentrated solution by hydrosulphuric acid, and subsequent addition of a little aqua ammoniæ.

Crystallized potassium and sodium tartrate, being in appearance somewhat similar to crystallized borax and alum, and therefore liable to incidental mistake, may readily be distinguished from either of these substances, in addition to its physical characters, by its taste, by its neutral reaction—alum being sour, borax alkaline, and by the black alkaline fuse upon incineration, while both borax and alum swell up to a porous mass, and yield a white or colorless fuse.

POTASSII SULPHAS.

POTASSIUM SEU KALIUM SULPHURICUM.

Sulphate of Potassium. Potassium Sulphate.

Hard, colorless, transparent, short, six-sided prisms, terminated by six-sided pyramids, or a white, granular powder, anhydrous, and permanent in the air, and having a spec. grav. of 2.66; when heated, the crystals decrepitate strongly; the salt imparts a purple color to the flame.

Potassium sulphate is soluble in about 12 parts of cold, and 4 parts of boiling, water, but insoluble in alcohol; the solution has a saline bitter taste, is neutral, and forms white precipitates with tartaric acid or sodium bitartrate, and with solutions of salts of calcium, barium, or lead.

Examination:

Metallic impurities are detected in the warm aqueous solution, by a dark coloration or turbidity, with an excess of hydrosulphuric acid, and subsequent acidulation with hydrochloric acid. Potassium ferrocyanide should cause neither a blue (*iron*) nor a reddish (*copper*) reaction in the slightly acidulated solution.

Sodium sulphate is indicated by a greater degree of solubility in cold water than that above stated; one drachm of the powdered salt, when dissolved in one fluidounce of boiling water, must, on cooling, give a crystalline deposit; otherwise sodium sulphate, or the admixture of more soluble salts, is indicated; in this case the diluted solution may be acidulated with nitric acid, and tested by means of argentic nitrate for chloride.

POTASSII SULPHURETUM.

POTASSII SULPHIDUM. POTASSIUM SEU KALIUM SULPHURATUM. HEPAR SULPHURIS.

Sulphuret of Potassium. Potassium Sulphide.

Solid, fused fragments, of a yellowish-brown color, when freshly prepared or recently broken; on exposure to the air they assume a greenish appearance, and finally become of a dirty white, in consequence of gradual decomposition by the action of atmospheric moisture and oxygen, the sulphides being successively converted, with the evolution of hydrogen sulphide, into hyposulphite, sulphite, and ultimately sulphate. When moistened with acids, potassium sulphide evolves hydrogen sulphide.

Potassium sulphide, which is usually a variable mixture of higher potassium sulphides with hyposulphite, sulphite, and sulphate, and with undecomposed carbonate, dissolves in 3 to 4 parts of water, and also in alcohol, leaving behind in the latter case the oxygen salts; these solutions have an orange-yellow color, a nauseous, alkaline, bitter taste, and the odor of hydrosulphuric acid, which is abundantly evolved, with the separation of sulphur, upon the addition of acids; they precipitate metallic sulphides from the solutions of most of the metallic salts.

POTASSII TARTRAS.

POTASSIUM SEU KALIUM TARTARICUM.

Tartrate of Potassium. Potassium Tartrate.

Colorless, semi-transparent, irregular, six-sided prisms, with dihedral summits, or a white, granular powder, slightly deliquescent. When heated in a dry test-tube, potassium tartrate melts, swells up, and is decomposed with the evolution of empyreumatic vapors, leaving behind a mixture of carbon and potassium carbonate, which effervesces with acids, and imparts a red color to the flame.

Potassium tartrate is soluble in a little less than its own weight of cold, and in half its weight of boiling, water, yielding a neutral solution, of a mild saline taste; it is but sparingly soluble in alcohol; its aqueous solution is decomposed by most acids and acidulous salts, forming, if not too dilute, a white, granular deposit of bitartrate.

For *Volumetric Estimation, see* page 58.

Examination:

Potassium and Sodium Tartrate.—One drachm of the salt is shaken with 1½ drachm of cold water; a clear and complete solution must take place, an incomplete solution may indicate an admixture of Rochelle salts. This may be ascertained by reducing about one drachm of the potassium tartrate, by ignition in a porcelain crucible, to carbonate, and then testing the filtered solution of the residue by mixing it with an equal volume of solution of potassium antimoniate; the occurrence of a white, crystalline deposit, at once or after several hours' standing, would prove an adulteration with potassium and sodium tartrate.

Bicarbonate, *carbonate*, and *bitartrate*, are recognized in the solution of the salt, the two former by a white precipitate with calcium chloride in the cold, and by an alkaline reaction upon turmeric-paper; the last by an acid reaction upon blue litmus-paper.

Sulphate and *chloride* may be detected in the dilute solution of the salt, when it is slightly acidulated with diluted nitric acid, and then tested, in separate portions, with argentic nitrate for chloride, and with barium nitrate for sulphate.

Metallic impurities are recognized by mixing a concentrated solution of the salt with two or three times its volume of hydrosulphuric acid, and by subsequent acidulation with acetic acid. Should a white, granular deposit of potassium bitartrate occur, the liquid is diluted with boiling water, or is warmed so as to redissolve the precipitate. Any reaction with the hydrosulphuric acid, before or after acidulation, would indicate metallic impurities.

QUINIA.

CHININUM. CHINIUM.

Quinine. Quinia.

A white, granular powder, or small, needle-like crystals, containing 14.28 per cent. of water of crystallization; permanent in the air; when heated, quinia fuses at about 196° C., and upon cooling solidifies into a resin-like mass, yellow, translucent, and friable; at a higher temperature it is wholly dissipated.

Quinia dissolves in about 1,200 parts of cold, and 250 parts of boiling, water, and in about 2 parts of boiling alcohol, and in 60 parts of ether; * it is also soluble in chloroform, and in benzol; its solutions have a bitter taste, and a feeble alkaline

* The solubility of quinia in ether differs according to the form of the alkaloid: it requires about 60 parts when in the crystallized state, but only about half that quantity when it is amorphous; for instance, when one grain of quinia sulphate

reaction, and neutralize acids. Quinia is freely soluble in diluted acids; these solutions exhibit an azure-blue fluorescence, caused by a change of the refrangibility of the invisible chemical rays. Concentrated sulphuric and nitric acids dissolve quinia without color.

Solutions of quinia and its salts are precipitated by the alkaline hydrates, carbonates, and bicarbonates, by calcium hydrate, by tannic and picric acid, by potassium ferrocyanide, and by potassio-mercuric iodide. The precipitates with calcium, potassium, sodium, and ammonium hydrates, are redissolved by a great excess of the precipitant.

Solutions of quinia and its salts render an emerald-green solution when dissolved in, or mixed with, chlorine-water, and afterward with an excess of aqua ammoniæ; the green color passes into purple upon the subsequent addition of potassium ferrocyanide; this characteristic reaction is most strikingly exhibited when, to the solution of quinia in chlorine-water, the solution of potassium ferrocyanide is first added, and subsequently the aqua ammoniæ.

When a solution of quinia or its salts in acidulated water is precipitated with ammonium hydrate, and the turbid mixture is then divided into three portions in as many test-tubes, and these shaken severally with a little ether, chloroform, and benzol, the precipitate will be dissolved, and the liquids subside into two clear, colorless strata in each test-tube.

Estimation of the Alkaloids in Cinchona-Barks:

The therapeutical value of the cinchona-barks is due to the alkaloids contained in the bark, and of which the principal ones are *quinia*, *quinidia*, *cinchonia*, and *cinchonidia*. The estimation of the commercial value, therefore, depends upon the determination of the quantity of these alkaloids, and in particular of the first one, in a known weight of the bark; and, from among the numerous methods employed for this purpose, the following ones are simple and expeditious:

is dissolved in about 15 drops of acidulated water, and subsequently precipitated with aqua ammoniæ, 10 grains of ether suffice to redissolve the alkaloid; while at least 35 grains of ether will be required for solution, when the precipitate is allowed to stand for 24 hours, and become crystalline, previous to the addition of the ether.

I. From a large number of pieces of the bark, small fragments are cut and reduced to a fine powder, so as to represent as nearly as possible the average of the bark to be examined; 10 drachms of the powder are mixed with an equal bulk of coarsely-powdered quartz or glass, and the mixture introduced into a comparatively narrow glass percolator, whose lower end is drawn into a point, and loosely closed with some cotton; the powder is covered with a layer of glass or quartz, and, when the percolator is placed in a perpendicular position, water, acidulated with about one per cent. of concentrated sulphuric acid, is introduced until it has penetrated the column and commences to drop out, then the upper orifice is corked, and the whole allowed to stand for several hours, when the cork is removed, and the percolation with the acidulated water is continued, until a few drops of the menstruum, received in a test-tube, cease to have a bitter taste, or to become turbid on addition of aqua ammoniæ. When this point is reached, and the bark completely exhausted, the entire liquid is poured into a porcelain dish, and three drachms of powdered, calcined marble are added, and the mixture evaporated to dryness on a water-bath. The dry residue is triturated, and introduced into a flask, and repeatedly and completely extracted with strong alcohol. The filtered alcoholic solution is evaporated in a small, tared beaker-glass at a gentle heat, and, when dry, the beaker is allowed to stand for several hours at a temperature near 80° C.; it is finally weighed, and, when the tare is subtracted, the difference answers approximately to the quantity of cinchona alkaloids contained in the 10 drachms of the bark.

When a quantitative determination of the quinia, quinidia, and cinchonia is desired, the alkaloids contained in the beaker are exhausted with ether, which dissolves the two former, and leaves the latter behind. The ethereal solution is then mixed with a little alcohol, acidulated with hydrochloric acid, and completely precipitated with platinic chloride; the ensuing precipitate of quinia- and quinidia-platinic chloride is collected upon a double filter, both of the same paper and size, is washed with alcohol, dried at 80° C., and then weighed; the second filter serving as the tare of the one containing the precipitate. The weight of the platinic precipitate is divided by

2.274, and the quotient gives the quantity of quinia and quinidia contained in 10 drachms of the cinchona-bark.

The residue remaining undissolved in ether is dissolved in some alcohol, one drop of hydrochloric acid is added, and it is then completely precipitated with platinic chloride; the precipitate of cinchonia-platinic chloride is washed upon a double filter with alcohol, is dried at 80° C., and weighed as the preceding one. The weight is divided by 2.34, the quotient gives the quantity of the cinchonia contained in 10 drachms of the cinchona-bark.

The platinum may be recovered by igniting the precipitates in an open crucible of platinum or porcelain.

II. The British Pharmacopœia gives the following process for the estimation of the quinia in cinchona-barks:

"Boil 100 grains of the bark, reduced to a fine powder, for a quarter of an hour in a fluidounce of distilled water, acidulated with 10 drops of hydrochloric acid, and allow it to macerate for 24 hours. Transfer the whole to a small percolator, and, after the fluid has ceased to drop, add at intervals about an ounce and a half of similarly acidulated water, or until the fluid which passes through is free from color. Add to the percolated fluid, solution of plumbic subacetate, until nearly the whole of the coloring-matter has been removed, taking care that the fluid remains acid in reaction. Filter, and wash with a little distilled water. To the filtrate add about 35 grains of potassium hydrate, or as much as will cause the precipitate, which is at first formed, to be nearly redissolved, and afterward 6 fluid-drachms of ether. Then shake briskly, and, having removed the ether, repeat the process twice with 3 fluid-drachms of ether, or until a drop of the ether employed leaves, on evaporation, scarcely any perceptible residue. Lastly, evaporate the mixed ethereal solutions in a capsule. The residue, which consists of nearly pure quinia, when dry, should weigh not less than two grains (2 per cent.), and should be readily soluble in diluted sulphuric acid."

III. Another method * is the following:

* This method, suggested by P. Carles, in 1869, differs from similar ones mainly by the employment of chloroform for exhaustion, instead of ethylic or amylic alcohol, of ether, or of benzol, and by the separation of the quinia sulphate from a solu-

One thousand grains of a good average sample of the powdered and dried bark are intimately mixed with from 300 to 400 grains of slacked lime, mixed with about 1,750 grains of water, and the pasty mixture of the bark and lime dried at a gentle heat; the cakes thus formed are reduced to a coarse powder, and pressed into a small percolator, with stop-cock and glass stopper; chloroform is then gradually poured on the contents until they are moistened throughout; stop-cock and glass stopper being closed, the whole is allowed to stand for 12 hours, when the bark is completely exhausted by successive slow percolation with chloroform. About 7,000 grains of chloroform will be required for the exhaustion of the bark, which may be ascertained by evaporating a few drops of the last percolate upon a watch-glass, and by dissolving the residue with a few drops of chlorine-water and one drop of diluted hydrochloric acid, and by adding the solution to a little aqua ammoniæ in a small test-tube.

The bark is exhausted when the occurrence of a greenish coloration with this test ceases. When this is the case, the chloroform is displaced by water poured upon the percolator, and, as soon as chloroform ceases to drop off, the entire percolate is evaporated upon a water-bath, until a dry residue is left. If desired to save the chloroform, especially when the test is performed on a larger scale, it may be distilled off in a small retort with a Liebig cooler, from a water-bath: the distillation should not, however, be carried on to dryness, but the remainder of the fluid is to be evaporated to dryness in a porcelain capsule. The residue consists of the alkaloids mixed with about their own weight of resinous and such other matters of the bark as are soluble in chloroform; it is treated several times with so much cold diluted sulphuric acid (1 part concentrated acid to 10 parts of water) as is required for solution of the alkaloids; about 500 to 600 fluid-grains of the diluted acid will be sufficient. The solution is passed through a small filter, previously moistened, and this is washed with a little cold water; the colorless filtrate is heated in a flask to near boiling, and, when

tion of ammonium sulphate, wherein it is said to be insoluble. It has the advantage of being simple in operation, and of occasioning the least loss, and giving the fullest yield of quinia, with an almost complete exclusion of cinchonia.

hot, so much aqua ammoniæ is added as nearly to neutralize the acid solution. This is then allowed to cool slowly. As quinia sulphate is insoluble in solution of ammonium sulphate, all the quinia sulphate will deposit in a crystalline crust, together, probably, with a part of the quinidia sulphate; the menstruum is then poured off into another beaker, and the crystalline mass broken and washed with a few drops of cold water, either in the beaker or upon a tared filter, after which it is dried at about 80° C.

If the mother-liquor is found to be very acid, it has to be neutralized with aqua ammoniæ, and once more allowed to stand and cool for crystallization; if this has taken place, the crystals are collected upon a tared filter, washed with cold water, and also dried.

The entire yield of dried crystals is then weighed, and the number of grains, after having added 11 per cent. to make up for the water of crystallization, contained in quinia sulphate, and lost on drying, when divided by 10, gives at once the percentage of quinia sulphate obtainable from the cinchona-bark.

IV. An adequate and perhaps the most reliable method for the estimation of the value of commercial cinchona-bark, designed for the manufacture of Quiniæ Sulphas, is the preparation of quinia sulphate on a small scale by the process of the U. S. Pharmacopœia:

Twelve ounces of the bark, in coarse powder, are thoroughly mixed with 4 pints of water and half an ounce of hydrochloric acid of 1.160 spec. grav.; the mixture is allowed to macerate for twelve hours, and is then gently boiled in a porcelain capsule for about one hour; it is then strained through muslin, and the residue is boiled twice successively, with half of a mixture consisting of 3 pints of water and 3 drachms of hydrochloric acid, and strained; the powder is finally submitted in the muslin to a gentle pressure, and this is repeated again after it has been moistened with about 6 ounces of boiling water. The obtained liquids are mixed in a porcelain capsule, and heated; now, while hot, a mixture of 10 drachms of lime, previously slacked with 8 ounces of water, is gradually added, with constant stirring with a glass rod, until the solution of the lime, and the corresponding formation of a precipitate of quinia, cease, or until the liquid turns turmeric-paper brown. The precipitate is collected upon a filter, and washed with distilled water, until this ceases to cause a turbidity, when a few drops are allowed to fall into a solution of argentic nitrate, acidulated with nitric acid; the precipitate is then dried, and, when detached from the filter, reduced to a fine powder, which is successively exhausted with decreasing quantities of boiling alcohol, separating the alcohol from the precipitate by decantation, which operation is repeated until the alcohol is no longer rendered bitter. The obtained alcoholic solutions are mixed, and are concentrated to a syrupy consistence, either by evaporation or by distillation; the remainder is then transferred and rinsed into a flask, with 12 ounces of boiling water; it is then heated to boiling, and so much diluted sulphuric acid is added as just completely to dissolve the quinia. Then half an ounce of common animal charcoal is added, or so much as to leave the liquid of a faintly acid reaction upon blue litmus-paper; after a few minutes' gentle stirring, the mixture is filtered while hot, and the residue washed on the filter with a little boiling water; the entire filtrate is then set aside to crystallize. The crystals are separated from the liquid, which is allowed to drop off as much as possible, and they are then dried in a porcelain capsule, covered with bibulous paper, at a heat not exceeding 50° C., taking care to avoid the efflorescence of the dry crystals. The remaining liquid is evaporated at a boiling heat to nearly two-thirds of its volume, and is then allowed to crystallize again. These crystals are separated and dried as before. The mother-water may be made to yield an additional quantity of quinia sulphate by precipitating the quinia with aqua ammoniæ, and treating the precipitated alkaloid as before with proportionate quantities of distilled water, sulphuric acid, and animal charcoal.

The entire yield of quinia sulphate, when dry, represents the available quantity of the salt obtained from 12 ounces of the bark, which number has to be multiplied by 8.33 in order to express the percentage value of the bark in quinia sulphate.

QUINIÆ HYDROCHLORAS.

QUINIÆ MURIAS. CHININUM SEU CHINIUM HYDROCHLORICUM SEU HYDROCHLORATUM.

Hydrochlorate of Quinia or Quinine. Quinia Hydrochloride.

White, silky needles, or a crystalline powder; when exposed to heat, the salt fuses, and finally burns away without residue. Quinia hydrochloride is soluble in about 20 parts of cold water, in from two to three parts of alcohol, and only slightly in ether; but it is freely soluble in acidulated water, and in diluted as well as in concentrated acids, without change of color; its solutions are neutral, have a bitter taste, and yield an emerald-green coloration with chlorine-water, on subsequent addition of an excess of aqua ammoniæ; the green becomes purple with potassium ferrocyanide; the aqueous solution, acidulated with nitric acid, yields a white, curdy precipitate with argentic nitrate, soluble in an excess of ammonium hydrate, and a white precipitate with aqua ammoniæ, which, however, is dissolved by ether, alcohol, chloroform, or benzol, when added and agitated with the liquid.

The examination of quinia hydrochloride for quality and purity is the same as that of quinia sulphate, as described on pages 321–324.

In reference to the mode of its preparation, its aqueous solution needs, moreover, to be examined for contamination with quinia sulphate and with barium chloride, by acidulation with nitric acid, and subsequent testing in separate portions with barium nitrate for the former, and with diluted sulphuric acid for the latter. Either of these impurities, of course, excludes the presence of the other one, without impairing the solubility of the salt.

In the application of the chemico-microscopic test (page 323), the concentrated aqueous solution of the quinia hydrochloride may be added to the solution of potassium sulphocyanide without the addition of acid.

QUINIÆ SULPHAS.

CHININUM SEU CHINIUM SULPHURICUM.

Sulphate of Quinia or Quinine. Quinia Sulphate.

Fine, silky, slightly flexible, snow-white needles, interlaced among one another, or grouped in small, starlike tufts; the crystals contain seven molecules of water of crystallization, five of which (about 11 per cent. by weight) evaporate, with a slight efflorescence of the salt, when exposed to a warm and dry air; the remaining two molecules separate at about 116° C., the fusing-point of quinia sulphate, when it still retains one molecule of water of constitution; at a stronger heat, the fused salt assumes a deep red color, and finally, when heated in the open air, burns slowly and wholly away.

Quinia sulphate is but sparingly soluble in cold water, one part requiring 790 parts of cold, and 33 parts of boiling, water; it dissolves in 60 parts of alcohol, and is also soluble in benzol, but only sparingly in ether and in chloroform. Dilute as well as strong acids will dissolve it freely, forming colorless, bitter solutions, which, when diluted, exhibit a blue fluorescence.

Solutions of quinia sulphate are precipitated by the alkaline hydrates, carbonates, and bicarbonates, by lime-water, by tannic and picric acids, by potassium ferrocyanide, and by potassio-mercuric iodide. The precipitates with calcium hydrate and with the alkaline hydrates are soluble in an abundance of the precipitant.

Like all quinia salts, quinia sulphate renders an emerald-green reaction when dissolved in, or when its solution is mixed with, chlorine-water, and afterward with an excess of aqua ammoniæ; the green color passes into purple upon the subsequent addition of a few drops of solution of potassium ferrocyanide, which reaction is especially striking when, to the solution of the quinia sulphate in water, the solution of potassium ferrocyanide is first added, and subsequently the aqua ammoniæ.

One hundred grains of quinia sulphate, dissolved in water acidulated with hydrochloric acid, yield, upon complete precipitation with barium chloride, a precipitate of barium sulphate, which, when collected, washed, and dried upon a filter, and completely incinerated by ignition, should weigh 26.6 grains.

Examination:

Mineral admixtures are detected by a residue left after igniting a little of the salt upon platinum-foil, or after dissolving about five grains of the salt in four drachms of boiling alcohol.

Ammonium salts are recognized by the odor of ammonia, and by the formation of white vapors from a glass rod moistened with acetic acid, when held in the orifice of the small test-tube, wherein a few grains of the quinia sulphate are heated with strong solution of potassium hydrate.

Chlorides and *hydrochlorides* may be recognized in the dilute solution of the salt in water, acidulated with nitric acid, by a white, curdy precipitate with argentic nitrate.

Stearic acid may be detected in the above-described alcoholic solution, by adding an equal volume of water; the liquid becomes turbid, but, on warming it gently, by dipping the test-tube in hot water, it becomes transparent again; the appearance of an oily layer on the surface would indicate the above fatty acid.

Salicin, *sugar*, and *mannite*, may be detected in the solution of the preceding test, if free from fatty substances, by mixing it, in a porcelain capsule, with about five grains of barium carbonate, and evaporating the whole to dryness with constant stirring; the residue is triturated with a little water, and transferred upon a moist filter; the obtained filtrate is evaporated at a gentle heat, upon a watch-glass, and must leave no residue, or only a very small one; if a residue remains, it is divided, and placed upon two watch-glasses, with one drop of water upon each, and is again dried up at a gentle heat; then, upon the one glass, a small drop of concentrated sulphuric acid

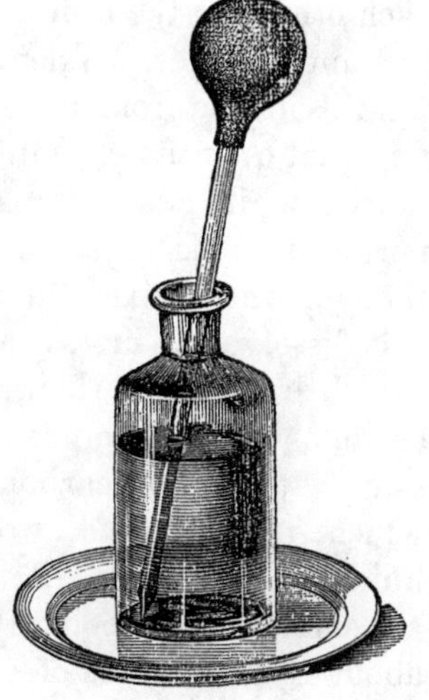

Fig. 83.

is allowed to fall from a glass rod or from a small pipette (Fig. 83); a red color will be produced if *salicin* is present, a black one if *sugar;* mannite remains unchanged, and may be detected on the second watch-glass, by a few drops of alcohol, which dissolve the *mannite*, and leave it behind in small, acicular crystals upon spontaneous evaporation.

Since quinia sulphate dissolves without apparent change in strong sulphuric acid, even when gently warmed, this test may be directly applied for the detection of admixtures of sugar, mannite, or fatty acids, which will produce a black coloration; a red coloration might be indicative of the presence of salicin, but, since many other compounds produce a similar reaction, the following additional test may be employed for salicin:

Four grains of the quinia sulphate are dissolved, in a test-tube, in one drachm of water and four drops of concentrated hydrochloric acid; the solution is boiled for a few minutes, when, if *salicin* be present, a white turbidity caused by the formation of saliretin will take place.

Cinchonia, Cinchonidia, and Quinidia.—These associate alkaloids of quinia are distinguished from the latter, and from each other, by the difference in their solubility in aqua ammoniæ and in ether. When freshly precipitated, quinidia requires for re-solution from 10 to 11 times as much aqua ammoniæ as an equal quantity of quinia; cinchonidia from 12 to 13 times as much; while cinchonia is not dissolved by a much larger proportion than is required by either of the other alkaloids, and, though, when mixed, in a very small proportion, with quinia, it is dissolved at first, it afterward separates on standing.

With regard to their solubility in ether, the alkaloids differ as follows: Thirty grains of quinia sulphate, when precipitated with sodium carbonate, yield about 22 grains of alkaloid, which, if all quinia, would require for solution about four fluidounces of ether; if all quinidia, about six fluidounces; if all cinchonidia, about 10 fluidounces; and, if all cinchonia, about 40 fluidounces of ether.

Hereupon the three following tests are based:

1. Thirty and two-thirds grains (two grammes) of the quinia sulphate are mixed and agitated with five fluid-drachms (20

cubic centimetres) of distilled water, at a temperature below 15° C., and the mixture allowed to stand for half an hour, at a temperature not exceeding 15° C.; it is then filtered, and $1\frac{1}{3}$ fluid-drachm (five cubic centimetres) of the filtrate are placed in a long, narrow test-tube, and $1\frac{7}{8}$ fluid-drachm (seven cubic centimetres) of aqua ammoniæ added, taking care that the two fluids do not mix; the tube is then closed, with the finger or a cork, and gently reversed, without undue agitation, when the liquid should, after a while, become clear and transparent. Otherwise an unusual quantity of the above alkaloids is indicated.

2. Ten grains of the quinia sulphate are agitated, in a narrow test-tube, with a mixture consisting of one drachm by weight of pure ether and 10 drops of aqua ammoniæ, and are then allowed to rest; the liquid should, after a while, form two separate, colorless, and transparent layers, without any white or crystalline matter floating at the line of contact between the two strata; such an appearance would also indicate the presence of the above associate alkaloids of quinia.

3. Ten grains of the quinia sulphate are dissolved, in a test-tube, in one fluid-drachm of distilled water, and 10 drops or less of diluted sulphuric acid; then $2\frac{1}{2}$ fluid-drachms of ether, 3 drops of alcohol, and 40 drops of a solution of potassium hydrate (1 part to 12 parts of water) are added; the test-tube is corked, agitated, and allowed to stand for 12 hours. When pure, the two fluid strata, and the line of their contact, will be clear and transparent; but, if even traces of cinchonia and cinchonidia are present, they will be seen floating as a white cloud at the line of separation between the two strata; when cinchonidia is the only impurity, it will appear at the line of contact as an oily-like stratum, consisting, when magnified, of minute amorphous particles, while cinchonia appears crystalline.

Another and ready method of distinguishing the three principal associate alkaloids, in quinia sulphate, is *Stoddard's chemico-microscopic test*: Five grains of the quinia sulphate are dissolved in two drachms of distilled water, acidulated with four drops of diluted sulphuric acid; of this solution, one small drop is placed upon the glass slip, and about half a drop of solution of potassium sulphocyanide is put by its side; both are then covered with one piece of thin glass, which will cause the drops to touch. At the line of junction, an immediate precipitation takes place, and the crystals of the different alkaloids may be recognized, under the microscope, at a magnifying

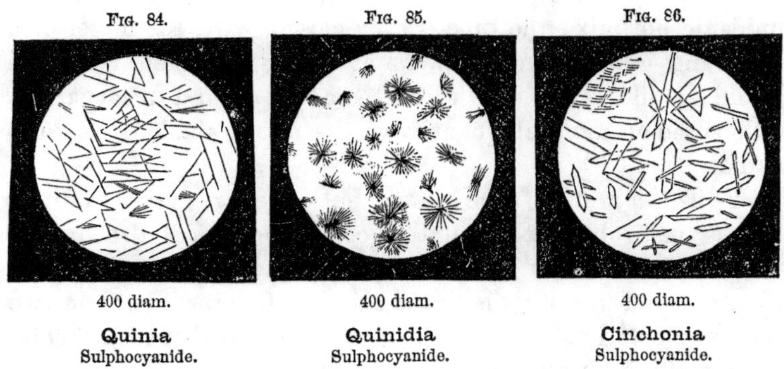

Fig. 84. 400 diam. **Quinia** Sulphocyanide.

Fig. 85. 400 diam. **Quinidia** Sulphocyanide.

Fig. 86. 400 diam. **Cinchonia** Sulphocyanide.

power of about 60 diameters. The particles arrange themselves into groups, quinia salt forming long, slender needles (Fig. 84), quinidia, radiate crystalline needles (Fig. 85), and cinchonia, comparatively large prisms (Fig. 86).

QUINIÆ TANNAS.

CHININUM SEU CHINIUM TANNICUM.

Tannate of Quinia or Quinine. Quinia Tannate.

A yellowish-white, amorphous powder, which, when heated, becomes brown, fuses, and is wholly dissipated; it is only sparingly soluble in cold water, requiring 453 parts of it, but dissolves in 21 parts of boiling water; addition of acids increases the solubility to some extent; it is also soluble in alcohol; its aqueous solution has an astringent, bitter taste, and a feebly acid reaction; it is precipitated by metallic salts, and becomes black upon the addition of a few drops of solution of ferric chloride. When 10 grains of quinia tannate are mixed with

10 grains of zinc oxide and a little water, and the mixture allowed to stand for several hours, in a warm place, then agitated with about two fluid-drachms of alcohol, and after a while filtered, and the filtrate evaporated to dryness, at a gentle heat, in a beaker-glass, a residue is obtained, which, when dissolved in a little chlorine-water, forms a solution which renders the characteristic emerald-green reactions of quinia with aqua ammoniæ, passing to purple upon addition of potassium ferrocyanide.

Examination:

Admixtures of *tannic* or *gallic acid*, *sugar*, *mannite*, or *dextrin*, may be recognized by their ready solubility in cold water in comparison with that of quinia tannate.

Starch is detected by a blue color, when one drop of iodinized solution of potassium iodide is added to a little of the quinia tannate shaken with some boiling water.

QUINIÆ VALERIANAS.

CHININUM SEU CHINIUM VALERIANICUM.

Valerianate of Quinia or Quinine. Quinia Valerianate.

Colorless, rhomboidal plates, of a pearly lustre and a faint odor of valerianic acid, permanent in the air; they fuse at a low temperature, and are entirely dissipated at a stronger heat, emitting white, inflammable vapors.

Quinia valerianate is soluble in six parts of cold, and one part of boiling, alcohol, in 110 parts of cold, and 40 parts of boiling, water, but only sparingly in ether; diluted acids dissolve it freely, and strong sulphuric acid does so without color, if heat is not applied; it is also soluble in chlorine-water, forming a solution which renders the emerald-green reaction of quinia with ammonium hydrate.

Examination:

Stearic acid, *sugar*, and *salicin*, are detected by agitating some of the quinia valerianate with strong sulphuric acid, in a test-tube; a black coloration would indicate one or both of the two former; a red one, salicin. In the case of a black reaction, a special test for *salicin* has to be made; a little of the valerianate is agitated with cold water, the filtrate is then evaporated at a gentle heat, to a small bulk, and this is strongly acidulated with a few drops of concentrated hydrochloric acid, and heated; a white turbidity, taking place after a while, would indicate salicin.

Quinia hydrochloride and *sulphate* may be detected, in the filtered aqueous solution of the salt, acidulated with a few drops of nitric acid, by testing portions of it with argentic nitrate for the former, and with barium nitrate for the latter. They will be indicated by a white precipitate with the respective reagent.

Zinc Valerianate or Acetate.—The absence of these or any other mineral salts, not readily volatilizable, may be ascertained by exposing the salt to a red heat, upon platinum-foil, whereby the organic matter is completely dissipated, leaving metallic oxides or carbonates behind, if any be present; if a residue remains which appears straw-yellow while hot and white when cold, it may be examined for zinc oxide by scraping it off carefully, and dissolving it, in a test-tube, in a few drops of diluted nitric acid; the solution is then mixed with twice its bulk of hydrosulphuric acid; if a precipitate is formed, it is collected upon a filter, and examined for the first group of metallic oxides by the method described on pages 39 and 41; the filtrate is over-saturated with ammonium hydrate; a white precipitate, soluble in liquor potassæ, would confirm the presence of *zinc oxide.*

QUINOIDINUM.

CHINOIDINUM.

Quinoidine.

A brittle, resin-like mass, of a deep brown color, a glassy, conchoidal fracture, and a peculiar aromatic odor; it becomes soft and tough at a moderate temperature, and melts like a resin when warmed; it is almost insoluble in water, and only partly soluble in chloroform and in glycerin, but freely soluble in diluted acids, in alcohol, and in ether, forming dark brown solutions, of an aromatic, bitter taste and odor; the alcoholic and ethereal solutions are precipitated by water, and the acid aqueous solution becomes green when first mixed with sufficient chlorine-water to decolorize it, and subsequently with an excess of aqua ammoniæ.

Examination:

Gum-Resins.—About 10 grains of the triturated quinoidine are agitated, in a large test-tube, with half an ounce of water; the mixture is then heated to boiling, with constant agitation; when cool, the water must be nearly colorless, and remain so when about 10 grains of dry potassium hydrate are dissolved in the decanted water, and the whole heated; if a brown coloration takes place, in either of these tests, gum-resins (aloes) or other soluble admixtures (liquorice, glucose, dextrin, etc.) are indicated.

Resins may be detected in the quinoidine remaining undissolved in the preceding test, by dissolving it, with the aid of heat, in about two drachms of diluted sulphuric acid; a complete or almost complete solution must take place, otherwise an admixture of resins, insoluble in diluted acids, is indicated.

As a confirmatory test for gum-resins, liquorice, glucose, etc., a few drops of the obtained solution may be allowed to fall into alcohol; they must form a clear solution; an ensuing turbidity would establish the presence of such admixtures.

Copper may be sought for, in the sulphuric-acid solution of the preceding test, by diluting it with twice its volume of water, and immersing in it, for several hours, a perfectly clean and bright iron spatula or the blade of a knife; if any copper be present, the iron will receive a slight coat of metallic copper.

Mineral admixtures are indicated by a residue upon complete exhaustion of the quinoidine with alcohol, and upon incineration.

SANTONINUM.

Santonin.

Small, flat, rhombic prisms, transparent, without odor or color, and of a slightly bitter taste; they fuse at 170° C., and assume, on cooling, the form of a crystalline mass; at a stronger heat, they volatilize in dense, white, irritating, inflammable vapors, which condense unaltered on cooling, forming a white crystalline sublimate; at a red heat, with free access of air, they burn away without residue.

Santonin is permanent in the air, and assumes a straw-yellow color, apparently without decomposition, when exposed to the solar light. When moistened with concentrated sulphuric acid, in a dry test-tube, it remains unchanged and colorless for a while (evidence of the absence of salicin, which at once becomes scarlet red); the mixture does not assume a bluish color upon the addition of a little powdered potassium bichromate (evidence of the absence of strychnia *), nor is it acted upon by concentrated nitric acid.

Santonin is almost insoluble in cold as well as in acidulated water, and also in dilute acids; the filtered liquid has only a feebly bitter taste, and renders no precipitate with tannic acid, nor with trinitrophenol (picric acid), either before or after the addition of a little solution of sodium acetate (further evidence of the absence of salicin, and of cinchonia, and other bitter alkaloids). Boiling water dissolves $\frac{1}{250}$th part of santonin. It is, however, readily soluble in diluted solutions of the alkaline hydrates, and in water containing them, but is reprecipitated by acids and acidulous salts. Santonin is also soluble in 43 parts of cold, and 3 parts of boiling, alcohol, in 75 parts of cold, and 2 parts of boiling, ether, in 3 parts of chloroform (distinction from cinchonia, which is almost insoluble in chloroform), and

* Santonin and strychnia have some similarity in their appearance, and this fact has repeatedly been the cause of incidental mistakes and sad accidents. They may, however, at once be distinguished, besides their difference in taste, by the solubility of *strychnia* in diluted acids, by its insolubility in ether and in liquor potassæ, and by its reaction with concentrated sulphuric acid, in which it dissolves without color, but produces, upon addition of a minute crystal of potassium bichromate, a bluish-violet tinge, which successively changes to violet, to red, and finally to yellow.

more or less freely in benzol, and in essential and fatty oils. The alcoholic solution burns with a pale yellow flame (evidence of the absence of an adulteration with boracic acid), and becomes transiently carmine-red upon the addition of concentrated solution of potassium hydrate. All the solutions of santonin have a bitter taste and a neutral reaction on test-papers, although it acts as an acid, forming crystallizable and soluble salts with the alkalies.

SODII ACETAS.

SODIUM SEU NATRIUM ACETICUM.

Acetate of Sodium. Sodium Acetate.

Large, colorless, transparent, striated oblique-rhombic prisms, containing three molecules (39.49 per cent.) of water of crystallization, which gradually evaporate upon exposure to a dry, warm air, leaving the anhydrous salt as a white powder. When heated, in a dry test-tube, the crystals undergo aqueous fusion at 288° C., and, upon continued heating, they give out their water of crystallization, and are decomposed, with the evolution of empyreumatic, inflammable fumes, leaving a black residue of carbon and sodium carbonate, which imparts to the flame a yellow color, blues moistened red litmus-paper, and effervesces with acids.

Sodium acetate is soluble in from two to three parts of cold, and in less than its own weight of boiling, water; it is also soluble to some extent in alcohol; its aqueous solution is neutral or nearly so, has a cooling, saline taste, is not precipitated when dropped into strong alcohol, nor when mixed with a diluted solution of sodium carbonate, or with a concentrated solution of sodium bitartrate; it assumes a red color upon the addition of a few drops of solution of a ferric salt, evolves the vapor of acetic acid, when warmed with concentrated sulphuric acid, and that of acetic ether, when heated with a mixture of alcohol and sulphuric acid.

For *Volumetric Estimation*, *see* page 58.

Examination:

Sodium chloride and *sulphate* are detected, in the solution of sodium acetate, acidulated with a few drops of diluted nitric acid, by testing it, in separate portions, with argentic nitrate and barium nitrate; a white precipitate with the first reagent would indicate *chloride*, and with the second one, *sulphate*.

Metallic impurities are detected by mixing a concentrated solution of the salt with an equal volume of hydrosulphuric acid, and by subsequent addition of a little aqua ammoniæ.

SODII ARSENIAS.

SODIUM SEU NATRIUM ARSENICUM.

Arseniate of Sodium. Sodium Arseniate.

Colorless, transparent, prismatic crystals, containing 14 molecules (40.38 per cent.) of water of crystallization, 10 of which are slowly given off at common temperatures, and in dry air, leaving a white powder still containing four molecules of water; these, however, are driven out at 148° C., when the salt fuses.

Sodium arseniate dissolves in about three parts of cold water, forming an alkaline solution, of a mild, feebly saline taste, which gives white precipitates with barium and calcium salts, and with magnesium and zinc sulphates, and a brick-red one with argentic nitrate, all of which are soluble in nitric acid; it suffers no alteration by hydrosulphuric acid, either in its alkaline solution or when this is acidulated with acids; the latter mixture, however, becomes turbid upon warming, separating white sulphur first, and yellow arsenic disulphide afterward. Fused upon charcoal, before the blow-pipe, sodium arseniate gives the garlic-like odor of arsenic, and imparts a yellow color to the flame; heated, in a narrow tube, with a little potassium cyanide, it forms a metallic mirror.

SODII BIBORAS.

BORAX. SODIUM SEU NATRIUM BIBORICUM.

Biborate of Sodium. Borax. Sodium Biborate.

Colorless, transparent, hard, six-sided prisms, containing 10 molecules (30.82 per cent.) of water of crystallization, ordinarily permanent, but slightly efflorescent in dry and warm air; when heated, they undergo aqueous fusion, give off the water, swell up, and form a white porous mass, which fuses at a red heat into a glass, which is a powerful solvent for the metallic oxides, forming colored fluxes. When powdered borax is mixed in a porcelain capsule with diluted sulphuric acid and subsequently with alcohol, and the mixture ignited, the alcohol burns with a greenish flame.

Sodium biborate is soluble in 12 to 15 parts of cold, and two parts of boiling, water, and, with the aid of heat, in its own weight of glycerin, but insoluble in alcohol; its aqueous solution has an alkaline, sweetish taste, and an alkaline reaction upon litmus and especially upon turmeric paper; it forms precipitates of insoluble or sparingly soluble borates with the solutions of most earthy and metallic salts, and acts upon salts of gold, silver, mercury, and others, almost like potassium hydrate, precipitating their hydrates.

When added to mucilage of gum-arabic or Iceland-moss, or to other similar vegetable mucilages, solution of sodium biborate thickens them considerably, unless they contain an addition of grape or cane sugar.

For *Volumetric Estimation*, *see* page 58.

Examination:

One drachm of the powdered borax, when dissolved in two ounces of warm water, should yield a complete and clear solution, remaining so after cooling; this solution may serve for the following tests:

Sodium carbonate is indicated by effervescence, or the rise of gas-bubbles, when a portion of the solution is added to concentrated hydrochloric acid.

Alum is indicated by a white precipitate with solution of sodium carbonate.

Chloride and *sulphate* may be detected in the solution, after dilution with three times its volume of water and acidulation with diluted nitric acid, by the formation of white precipitates when tested, in separate portions, with argentic nitrate for chloride, and with barium nitrate for sulphate.

Phosphate may be detected by a white granular precipitate, when the solution is mixed with nearly an equal volume of solution of ammoniated magnesium sulphate.

In order to ascertain the absence of arsenic acids, which would render the same reaction, the precipitate may be collected, washed, and dried, and then tested by heating a portion of it, mixed with a little exsiccated sodium carbonate, upon charcoal, and another portion, with a little potassium cyanide, in a narrow glass tube; a garlic-like odor in the first test, and a metallic mirror in the second, would indicate an incidental contamination with an arsenite or arsenate.

Metallic impurities are detected by mixing the solution of the salt with an equal volume of hydrosulphuric acid, and by subsequent acidulation with diluted hydrochloric acid.

SODII BICARBONAS.

SODIUM SEU NATRIUM BICARBONICUM.

Bicarbonate of Sodium. Sodium Bicarbonate.

White, opaque masses, made up of aggregations of small, oblique, four-sided plates or irregular scales, of a white appearance, in consequence of a slight efflorescence of the surfaces of the crystals, or a snow-white powder. The salt contains two molecules (10.65 per cent.) of water of crystallization, and is permanent in the air. Exposed to a strong heat, sodium bicarbonate loses the water of crystallization and one molecule of carbonic acid, amounting together to 36.8 per cent. by weight, and anhydrous sodium carbonate remains behind.

Sodium bicarbonate is soluble in 13 to 14 parts of water at 15.5° C., and in 10 parts at 24° C., forming a solution of a mild, alkaline taste and reaction; when heated, effervescence takes place, and at the boiling-point of the solution the salt is converted into sodium carbonate and sesquicarbonate. Solution of sodium bicarbonate affords no precipitate upon the admixture of a concentrated solution of sodium bitartrate or of tartaric acid.

One drachm of crystallized sodium bicarbonate saturates 49.98 grains of citric acid, and 53.55 grains of tartaric acid.

For *Volumetric Estimation*, *see* page 58.

Examination:

Sodium mono-carbonate may be recognized in the solution of the salt by a white precipitate, when tested with magnesium sulphate.

A confirmatory test is to mix a solution of 4½ grains (0.3 gramme) of mercuric chloride in 1½ drachm of water, with a solution of half a drachm (2.0 grammes) of the sodium bicarbonate in 1 ounce of cold water, and to allow the mixture to stand for about three minutes, when only a slight white turbidity should have occurred; a reddish-brown deposit would confirm the presence of mono-carbonate.

Sodium chloride and *sulphate* are detected in the solution of the salt, when over-saturated with diluted nitric acid, by testing it in separate portions, with argentic nitrate for chloride, and with barium nitrate for sulphate. If a precipitate is formed with the argentic nitrate, which gradually turns gray or grayish black, the presence of sodium *hyposulphite* is also indicated.

Metals.—About ten grains of the powdered sodium bicarbonate are dissolved in about one ounce of hydrosulphuric acid, and, when solution has taken place, this is over-saturated with diluted nitric acid; a dark turbidity, occurring either before or after acidulation, would indicate metallic impurities; these, if considerable in amount, may be obtained as a precipitate from a larger quantity of the salt, and the nature of the metals determined by the methods described on pages 41 and 42. The occurrence of a slight white turbidity, upon the addition of the acid, would be due to the presence of traces of sodium hyposulphite.

SODII CARBONAS.

SODIUM SEU NATRIUM CARBONICUM.

Carbonate of Sodium. Sodium Carbonate.

Large, colorless, transparent, oblique-rhombic prisms, containing 10 molecules (62.85 per cent.) of water of crystallization; they effloresce in dry air, lose their water, and crumble to a white powder; when exposed to heat, they undergo aqueous fusion, and, after the evaporation of the water, the anhydrous salt fuses at a red heat without undergoing further change.

Crystallized sodium carbonate dissolves in two parts of cold, and in less than half its weight of boiling, water; or in other words, 100 parts of water dissolve, at 14° C., 60.4 parts, at 36° C., 833 parts, and at 104° C. (the boiling-point of the saturated solution), 445 parts, of crystallized sodium carbonate.* The salt is insoluble in alcohol. Its aqueous solution has a strong alkaline taste and reaction; dropped into solution of tartaric acid, it produces no precipitation; it effervesces with acids and acidulous salts, and decomposes the soluble salts of the earthy and heavy metals, forming, with most of them, insoluble or sparingly soluble carbonates or hydrates.

One drachm of crystallized sodium carbonates saturates 29.34 grains of citric, and 31.47 grains of tartaric, acid.

For *Volumetric Estimation, see* page 58.

Examination:

Sodium Chloride, Sulphate, and Hyposulphite.—A solution of the sodium carbonate is over-saturated with diluted nitric acid, and is subsequently tested in separate portions, with barium nitrate for sulphate, and with argentic nitrate for chloride; if the latter reagent causes a white precipitate which soon turns more or less gray or grayish black, it would also indicate the presence of sodium hyposulphite.

* Sodium carbonate, with ten molecules of water of crystallization, is altered in its solution, at near the boiling-point, into a salt with only one molecule of water of crystallization, which is less soluble, and gives rise to the anomaly in the solubility of sodium carbonate. A similar instance is met with in sodium sulphate and several other salts.

Arsenic.—About one ounce of the crystallized salt is dissolved in 4 ounces of water, the solution is slightly over-saturated with concentrated hydrochloric acid, is filtered, and then is warmed to about 60 to 70° C.; while still warm, hydrosulphuric-acid gas is allowed to pass into the solution until it is nearly cooled, the flask is then corked, and allowed to stand for 12 hours, when a flocculent yellow precipitate would indicate the presence of arseniate.

As a confirmatory test, about 10 grains of the powdered salt are dissolved in half a fluidounce of concentrated hydrochloric acid, and then 20 drops of concentrated solution of stannous chloride and subsequently about one fluid-drachm of concentrated sulphuric acid are added, and the whole, if necessary, is heated nearly to boiling. The formation of a brown turbidity would verify the presence of arsenic.

SODII CHLORIDUM.

SODIUM SEU NATRIUM CHLORATUM.

Chloride of Sodium. Common Salt. Sodium Chloride.

Anhydrous, colorless, transparent, cubical crystals, often agglomerated into hollow, quadrangular pyramids, or a white, granular powder, having a spec. grav. of 2.15; permanent in the air, but slightly deliquescent when containing traces of magnesium and calcium chlorides. When exposed to heat, sodium chloride decrepitates, from the presence of interstitial moisture, melts at a red heat, and volatilizes with partial decomposition at a high temperature. It imparts a yellow color to the flame, and evolves vapors of hydrochloric acid, when heated with strong sulphuric acid.

Sodium chloride is almost equally soluble in water at all temperatures: 100 parts of water dissolve at 0° C. 35.52 parts, at 14° C. 35.87 parts, at 25° C. 36.13 parts, at 40° C. 36.64 parts, at 80° C. 38.22 parts, at 100° C. 39.61 parts, and at 110° C., the boiling-point of the saturated solution, 40.35 parts, of the salt; it is also soluble in glycerin, but not perceptibly soluble in absolute alcohol, in ether, or in chloroform, but its solubility in alcohol increases with the quantity of water contained therein. Its aqueous solution is neutral, and remains colorless upon the addition of chlorine-water (distinction from the alkaline bromides); it forms white precipitates with the solutions of those metallic salts whose chlorides are quite or almost insoluble in water—for instance, with the salts of silver, bismuth, and lead, and with the subsalts of mercury.

Examination:

Magnesium and *calcium chlorides* are detected in the solution of sodium chloride by a white turbidity taking place upon the addition of a diluted solution of sodium carbonate. They may be distinguished by completely precipitating a warm dilute solution of sodium chloride, after the addition of a few drops of aqua ammoniæ, with ammonium oxalate; if this causes a precipitate, it eliminates only the calcium without acting upon the magnesium salt; after a while the liquid is passed through a filter, and the filtrate tested, with a few drops of sodium phosphate, for magnesium salts, which, when present, will give rise to the formation of a white precipitate, soluble upon the addition of acetic acid.

Alkaline and *earthy sulphates* are recognized in the dilute solution, acidulated with hydrochloric acid, by a white precipitate with barium chloride.

Metallic impurities may be detected by the occurrence of a dark coloration or precipitate, when the solution of the salt is mixed with an equal volume of hydrosulphuric acid, and subsequently over-saturated with aqua ammoniæ.

Iodides and Bromides.—A glass tube of about 20 inches in length, and one-half to three-fourths of an inch wide, which is drawn out at its lower end into a narrow aperture, is filled to about 16 inches with the powdered salt, and this percolated with about 4 ounces of alcohol, of a spec. grav. between 0.834 and 0.864, and the obtained percolate evaporated to dryness at a gentle heat. The residue thus obtained is dissolved in a little water, the solution slightly acidulated with diluted sulphuric acid, and a little mucilage of starch is first added, and subsequently a little chlorine-water, drop by drop, and with gentle stirring with a glass rod. The presence of even minute traces of iodine will cause a bluish coloration of the fluid; when iodine alone is present, the blue color will gradually become purple, and decrease, until it finally disappears; but, when bromine is also present, the blue color will not turn purple, but brownish, then orange, and finally yellow.

Nitrate.—To a little strong sulphuric acid is added part of a drop of sulphuric-acid solution of indigo, so as to communicate to the acid a bluish tint. An equal volume of concentrated solution of the salt is then added, keeping the mixture cool; if nitrate be present, the blue tint will disappear.

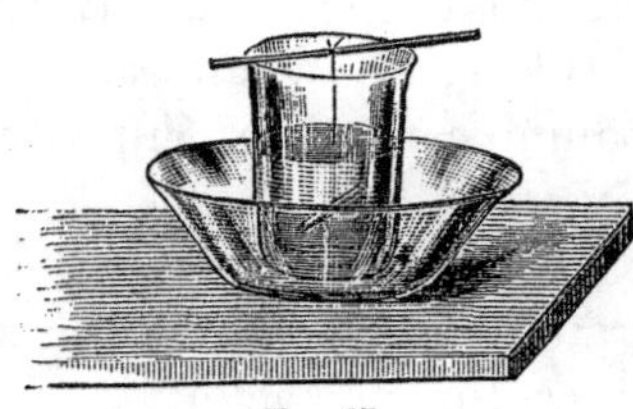

FIG. 87.

A confirmatory and still more sensitive test is to dip a bright zinc rod into a test-tube, or to suspend it in a small beaker (Fig. 87) containing a little diluted sulphuric-acid mucilage of starch, to which one drop of solution of pure potassium iodide, and subsequently twice the bulk of the liquid of a solution of the salt, has been added; if nitrate be present, a bluish coloration of the liquid will issue from the zinc.

Moisture.—When the quantity of water in sodium chloride has to be estimated, 10 scruples of the salt are weighed into a tared, dry beaker, or porcelain dish; this is covered with bibulous paper, and allowed to stand in a warm, dry place, at nearly 100° C., for several days, or until no more loss of weight takes place; the whole is then weighed, and the loss of weight, multiplied by 10, indicates the percentage of moisture contained in the salt.

SODII HYDRAS.

SODA. SODIUM SEU NATRIUM HYDRICUM. NATRIUM CAUSTICUM.

Soda. Sodium Hydrate.

Hard, white, fusible masses, in flat, tabular fragments or sticks, of a fibrous fracture, or a coarse, white powder; very deliquescent, but solidifying again after a time, in consequence of the absorption of carbonic-acid gas, and the formation of sodium carbonate. Heated upon the looped end of a platinum wire, it imparts a yellow color to the flame.

Sodium hydrate is very soluble in water, with the evolution of heat; it is also freely soluble in alcohol; its solutions are highly alkaline and caustic, and act destructively upon animal tissues. Dropped into a diluted solution of plumbic acetate, it causes a white turbidity, which disappears again upon continued addition of the caustic solution, without leaving a black residue (evidence of the absence of sodium sulphide). When the concentrated aqueous solution is dropped into strong alcohol, no precipitate should take place, as its appearance would indicate the presence of sodium carbonate, sulphate, chloride, or other salts, less soluble in alcohol.

Sodium hydrate may readily be distinguished from potassium hydrate, by dropping concentrated solutions of the salts into solution of tartaric acid, taking care that the acidity of the solution remains prevalent; sodium hydrate will yield no precipitate unless containing potassium hydrate to a considerable extent, while potassium hydrate forms a white, granular precipitate.

For *Volumetric Estimation*, *see* page 58.

Examination:

Sodium carbonate may be detected in the hydrate by effervescence, or by the formation of gas-bubbles, when a small piece of the hydrate is thrown into acetic acid, and by the occurrence of a white turbidity upon the admixture of an equal volume of lime-water with the aqueous solution of the salt.

Chloride and *sulphate* are detected in the diluted solution, when over-saturated with diluted nitric acid, and tested in separate portions, with argentic nitrate for the former, and with barium nitrate for the latter.

Silicic acid and *aluminium salts* may be detected by over-saturating the dilute solution of sodium hydrate with an excess of diluted nitric acid, and subsequently evaporating to dryness; the residue is dissolved in warm water, and should be wholly soluble; an insoluble residue would indicate silicic acid; the solution is filtered, and the filtrate tested with aqua ammoniæ;

the formation of a white, gelatinous precipitate would indicate aluminium salts.

SODII HYPOPHOSPHIS.

SODIUM SEU NATRIUM HYPOPHOSPHOROSUM.

Hypophosphite of Sodium. Sodium Hypophosphite.

Small, colorless, transparent, rectangular tables, of a pearly lustre, or a white, granular powder, which, when heated in a dry test-tube, will first give off moisture, and subsequently phosphoretted hydrogen, burning with a bright light; a residue of sodium pyrophosphate, reddened by traces of red amorphous phosphorus, is left behind.

Sodium hypophosphite is deliquescent if not dry, or in a moist atmosphere; it is readily soluble in water, and in absolute alcohol (distinction from calcium hypophosphite), but insoluble in pure ether; its aqueous solution has a slightly alkaline reaction, and is gradually oxidized on exposure, especially when warm; it affords, when much diluted, no reaction with diluted sulphuric acid, or with calcium chloride (distinction from soluble phosphates and phosphites), nor does it render a precipitate with oxalic acid (further evidence of the absence of calcium salts); it acts as a powerful reducing agent, and forms a white precipitate with argentic nitrate, which quickly turns brown, and is converted into metallic silver.

SODII HYPOSULPHIS.

SODIUM SEU NATRIUM HYPOSULPHUROSUM SEU SUBSULPHUROSUM.

Hyposulphite of Sodium. Sodium Hyposulphite. Sodium Dithionate.

Large, colorless, transparent, right-rhombic, striated prisms, containing five molecules (41.7 per cent.) of water of crystallization; permanent at ordinary temperatures, but efflorescent in dry and warm air; when exposed to heat, the crystals undergo aqueous fusion, and then dry up again into a white mass, which, when the heat is increased, are decomposed, with the evolution of vapors of sulphurous acid and sulphur, which take fire, and burn away, leaving behind a residue of neutral sodium sulphate, containing a little sulphide.

Sodium hyposulphite is soluble in 1½ part of cold water, but insoluble in alcohol; its solution has a cooling and afterward a bitter taste, and a feebly alkaline reaction; on exposure to the air, it is gradually decomposed, the hyposulphite being converted into sulphur and sodium sulphite, which latter salt, on the exposure of the solution to the air, is further decomposed into sulphur and sodium sulphate; when dropped into diluted hydrochloric, nitric, or sulphuric acid, solution of sodium hyposulphite gradually becomes turbid, sulphur being precipitated, and sulphurous acid disengaged.

With solution of barium chloride, a concentrated solution of sodium hyposulphite forms a white precipitate, which dissolves, however, upon sufficient dilution with water (evidence of the absence of sodium sulphate); when dropped into dilute solution of argentic nitrate, a white precipitate is formed, which soon turns yellow, and finally black; when, however, on the other hand, the argentic solution is dropped into the solution of sodium hyposulphite, the ensuing white precipitate of argentic hyposulphite is redissolved upon agitation, and the solution remains clear as long as sodium hyposulphite is in excess.

When iodine, either alone or dissolved in alcohol, is added to solution of sodium hyposulphite, it is immediately decolorized, sodium iodide and tetrathionate being formed; this process takes place in the proportion, approximately, of one part (127) of iodine to two parts (248) of crystallized sodium hyposulphite; a solution in these proportions dissolves iodine readily, with a brown color, but it is decolorized again upon the restoration of those proportions by the addition of sodium hyposulphite.

Solution of sodium hyposulphite is a solvent for several otherwise insoluble compounds, as argentic oxide, argentic iodide, bromide, and chloride, plumbic iodide, plumbic and calcium sulphates, etc.

For *Volumetric Estimation, see* page 63.

Examination:

Sodium sulphate is detected by the occurrence of a white precipitate, when the dilute solution of the salt is tested with barium nitrate.

Sodium carbonate is indicated by effervescence when a concentrated solution of the salt is dropped into diluted acetic or hydrochloric acid.

SODII IODIDUM.

SODIUM SEU NATRIUM IODATUM.

Iodide of Sodium. Sodium Iodide.

A white, granular, deliquescent powder, or colorless, striated, oblique-rhombic prisms, with four molecules of water of crystallization, when crystallized at temperatures below 30° C., and anhydrous cubes when obtained at temperatures above that point. When exposed to heat, sodium iodide fuses, and is volatilized at a higher temperature, giving off part of its iodine; it is decomposed at a strong heat; when a few grains of the salt are heated in a test-tube, either in concentrated sulphuric acid, or with a little potassium bisulphate, violet-colored vapors of iodine are evolved; when a few grains are dissolved in chlorine-water, the latter assumes a brown color, and, when shaken with some chloroform or carbon bisulphide, and subsequently with a little water added, imparts a fine purple color to the chloroform or carbon bisulphide.

Sodium iodide is freely soluble in water, in glycerin, and in alcohol; 100 parts of water, at 14° C., dissolve 173 parts of the salt; the solution has an acrid, saline taste, and a feeble, alkaline reaction; it gives no precipitate with tartaric acid, with sodium bitartrate, or with sodium carbonate, but forms a yellowish one with argentic nitrate, insoluble in diluted nitric acid, and but very sparingly soluble in ammonium hydrate, and a vermilion-red one with mercuric chloride, soluble in excess of either reagent.

Examination:

Impurities and *admixtures*, less soluble in alcohol, are indicated by a white turbidity or granular deposit, when a saturated aqueous solution of the salt is dropped into an excess of alcohol fortius.

Potassium salts are indicated by a white, crystalline precipitate in the concentrated aqueous solution, when added to a strong solution of sodium bitartrate.

Chloride and *bromide* may be detected by completely precipitating a warm aqueous solution of the salt, acidulated with a few drops of nitric acid, with argentic nitrate; the precipitate is separated from the menstruum, as much as practicable, by decantation, is washed, and then agitated with a little dilute aqua ammoniæ; the liquid is then filtered, and the filtrate oversaturated with nitric acid; an opalescence of the liquid will take place; a precipitate would indicate chloride or bromide. If a precipitate be formed, it may be collected upon a filter, washed,

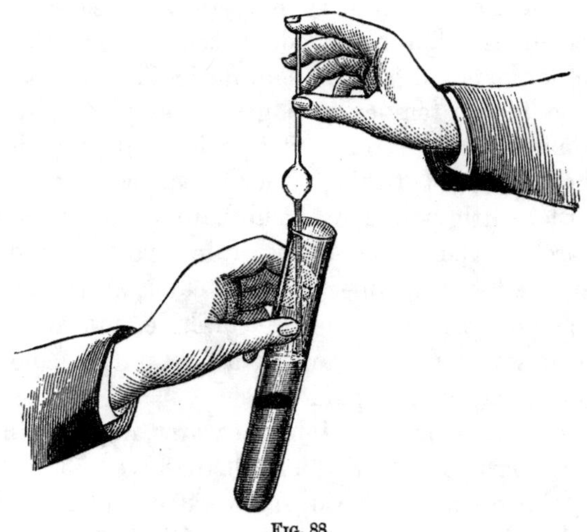

Fig. 88.

and subsequently removed into a test-tube, by piercing the filter, and rinsing the precipitate into the tube by means of a washing-bottle; after subsiding, the supernatant water is drawn off, and the precipitate agitated with chlorine-water; if argentic chloride, it remains unchanged; if bromide, the chlorine-water will assume a yellowish or reddish color, which, on agitation with chloroform, will be transferred to the latter.

Carbonate may be detected, in the aqueous solution, by a white turbidity when mixed with twice or thrice its volume of lime-water.

Sodium nitrate may be detected, when 10 grains of the sodium iodide and 12 grains of argentic sulphate are triturated together, with a little water; the mixture is then transferred to a small, tared, moist filter, and the filtrate mixed with an equal volume of concentrated solution of ferrous sulphate, and subsequently placed upon concentrated sulphuric acid (Fig. 88); a dark coloration of the line of junction of the two liquids would indicate nitrate.

The precipitate upon the filter is washed with tepid water, until this, when allowed to drop into dilute hydrochloric acid, ceases to produce a turbidity; the precipitate is then washed with a little greatly-diluted aqua ammoniæ, and subsequently dried at a temperature not exceeding 90° C.; it must weigh not less than 15 grains, if the sodium iodide employed was pure.

For *Volumetric Estimation, see* page 62.

SODII NITRAS.

SODIUM SEU NATRIUM NITRICUM.

Nitrate of Sodium. Chili Salpetre. Sodium Nitrate.

Anhydrous, colorless, transparent, obtuse-rhombohedral crystals, slightly deliquescent, and generally of a moist appearance; when thrown upon burning coals, the salt deflagrates with an orange-yellow flame.

Sodium nitrate is freely soluble in water, but only sparingly in alcohol; the aqueous solution, saturated at 18.5° C., contains 46.81 per cent., and saturated at its boiling-point (121° C.), nearly 70 per cent., of the salt; the solution is neutral, has a sharp, cooling, saline taste, and is not acted upon, when diluted, by reagents.

Its concentrated solution may readily be distinguished from that of potassium nitrate by not being acted upon by a solution of sodium bitartrate, which gives a white, granular precipitate with potassium nitrate.

Examination:

Salts of the earthy oxides are detected in the solution of sodium nitrate, by a white turbidity, upon the addition of solution of sodium carbonate.

Chlorides and *sulphates* are detected in the diluted solution, after acidulation with diluted nitric acid, by white precipitates when tested, in separate portions, with argentic nitrate for chloride, and with barium nitrate for sulphate.

Metals are detected by a dark coloration or precipitate, when the concentrated solution of the salt is mixed with an equal volume of hydrosulphuric acid.

Sodium Iodide and Iodate.—About two drachms of the salt are dissolved in one ounce of water, and the solution is equally divided into two test-tubes, to one of which a little sulphuric-acid mucilage of starch is added; an ensuing blue coloration would indicate either sodium iodide and iodate or sodium iodide and nitrite. When no blue reaction takes place, to the second part of the solution is also added a little sulphuric-acid mucilage of starch, and it is divided into two portions, and tested severally with one drop of solution of potassium nitrite, in the one, and in the other by adding, drop by drop, a little hydrosulphuric acid. An ensuing blue reaction, in the first instance, would indicate the presence of sodium *iodide;* a reddish, purple, or violet coloration, in the second one, sodium *iodate.*

Another test for iodide and iodate is to add, to a concentrated aqueous solution of the sodium nitrate, some chlorine-water, and to agitate the mixture afterward with carbon bisulphide; the latter, when it has subsided, is separated by means of a pipette, and agitated with a little powdered zinc and a few drops of diluted sulphuric acid; a purple coloration of the carbon bisulphide, either at once or upon the latter treatment, would confirm the presence of iodine compounds.

SODII PHOSPHAS.

SODIUM SEU NATRIUM PHOSPHORICUM.

Phosphate of Sodium. Tri-basic Sodium Phosphate.

Colorless, transparent, oblique-rhombic prisms, terminated by four converging planes, containing 24 molecules (60.10 per cent.) of water of crystallization; they are readily efflorescent, losing, when exposed to the air at common temperatures, 10 molecules (22.25 per cent.) of water, and becoming opaque; when heated to near 100° C., the rest of the water of crystallization evaporates. When exposed to a strong heat, sodium phosphate first undergoes aqueous fusion, and afterward melts at a red heat into a limpid glass of sodium pyrophosphate, which becomes opaque on cooling.

Sodium phosphate dissolves in 6 parts of cold, and in 2 parts of boiling, water, but is insoluble in alcohol. Its solution has a cooling, saline taste, a faintly alkaline reaction, affords no effervescence upon addition of an acid, and gives, with solution of completely neutral (fused) argentic nitrate, a bright-yellow precipitate, soluble in both ammonium hydrate and nitric acid; the ammoniacal solution remains unchanged, when the test-tube, wherein it is contained, is immersed in boiling water (distinction from the similar argentic arsenite, whose ammoniacal solution deposits metallic silver upon the walls of the test-tube upon warming). With solution of ammoniated magnesium sulphate, sodium phosphate gives a white precipitate, insoluble in an excess of the salt as well as of the reagent.

Examination:

Sodium carbonate is detected by effervescence, upon the addition of hydrochloric acid to the concentrated solution of the salt.

Sulphates and *chlorides* are detected in the diluted solution, strongly acidulated with nitric acid, when tested in separate portions, with barium nitrate for sulphate, and with argentic nitrate for chloride.

Metallic Impurities.—About half an ounce of the salt is dissolved in one ounce of boiling water, and the solution added to about 3 ounces of hydrosulphuric acid in a flask; about one drachm of diluted hydrochloric acid is added, and the flask corked and allowed to stand in a warm place for about 12 hours. A yellow precipitate would indicate arsenic, a dark one metallic impurities.

As a confirmatory test, or if the presence of other metals requires a special test for arsenic, 20 grains of the sodium phosphate are dissolved in about one drachm of boiling water in a wide test-tube; half a fluidounce of pure, concentrated hydrochloric acid is then added, and a strip or roll of bright copper-foil completely immersed in the fluid; the tube is then dipped into boiling water, and allowed to stand therein for half an hour. The copper must remain bright; a grayish or grayish-black coating of the copper would be evidence of the presence of *arsenic.*

SODII PYROPHOSPHAS.

SODIUM SEU NATRIUM PYROPHOSPHORICUM.

Pyrophosphate of Sodium. Sodium Pyro- or Tetraphosphate.

Colorless, transparent, brilliant, rhomboidal plates, or a white granular powder, containing ten molecules of water of crystallization, permanent in the air; when exposed to heat, the salt gives off its water of crystallization, fuses, and, on cooling, concretes to a crystalline semi-transparent mass.

Sodium pyrophosphate dissolves in ten parts of cold water, forming a slightly alkaline solution, which renders a white precipitate and a neutral menstruum with argentic nitrate, and a white granular precipitate with ammoniated magnesium sulphate, insoluble in an excess of either the reagent or the salt. When solution of sodium pyrophosphate is boiled for some time, the salt returns to the state of the tri-basic phosphate, and regains the properties, and renders the reactions, of tri-basic sodium phosphate (page 345).

Examination:

Carbonate is detected in the solution of the salt, by effervescence upon the addition of a little concentrated nitric acid.

Sulphate and *chloride* may be detected in the diluted solution, after acidulation with nitric acid, by white precipitates when tested in separate portions, with barium nitrate for the former, and with argentic nitrate for the latter.

Metallic impurities are detected by mixing the concentrated solution of the salt with an equal volume of hydrosulphuric acid, and subsequently acidulating with diluted nitric acid.

SODII SULPHAS.

SODIUM SEU NATRIUM SULPHURICUM.

Sulphate of Sodium. Glauber's Salt. Sodium Sulphate.

Colorless, transparent, six-sided, striated, oblique prisms, containing ten molecules (56 per cent.) of water of crystallization, efflorescent, and gradually losing water and crumbling to a white powder. Exposed to heat, the crystals undergo first aqueous, and subsequently igneous, fusion.

Sodium sulphate is very soluble in water; 100 parts of water at 0° C. dissolve 12 parts, at 18° C. 48 parts, at 25° C. 100 parts, and at 33° C. 322.6 parts, of the crystallized salt; above that temperature the salt passes into the anhydrous state, in which it is less soluble, and the solution then separates salt upon an increase of heat (*see* foot-note on page 334). The solution has a salty and feebly bitter taste, is neutral, remains unaltered with sodium carbonate as well as with sodium bitartrate, and gives a granular white precipitate with lime-water, and a copious white one with solutions of both barium and lead salts, which latter precipitates are insoluble in diluted acids.

Fifty grains of crystallized sodium sulphate, dissolved in two ounces of water acidulated with hydrochloric acid, give, when completely precipitated with barium chloride, a precipitate which, when collected upon a tared filter, washed, and dried, should weigh 72.2 grains.

Examination:

A solution of one part of the crystallized salt in four parts of water, tested with blue and with red litmus-paper, must not change the color of either.

Chloride may be detected in the diluted solution, acidulated with nitric acid, by a white turbidity or precipitate with argentic nitrate.

Ammonium sulphate may be recognized by the odor as well as by the rise of white vapors, when a little of the triturated salt is heated in a strong solution of potassium hydrate and a glass rod, moistened with acetic acid, is held in the orifice of the test-tube.

Magnesium and *calcium salts* are detected in the solution by a white precipitate with sodium carbonate; a reddish or brownish appearance of the precipitate would indicate metallic impurities (iron and manganese); the presence of *manganese salts* may be confirmed by a brown precipitate upon the addition of solution of chlorinated lime to the solution of the sodium sulphate, that of *iron* by a blue turbidity, when the solution of the salt is acidulated with hydrochloric acid and tested with potassium ferrocyanide.

Metals may further be detected in the diluted solution by introducing ammonium sulphydrate, and allowing the mixture to stand for a few hours; a white turbidity would indicate *zinc*, and a brownish-black one, *copper;* a greenish-black one would confirm the presence of *iron*, and a pale-reddish one, that of *manganese.*

If a test for arsenic is required, about 30 grains of the crystallized sodium sulphate are dissolved in one drachm of warm water in a wide test-tube; half a fluidounce of concentrated hydrochloric acid is added, and a strip or roll of bright copper-foil completely immersed in the fluid; the tube is then dipped into boiling water and allowed to stand in the water for half an hour. The copper must remain bright; a grayish-black coating would indicate *arsenic.*

Sulphite and *hyposulphite* may be detected in a solution of one part of the salt in three parts of water, by mixing it with one-third of its volume of concentrated hydrochloric acid, and heating it gently with a few grains of granular zinc; the presence of either of the above salts will give rise to the formation of hydrosulphuric acid, which may be recognized by placing a small bunch of cotton, moistened with solution of plumbic acetate, in the orifice of the tube, or by closing it with bibulous paper moistened with the plumbic solution (Fig. 89).

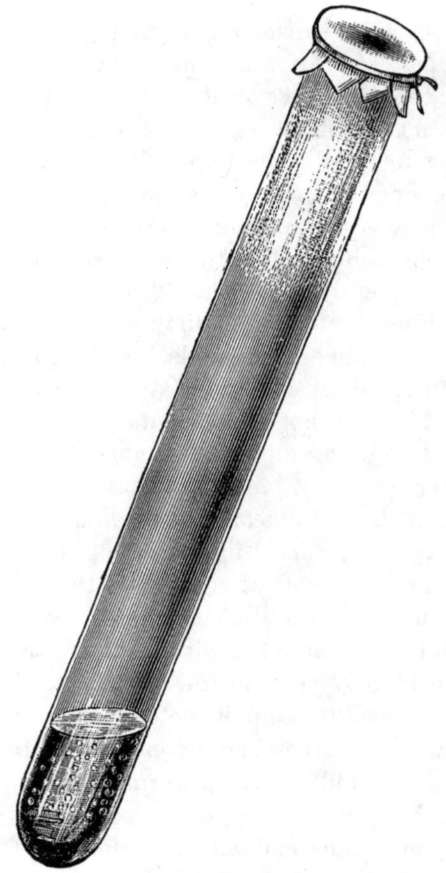

Fig. 89.

A black coloration of the solution would shŏw the presence of either or both the above salts.

SODII SULPHIS.

SODIUM SEU NATRIUM SULPHUROSUM.

Sulphite of Sodium. Sodium Sulphite.

Colorless, opaque, prismatic crystals, containing seven molecules (50 per cent.) of water of crystallization; on exposure to the air they effloresce somewhat, and the salt is gradually converted into sulphate, emitting a feeble odor of sulphurous acid. This liability to decomposition is retarded, and the salt made more permanent, by exsiccating it at a gentle heat, when it undergoes aqueous fusion, loses its water of crystallization, and becomes white. It is this form in which sodium sulphite is now frequently met with. When this salt is exposed to a strong red heat, it fuses to a dirty yellowish mass, consisting of sodium sulphate and sulphide, which may be separated by extracting the cold residue with strong alcohol, which dissolves the sulphide, but not the sulphate.

Crystallized sodium sulphite is soluble in 4 parts of cold, and in less than its own weight of boiling, water, but only sparingly in alcohol; its aqueous solution has a feeble alkaline reaction, and becomes turbid upon heating, but limpid again on cooling; on exposure of the solution to the air, the sulphite is gradually converted into sulphate with the separation of sulphur, as it is also by treatment with oxidizing agents, such as chlorine, hypochlorous acid, nitrous acid, etc. When acidulated, solution of sodium sulphite acts as a powerful reducing agent; it emits sulphurous acid upon the addition of strong acids, slowly when cold, freely on warming, and with the separation of sulphur; when this test is performed with hydrochloric or sulphuric acid, and with the addition of a little zinc, hydrosulphuric acid is evolved. With barium chloride or nitrate, solution of sodium sulphite forms white precipitates, soluble in diluted hydrochloric acid.

Examination:

Sodium sulphate may be detected in the diluted solution of the salt, by precipitating it with barium chloride, and by decantation of the supernatant menstruum after subsiding; the precipitate is once or twice washed with a little cold water, and then treated with diluted hydrochloric acid, wherein it should be wholly soluble; an insoluble residue will indicate the presence of sulphate.

SPIRITUS ÆTHERIS NITROSI.

SPIRITUS NITROSO-ÆTHEREUS. SPIRITUS NITRI DULCIS.

Sweet Spirit of Nitre. Spirit of Nitrous Ether. Alcoholic Solution of Ethyl Nitrite.

A colorless, or pale-yellow, volatile liquid, of a fragrant, ethereal odor, and sharp, aromatic, sweetish taste; its spec. grav. is 0.837 U. S. Pharm., 0.845 Brit. Pharm., and 0.840–0.850 Pharm. Germ., and it should contain between 4 and 5 per cent. of ethyl nitrite. It is inflammable, reddens blue litmus-paper not at all or only faintly, and assumes a dark color upon the addition of a few drops of a solution of ferrous chloride.

Spirit of nitrous ether is miscible with water, alcohol, chloroform, ether, carbon bisulphide, benzol, and essential and fatty oils. A portion of the spirit, in a test-tube half filled with it, plunged into water heated to 63° C., and held there until it has acquired that temperature, will boil distinctly on the addition of a few small pieces of glass.

Examination:

Aldehyde is indicated by a brown coloration of the spirit when agitated in a test-tube with a few fragments of fused potassium hydrate.

Acids.—Spirit of nitrous ether containing so much of free acid as to have a perceptible sour taste and an acid reaction upon blue litmus-paper, and to cause the rise of gas-bubbles from a few crystals of potassium bicarbonate when dropped into it, cannot be considered admissible for medicinal use.

Ethyl chloride may be detected by burning away a small quantity of the spirit upon a little water in a porcelain capsule, and by subsequently testing the water, after acidulation with a few drops of nitric acid, with a few drops of solution of argentic nitrate; the occurrence of a white turbidity would indicate the presence of ethyl chloride.

Methylic Alcohol.—About one ounce of the spirit is shaken with 30 to 40 grains of anhydrous (exsiccated) potassium carbonate; after subsiding, the supernatant spirit is decanted; about half a fluidounce of this dehydrated spirit is introduced into a small flask, or a test-tube of a proper size (Fig. 90), 2½

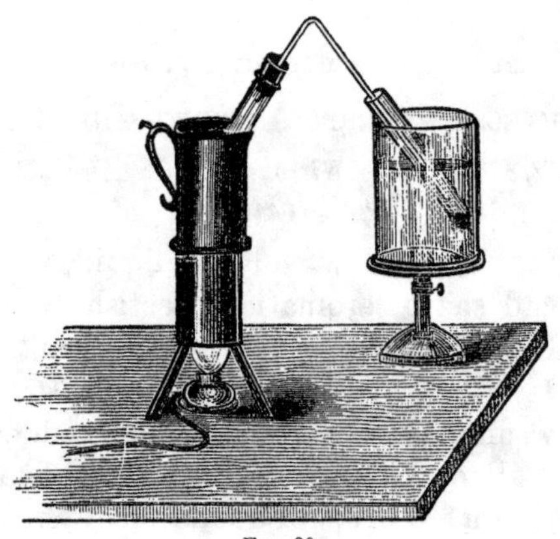

Fig. 90.

drachms of anhydrous calcium chloride in powder are added, and, after thoroughly mixing, the flask is connected with a condenser, and is then placed in a water-bath for distillation; this distillation is continued until one and a half fluid-drachm of distillate have been obtained. The test-tube is then removed from the water-bath, and, when cool, one drachm of water is added, and the distillation once more resumed until a little more than half a fluid-drachm of distillate is obtained. The latter distillate is mixed with half an ounce of water, wherein 30 grains of potassium bichromate and 30 drops of concentrated sulphuric acid have been dissolved. After having allowed the mixture to stand for a quarter of an hour, it also is submitted to distillation, until half an ounce of distillate is obtained; to this 30 grains of crystallized sodium carbonate are added, in a porcelain capsule, and the whole evaporated to half its bulk; it is then slightly over-saturated with acetic acid, filtered into a test-tube, and a solution of one grain of argentic nitrate in half a drachm of water acidulated with two drops of diluted acetic acid added, and the whole gently boiled for about two minutes. If the spirit is free from methylic alcohol, the solution darkens, and often assumes transiently a purplish tinge, but continues quite transparent, and the test-tube, after being rinsed out and filled with water, appears clean. But, if the spirit contains even traces of methylic alcohol, the liquid becomes at first brown, then almost black and opaque, and a film of silver is deposited on the tube, which appears brown by transmitted light. When only 3 to 4 per cent. of methylic alcohol are present, the film is sufficiently thick to form a brilliant metallic mirror.

Estimation of the Quantity of Ethyl Nitrite.—One hundred grains of the spirit of nitrous ether are macerated in a little corked flask for 12 hours, with occasional agitation, with 12 to 15 grains of fused potassium hydrate; the ethereal odor will then have disappeared, and the liquid is poured into a beaker, diluted with an equal bulk of water, and left in a warm place until the odor of alcohol has also disappeared. The remaining solution is then slightly acidulated with diluted sulphuric acid, and tested by delivering into it test solution of potassium permanganate, containing a certain and known quantity of the permanganate in each centimetre, or volume-unit, drop by drop, and with constant stirring, until the color of the permanganate ceases to be discharged (Fig. 91). The number of grains and parts of grains of potassium permanganate required for the test is readily calculated from the volume of the solution

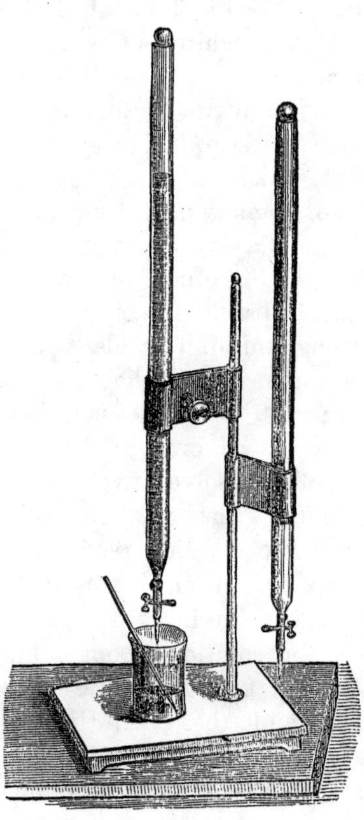

Fig. 91.

used; and this number, multiplied by 1.18, indicates at once the percentage of ethyl nitrite contained in the spirit.

STRYCHNIA.

STRYCHNINUM. STRYCHNIUM.

Strychnia. Strychnine.

Small, brilliant, octahedral crystals, colorless and transparent, or a white, crystalline powder, permanent in the air. Heated upon platinum-foil, strychnia melts, and spreads over the foil like melted resin; it partly volatilizes without decomposition; most of it, however, is decomposed, leaving a charred residue, which, at a stronger heat, is wholly dissipated (evidence of the absence of fixed admixtures).

Strychnia is almost insoluble in cold water, in absolute alcohol, and in ether, and only sparingly soluble in boiling water, but it dissolves freely in boiling alcohol, in chloroform, and in strong and dilute acids, as also, to some extent, in alcohol containing water; 100 parts of such alcohol, cold, of a spec. grav. of from 0.890 to 0.880, dissolve five parts of strychnia. The alcoholic solution has a feeble alkaline reaction upon test-paper, and an intensely bitter taste. One grain of strychnia requires for solution about 7 ounces of boiling, and 16 ounces of cold, water; and these solutions, even when greatly diluted, preserve the intensely bitter taste of strychnia.

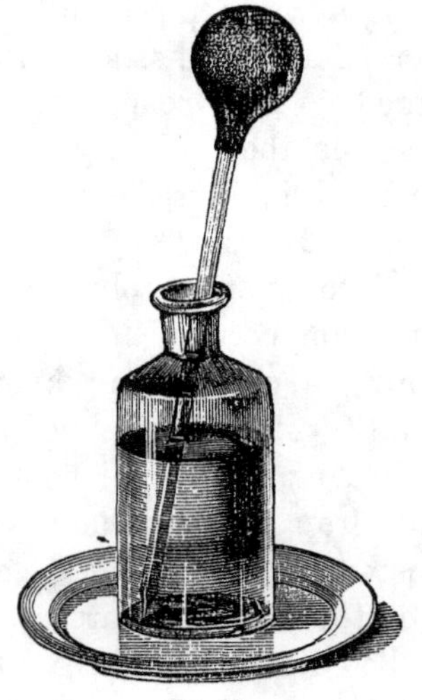

FIG. 92.

When a few drops of cold concentrated nitric acid are added, by means of a glass rod or a small pipette (Fig. 92), to a little strychnia, or its salts, on a watch-glass, it dissolves without any color, or with only a pale-greenish or yellow tint (distinction from brucia and morphia, and their salts, which give intensely red solutions). Strong sulphuric acid also dissolves strychnia and its salts without color (distinction from brucia, veratria, and salicin, which yield red or purple reactions); but, when a minute fragment of a crystal, or one drop of a solution of potassium bichromate or permanganate is added, the solution assumes at once a deep-violet or blue color, which successively changes from violet to red, and finally to green or yellow. This characteristic reaction * may also be performed upon a porcelain plate, or a glass-pane placed upon white paper, by allowing four drops of concentrated sulphuric acid to fall on the plate, two upon about half a grain of strychnia each, the third upon a small fragment of potassium bichromate, and the fourth upon a small fragment of a crystal of potassium permanganate; the drops with the two reagents are then drawn by a glass rod, each to one of the colorless drops containing the strychnia; the intense coloration will occur at once.

When a cold, saturated alcoholic solution of strychnia is mixed with about an equal volume of an alcoholic solution of ammonium sulphydrate, and the mixture is allowed to stand for 12 hours, long, brilliant, orange-red needles are formed, which are insoluble in water, alcohol, ether, and carbon bisulphide, but which are decolorized and decomposed when treated with concentrated sulphuric acid, with the formation of strychnia sulphate, and of an oily compound, which, in contact with water, is resolved into sulphur and hydrosulphuric acid. Strychnia only is known to render this reaction.

When about two grains of strychnia are agitated with five fluid-drachms of warm water, they will not wholly dissolve, but will do so at once upon the addition of a few drops of diluted sulphuric acid; this solution, when tested in separate portions, will yield precipitates with tannic acid, with potassio-mercuric iodide, and with iodinized potassium iodide; it will remain unaltered with potassium bicarbonate (distinction from the cinchona alkaloids), but it will yield a white precipitate with the

* Only aniline and its salts are known to render, with the same reagent, a similar reaction, which, however, is less of a violet, and more of a blue tint throughout, and which does not appear immediately.

alkaline hydrates, insoluble in an excess of the precipitant (further distinction from morphia), and also insoluble when agitated with ether, but soluble in chloroform.

Examination:

Incidental or fraudulent admixtures of other alkaloids are recognized by the above-described characteristics and reactions of strychnia.

Brucia and *salicin* are indicated by a red coloration with either concentrated nitric or sulphuric acid.

Brucia may also be recognized by its ready solubility in absolute alcohol (wherein strychnia is almost insoluble), and by the reaction of its solution in strong nitric acid with stannous chloride or sodium hyposulphite. While pure strychnia renders a pale-green or yellowish solution with strong nitric acid, this will appear more or less red, if brucia be present, and will assume, upon the addition of solution of stannous chloride or of sodium hyposulphite, a deep violet color, which will not disappear upon dilution with water; if the latter reagent has been employed, a white turbidity from the separation of sulphur will occur after a while.

Santonin is known by its insolubility in dilute acids, and by its property of becoming lemon-yellow when the sample, covered with a sheet of thin white paper, is exposed to solar light for one or two days.

Cinchona alkaloids may be detected by a white precipitate, when a solution of two grains of the strychnia in two drachms of water and three drops of diluted sulphuric acid is tested with solution of potassium bicarbonate.

Cinchonia may also be recognized by its insolubility in chloroform, remaining behind when a little powdered strychnia is exhausted with that solvent; its identity may then be ascertained by its properties, described on page 182.

STRYCHNIÆ NITRAS.

STRYCHNINUM SEU STRYCHNIUM NITRICUM.

Nitrate of Strychnia or Strychnine. Strychnia Nitrate.

Colorless, transparent, flexible needles of a silky lustre; when heated upon platinum-foil, they become slightly yellow, melt, and are finally wholly dissipated (evidence of the absence of fixed admixtures).

Strychnia nitrate is soluble in 60 parts of cold, and in three parts of boiling, water; it is also soluble in dilute alcohol and in chloroform, but only sparingly in strong alcohol, and almost insoluble in absolute alcohol, and in ether. Its solutions are neutral, and have an intensely bitter taste.

Strychnia nitrate answers to all the reactions of strychnia, and may be recognized by the characteristic coloration with strong sulphuric acid and potassium bichromate or permanganate, as described on page 355. The evidence of being a nitrate may readily be obtained by decolorization, when its solution is faintly blued with sulphuric-acid indigo-solution, and heated, or by dissolving a few grains of the salt, and an equal quantity

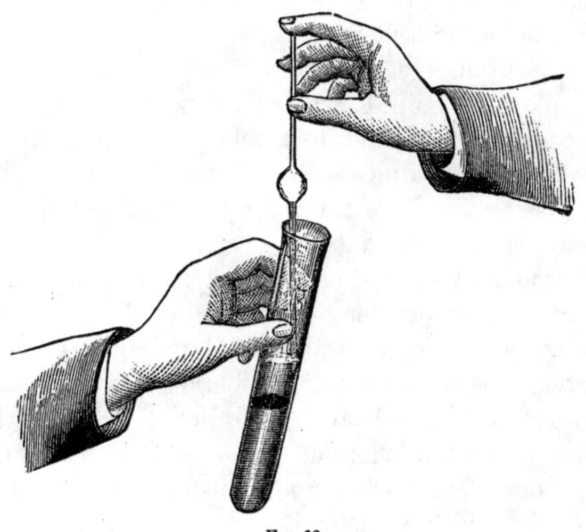

FIG. 93.

of ferrous sulphate, in a little diluted sulphuric acid, and placing the colorless solution upon concentrated sulphuric acid (Fig.

93); a dark-brown coloration will take place at the junction of the two liquids.

The methods for testing the purity of strychnia nitrate are the same as described with strychnia, on pages 355, 356. It needs only to be added that the salt should not emit ammoniacal odors, when heated with liquor potassæ, nor cause the rise of white vapors, when a glass rod, moistened with acetic acid, is held over the orifice of the test-tube.

STRYCHNIÆ SULPHAS.

STRYCHNINUM SEU STRYCHNIUM SULPHURICUM.

Sulphate of Strychnia or Strychnine. Strychnia Sulphate.

Fine, colorless, and transparent prismatic crystals, containing seven molecules (nearly 14 per cent.) of water of crystallization; slightly efflorescent on exposure to the air; when heated, they melt, and lose the water of crystallization; at a strong heat, they burn away without residue (evidence of the absence of fixed admixtures).

Strychnia sulphate is soluble in 42 parts of cold, and in about two parts of boiling, water; it also dissolves, in nearly the same proportions, in dilute alcohol, but only sparingly in strong alcohol, and is almost insoluble in absolute alcohol and in ether. Its solutions are intensely bitter; their deportment with reagents answers to that of strychnia, and they also render the characteristic reaction with sulphuric acid and potassium bichromate or permanganate; the evidence of being a sulphate may be obtained by the white precipitate which barium nitrate yields with a solution of strychnia sulphate in dilute nitric acid. When heated with liquor potassæ, the salt should not emit ammoniacal odors, nor should it cause the rise of white vapors when a glass rod, moistened with acetic acid, is held over the orifice of the test-tube.

The purity of the salt may be ascertained by the same tests as described with pure strychnia, on pages 355, 356.

SULPHUR PRÆCIPITATUM.

LAC SULPHURIS.

Precipitated Sulphur.

A fine, slightly coherent powder, of a pale yellowish or grayish color, without taste or smell, and free from grittiness, consisting, when seen under the microscope, of minute, opaque globules, without any admixture of crystalline matter. When thrown upon burning coal, or heated in an open vessel, precipitated sulphur first emits a little hydrosulphuric-acid gas, then fuses and burns wholly away at a stronger heat.

Precipitated sulphur is insoluble in the common solvents, but is dissolved readily and wholly in carbon bisulphide, and in strong, boiling solutions of potassium and sodium hydrates, and also more or less in benzol, in hot oil of turpentine, and other essential and fatty oils.

Examination:

Fixed admixtures are indicated by a white ash left behind upon complete dissipation of about 10 grains of the sulphur in an open porcelain crucible.

Calcium Sulphate.—About one drachm of the sulphur is triturated with about 6 fluid-drachms of tepid water, and the mixture agitated for a few minutes until cold, when it is filtered; the filtrate must not act upon test-paper, as an acid reaction would indicate long exposure to the air; nor must it leave any residue upon evaporation upon a watch-glass, which would indicate either insufficient washing, or an admixture of a soluble fixed compound; a white precipitate of the filtrate, when tested with ammonium oxalate in one portion, and with a few drops of nitric acid and barium nitrate in another portion, would indicate calcium sulphate.

Earthy Carbonates or Phosphates.—When the sulphur leaves a residue on incineration, or on solution in carbon bisulphide, about 20 grains of the sulphur are digested for several hours, with occasional agitation, with about 3 fluid-drachms of a mixture consisting of equal parts of concentrated hydrochloric acid and water; effervescence upon the addition of the acid would indicate the presence of carbonates. The mixture is then filtered, and one portion of the filtrate over-saturated with sodium carbonate; an ensuing white precipitate would indicate the presence of the above admixtures; the other portion is tested with ammonium molybdenate, a yellow coloration of the liquid, and after a while a yellow crystalline deposit, would indicate phosphates (if the sulphur be free from arsenic).

Any admixture, except *powdered resin* or *pitch*, which are recognized by a sooty flame when ignited, and by their solubility in strong alcohol or ether, may be quantitatively determined by remaining behind upon solution of a known quantity of the sulphur in carbon bisulphide, or, when calcium sulphate is the only admixture, by complete incineration of a weighed quantity of the sulphur in a tared porcelain crucible; the weight of the remaining anhydrous calcium sulphate, with one-fourth thereof added to make up for the loss of the water of crystallization, gives the amount of crystalline calcium sulphate present in the quantity of sulphur under examination.

Arsenic may be detected by triturating and digesting about 30 grains of the sulphur with half an ounce of solution of ammonium sesqui-carbonate, or aqua ammoniæ, in a corked test-tube, for about one hour, with occasional agitation; the liquid is then passed through a filter, and a portion of the filtrate over-

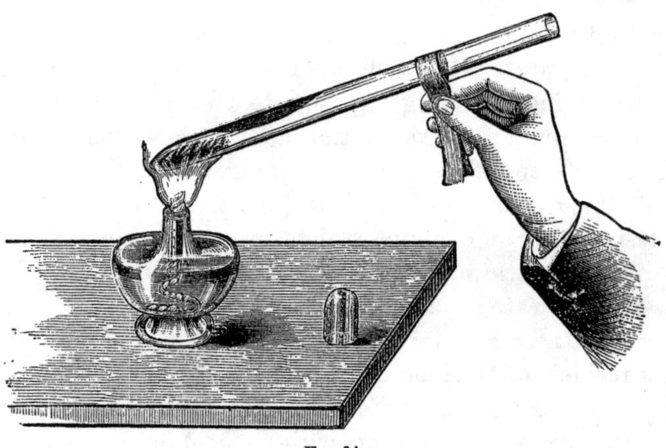

FIG. 94.

saturated with hydrochloric acid; the formation of a yellow precipitate would indicate arsenic; the rest of the filtrate is evaporated to dryness in a small porcelain capsule; if a residue remains, it is scraped off by triturating it with a few grains of powdered magnesite, or pumice-stone; a few grains of potassium cyanide are added, and the mixture gently warmed so as to expel moisture; it is then introduced into a reduction-tube and heated (Fig. 94); whereupon, the formation of a metallic mirror and the evolution of a garlic-like odor would further indicate arsenic.

An admixture of *starch* may be recognized by examination of the precipitated sulphur under the microscope, or by boiling about 10 grains of it with 2 fluid-drachms of water, and testing the liquid with one drop of iodinized potassium iodide. The occurrence of a blue coloration would show such an adulteration.

SULPHUR SUBLIMATUM.

FLORES SULPHURIS.

Sublimed Sulphur.

A gritty, yellow, tasteless, and odorless powder, consisting, when seen under the microscope, of a mixture of minute, smooth globules, and of rhombic-octahedral crystals. When heated in a dry tube, sulphur fuses at about 112° C., forming an amber-colored fluid, which, when heated to 240° C., becomes more and more thick and tenacious, until, beyond 260° C., it becomes thin and liquid again, and boils at 420° C., volatilizing in colorless vapors, which condense on cooling; when heated with free access of air, sulphur takes fire at about 270° C., and slowly burns away with a pale-blue flame, forming sulphurous-acid gas.

Sublimed sulphur is insoluble in water, and almost insoluble in alcohol and in ether; it dissolves to some extent in chloroform, and for the most part in carbon bisulphide,* in benzol,

* Both the *amorphous* (spec. grav. 1.95) and the *prismatic* (spec. grav. 1.982, fusing-point 120° C.) modifications of sulphur are almost insoluble in carbon bisulphide, while the *octahedral* form (spec. grav. 2.045, fusing-point 115° C.) is readily soluble therein. Both the former varieties pass into the octahedral form,

and in warm or boiling essential and fatty oils; it is wholly soluble in warm liquor potassæ or sodæ.

Commercial sublimed sulphur has generally an acid reaction upon moist test-paper, and contains traces of oxygen acids of sulphur, occasionally also of selenium, and frequently of sulphides of arsenic, all which impurities have to be eliminated from such sulphur as is intended for medicinal use (SULPHUR LOTUM, SULPHUR DEPURATUM); this is done by digesting the crude sublimed sulphur for a few days with very dilute aqua ammoniæ, or with a solution of ammonium sesqui-carbonate, and by subsequent thorough washing with water, and drying.

Examination:

Washed sulphur should not redden moist blue litmus-paper, nor affect the color of water which has been slightly blued with litmus-tincture, when agitated with a little of the sulphur. Warm water, when rubbed with the sulphur in a mortar, should render a filtrate which leaves no residue on evaporation on platinum-foil or on a watch-glass.

Mineral and *fixed admixtures* are recognized by remaining behind, either upon complete dissipation of the sulphur in a porcelain crucible, or upon solution of about 20 grains of the sulphur in a strong boiling solution of potassium or sodium hydrate.

Arsenic may be detected by digesting the sublimed sulphur for several hours, with about four times its weight of a dilute solution of ammonium sesqui-carbonate. This dissolves only the arsenic sulphides; they may be recognized by a yellow precipitate, either at once or after a while, when a portion of the filtrate is over-saturated with hydrochloric acid, and by a yellow residue upon evaporation of the filtrate on a water-bath, and also by the formation of a metallic mirror when this residue is heated in a dry reducing-tube with potassium cyanide (Fig. 94, page 360).

Selenium may be detected by heating to boiling a mixture

slowly at ordinary temperatures, and more rapidly at higher ones. Therefore, the older sublimed sulphur is, the more soluble it is in carbon bisulphide.

There are, however, minor varieties of both the amorphous and the crystalline modifications of sulphur, which appear to differ in their deportment with solvents, and thereby also to alter the solubility of sublimed sulphur in carbon bisulphide.

consisting of about one drachm of the sulphur, and a solution of half a drachm of potassium cyanide in one ounce of water; when cool, this mixture is filtered, and the filtrate over saturated with concentrated hydrochloric acid—taking care not to inhale the vapors of the evolved hydrocyanic acid; the solution is allowed to stand in a corked vial for about 24 hours. A *reddish* turbidity or deposit would indicate selenium. If the sulphur contains arsenic sulphides, they will give rise to the simultaneous formation of a yellow precipitate, which, however, will appear more or less reddish in hue when selenium is contained in the sulphur.

SULPHURIS IODIDUM.

SULPHUR IODIDUM. SULPHUR IODATUM.

Iodide of Sulphur. Sulphur Iodide.

Grayish-black, flat plates, or fragments of plates, of a radiated, crystalline fracture, and the odor of iodine. Sulphur iodide is an unstable compound, and is readily decomposed; on exposure to the air, iodine evaporates gradually; it is also decomposed by boiling water, and, when heated in the air, the iodine passes off in vapor, and is wholly expelled, leaving behind sulphur, which burns away at a strong heat with a pale blue flame.

Sulphur iodide is insoluble in water, but this takes up a trace of iodine; it is soluble in about 60 parts of glycerin; alcohol, as well as strong solutions of potassium iodide, or of potassium hydrate, deprives it completely of the iodine, leaving the sulphur behind; in this way, sulphur iodide may be examined by exhausting 50 grains of it with alcohol; only 10 grains of sulphur should remain behind, and, when this is divided into two portions, one of them should burn away at a strong heat, with the odor of sulphurous acid, and the other must be completely soluble in carbon bisulphide.

VERATRIA.

VERATRINUM. VERATRIUM.

Veratria. Veratrine.

A white or grayish-white coherent powder, or minute, efflorescent prismatic crystals, without smell, but exciting violent sneezing when admitted into the nostrils. Heated upon platinum-foil, veratria fuses into a yellow liquid, which, on cooling, solidifies to a transparent yellowish mass; at a stronger heat, it is charred, and burns wholly away.

Veratria is soluble in three parts of cold alcohol, and freely in acids, in ether, and in chloroform, less so in amylic alcohol, and only sparingly in cold, as well as in boiling, water, but readily upon the addition of acids. The solution in diluted acids has a persistent acrid, though not bitter, taste, causing a sensation of tingling, with numbness of the tongue. It has an alkaline reaction, and gives a white precipitate with tannic acid and with potassio-mercuric iodide, a brown one with iodinized solution of potassium iodide, and a white one with the alkaline hydrates, insoluble in an excess of the precipitant, but soluble in alcohol, ether, and chloroform; when concentrated sulphuric acid is added to the solution of veratria, so that heat is evolved, it assumes a purple coloration; concentrated hydrochloric acid produces the same color, though less intense, and not without heating the mixture.

Concentrated sulphuric acid dissolves veratria, with a yellow color, which successively becomes orange, purple, and deep red or violet; gentle heat accelerates this reaction; concentrated hydrochloric acid dissolves veratria without color; the solution becomes purple, however, upon heating. Concentrated nitric acid does not effect any coloration with veratria; nor does concentrated sulphuric acid, when diluted with one-third its bulk of water, produce any coloration, unless heated.

Examination:

Mineral or other *insoluble admixtures* may be readily detected by their insolubility in chloroform and in alcohol.

Brucia remains behind upon solution in ether;* it may

* The solubility of commercial veratria in ether varies, some kinds being less readily soluble, and the crystalline more so than the amorphous.

also be confirmed or recognized by dissolving a little of the veratria in some concentrated nitric acid, diluted with an equal part of water; veratria yields a colorless solution, which, however, will appear red when brucia is present; the red solution gives way to a yellow color upon heating the test-tube by dipping it in boiling water; this yellow, however, will turn deep violet upon the addition of a few grains of stannous chloride or sodium hyposulphite. None of these color-reactions will take place with pure veratria.

ZINCI ACETAS.

ZINCUM ACETICUM.

Acetate of Zinc. Zinc Acetate.

Colorless, translucent plates, or lamellar, rhombic prisms, of a pearly, unctuous lustre, flexible, and with a faint odor of acetic acid, which is freely evolved when the crystals are treated with sulphuric acid; they contain three molecules (22.88 per cent.) of water of crystallization, and are ordinarily permanent in the air, but efflorescent in air that is dry and warm. When heated upon charcoal, before the blow-pipe, zinc acetate undergoes aqueous fusion, solidifies again, after the evaporation of the water of crystallization, and emits vapors of acetic acid, and the products of the decomposition of the latter; finally zinc oxide is left behind, yellow while hot, white when cold. When this residue is moistened with one drop of solution of cobaltous nitrate, and heated to redness, it will appear green, after cooling.

Zinc acetate dissolves in 3 parts of cold, and in 1½ part of boiling, water, and in 30 parts of cold, and about 3 parts of boiling, alcohol; the aqueous solution has an astringent, metallic taste, and a slightly acid reaction, and gives only a white turbidity or an incomplete precipitation with hydrosulphuric acid, but a complete one with ammonium sulphydrate; it forms white precipitates with the alkaline hydrates and carbonates, of which those with the hydrates, and with ammonium carbonate, are redissolved by an excess of the precipitant, but these solutions are precipitated again, by boiling, if not too concentrated.

Solution of zinc acetate acquires a red color, upon the addition of a few drops of dilute solution of a ferric salt.

Examination:

Metallic and *earthy impurities* are indicated by a precipitate, when the aqueous solution, after acidulation with acetic acid, is mixed with an equal volume of hydrosulphuric acid, and allowed to stand for a few hours, and by a not quite white precipitate when this liquid is afterward slightly over-saturated with ammonium hydrate.

Magnesium salts may also be recognized, in the solution of zinc acetate, by mixing the latter with so much ammonium sesqui-carbonate as is required to redissolve the precipitate formed at first; one or two drops of phosphoric acid are then added, and will cause a white precipitate, if magnesia be present.

Alkaline salts may be recognized by a strong alkaline reaction, when about 10 grains of the salt are completely reduced upon charcoal, before the blow-pipe, and the residue tested with moist blue litmus-paper.

ZINCI CHLORIDUM.

ZINCUM CHLORATUM. ZINCUM MURIATICUM.

Chloride of Zinc. Zinc Chloride.

A colorless, coherent, granular powder, or colorless, opaque, fused rods or fragments of tablets, very deliquescent and caustic. When heated upon charcoal or platinum-foil, zinc chloride fuses at about 225° C., and volatilizes, emitting thick, white vapors, and leaving behind a slight coating, yellow when hot, white when cold.

Zinc chloride is soluble in water, glycerin, alcohol, and ether, giving more or less turbid, and slightly acid, solutions, which, however, become clear upon the addition of hydrochloric acid; the aqueous solution is precipitated by the alkaline hydrates and carbonates, forming white precipitates, which, with the exception of those with potassium or sodium carbonate, are soluble in an excess of the precipitant; these last solutions may be reprecipitated either by hydrosulphuric acid or by boiling. The solution of zinc chloride, acidulated with nitric acid, yields, when diluted with water, a curdy white precipitate with argentic nitrate, soluble in aqua ammoniæ, and it occasions white precipitates with liquids containing albumen or gluten.

Examination:

Salts.—About 10 grains of the zinc chloride are dissolved in 10 drops of water and two drops of concentrated hydrochloric acid; part of this solution is dropped into strong alcohol; an ensuing turbidity, or the separation of a granular salt, would indicate the admixture of salts insoluble or less soluble in alcohol.

Calcium chloride may be detected in the rest of the solution obtained in the preceding test, by precipitating it with ammonium sesqui-carbonate, and by subsequently redissolving the precipitate by the addition of an excess of the reagent; an incomplete solution would indicate calcium chloride.

Sulphate may be recognized by a white precipitate in the diluted solution, acidulated with hydrochloric acid, on testing with barium chloride.

Magnesium and *manganese chlorides* may be detected, in the ammoniacal solution of the preceding test, after filtering, if necessary, by testing it with a few drops of solution of sodium phosphate; the occurrence of a turbidity, not disappearing upon the addition of solution of ammonium sesqui-carbonate, would indicate magnesium salts, if the precipitate is white, and manganese salts, if it has a pale reddish color.

Ammonium chloride (ammonio-zinc chloride) may be detected by an ammoniacal odor, and by white vapors when a glass rod, moistened with acetic acid, is held in the orifice of the test-tube, wherein a few grains of the salt are heated with a strong solution of potassium hydrate.

Metallic impurities are detected by mixing a strong solution of the salt, rendered limpid by a few drops of hydrochloric acid, with twice its volume of hydrosulphuric acid; a white turbidity will occur, which must disappear upon further acidulation with hydrochloric acid, and warming; a colored precipitate would indicate metallic impurities (lead, copper, or cadmium). If a yellow precipitate takes place, *arsenic* is indicated, The presence of this substance may be confirmed by the occurrence of a brown precipitate, when a solution of about 10 grains of the zinc chloride in half an ounce of concentrated hydrochloric acid is heated to boiling with about 20 drops of concentrated solution of stannous chloride.

ZINCI OXIDUM.

ZINCUM OXYDATUM. FLORES ZINCI.

Oxide of Zinc. Zinc Oxide.

A white powder, soft and loose, inodorous and tasteless, remaining white when mixed with hydrosulphuric acid. When heated in a dry tube or a porcelain crucible, it neither fuses nor volatilizes, but assumes a lemon-yellow color, which disappears again on cooling; when the oxide is subsequently heated in a mixture of equal parts of acetic acid and water, it dissolves wholly and without effervescence. When moistened with one drop of solution of cobaltous nitrate, and heated in the flame of the blow-pipe, zinc oxide assumes a green color.

Zinc oxide is insoluble in water, glycerin, and alcohol, but soluble in diluted acids, forming colorless solutions, which, when acid, are not acted upon by hydrosulphuric acid or ammonium sulphydrate; when neutral, they are only incompletely precipitated by the former reagent, but completely by the latter; when alkaline, they are wholly precipitated by both reagents. The solutions of zinc oxide form white precipitates with the alkaline hydrates and carbonates, of which those with the former, and with ammonium carbonate, are soluble in an excess of the precipitant, but they are reprecipitated from these solutions, if not too concentrated, by boiling. Zinc oxide is, therefore, soluble in concentrated solutions of the alkaline hydrates (when frée from carbonates), and of ammonium carbonate.

Zinc oxide absorbs carbonic acid slowly from the atmosphere.

Examination:

Sulphates and Chlorides.—About one drachm of the zinc oxide is agitated for a few minutes with about one ounce of boiling water, and subsequently filtered; a few drops of the filtrate, evaporated upon platinum-foil, should leave no residue; nor should the filtrate, after the addition of a few drops of nitric acid, give any reaction with barium nitrate or with argentic nitrate.

Carbonates and Salts of Aluminium, and Alkaline Earthy Sulphates and Phosphates.—The oxide left on the filter in the preceding test is dissolved, with the aid of heat, in about half a fluidounce of acetic acid diluted with an equal volume of water; effervescence would indicate carbonates, and an insoluble residue, aluminium salts or calcium or barium sulphates (zinc oxide prepared in the dry way generally leaves a small gray residue, consisting of minute particles of metallic zinc, readily soluble in hydrochloric or nitric acid); the solution is filtered, if necessary, and is then slowly over-saturated with aqua ammoniæ; the ensuing white turbidity must disappear upon the addition of an excess of the reagent; a remaining turbidity would indicate earthy phosphates.

Calcium and Magnesium.—The ammoniacal solution of the preceding test is filtered, if necessary, and is tested in separate portions with ammonium oxalate for calcium, and with a few drops of diluted phosphoric acid for magnesium; a white turbidity, not disappearing upon addition of aqua ammoniæ, will indicate calcium with the first and magnesium with the latter reagent.

Arsenic and Cadmium.—The rest of the ammoniacal solution is mixed with twice its volume of hydrosulphuric acid; a white precipitate should occur, which disappears upon oversaturation with hydrochloric acid and gentle warming; a slight remaining precipitate, yellow or pale brown, would indicate arsenic or cadmium.

If such a reaction takes place, so much of the precipitate may be obtained as to collect it upon a small filter; it is washed with a little hydrosulphuric acid, and subsequently with water, and is then treated upon the filter with a few drops of a warm strong solution of ammonium sesqui-carbonate; arsenic sulphide is dissolved, and may be reprecipitated in the solution by over-saturating it with hydrochloric acid diluted with some hydrosulphuric acid. Cadmium sulphide remains undissolved upon the filter, and may be recognized by a red-brown coating of the coal, when heated with a little exsiccated sodium carbonate upon charcoal before the blow-pipe (Fig. 95).

Fig. 95.

As a confirmatory test for arsenic, or to detect a minute quantity of it, about 10 grains of the oxide may be dissolved in half an ounce of concentrated hydrochloric acid, and, after the addition of 20 drops of concentrated solution of stannous chloride, heated to boiling; a brown turbidity would confirm the presence of arsenic.

ZINCI SULPHAS.

ZINCUM SULPHURICUM.

Sulphate of Zinc. White Vitriol. Zinc Sulphate.

Colorless, transparent rhombic prisms, containing 7 molecules (43.89 per cent.) of water of crystallization, of which 6 molecules evaporate at common temperature in dry air, with efflorescence of the salt, while the remaining one does not separate below 220° C. When heated, zinc sulphate undergoes aqueous fusion, and becomes solid again after the evaporation of the water; at a stronger heat, it is decomposed, sulphurous acid and oxygen being evolved, and zinc oxide remaining behind; when this is moistened with one drop of solution of cobaltous nitrate, and heated again to redness, it assumes a green color (magnesium sulphate, when similarly treated, gives a reddish coloration, alum a blue one).

Zinc sulphate is readily soluble in water, 100 parts of which dissolve at 10° C. 138 parts, at 20° C. 161.5 parts, and at 100° C. 653.5 parts, of the crystallized salt; it is soluble in about three parts of glycerin, and in an excess of the solutions of the alkaline hydrates, but it is little soluble in strong, and not at all in absolute, alcohol; the aqueous solution reddens blue litmus-paper and has a metallic styptic taste, remains colorless with solution of tannic acid, and gives a copious white precipitate with highly diluted solution of barium chloride. Its deportment with reagents is the same as described on page 368.

Examination:

Metallic Impurities.—A concentrated solution of zinc sulphate is slightly acidulated with a few drops of diluted sulphuric acid, and added to about three times its volume of hydrosulphuric acid; no turbidity or coloration should take place; aqua ammoniæ is then added in a slight excess, and must produce a perfectly white precipitate; or otherwise metallic impurities (copper, lead, cadmium, iron, arsenic) are indicated.

The same test may serve to distinguish at once magnesium sulphate from zinc sulphate; these substances, being isomorphous and of a similar appearance, are liable to be taken one for the other; solution of magnesium sulphate is not acted upon by hydrosulphuric acid, nor by ammonium sulphydrate.

Alkaline and Earthy Sulphates.—A solution of 20 grains of zinc sulphate in about 3 drachms of water is mixed with a solution of 30 grains of crystallized plumbic acetate in half an ounce of water; after repeated agitation, the mixture is filtered, and the filtrate completely precipitated with hydrosulphuric acid; this mixture is also filtered, and the filtrate must leave no residue upon evaporation on platinum-foil; if any such remains, the presence of alkaline or earthy sulphates is indicated.

Magnesium and *aluminium sulphates* are indicated by a white insoluble residue, when the solution of the salt is precipitated by ammonium hydrate, and the precipitate redissolved in an excess of the precipitant.

ZINCI VALERIANAS.

ZINCUM VALERIANICUM.

Valerianate of Zinc. Zinc Valerianate.

Anhydrous, white, pearly lamellar crystals,* somewhat unctuous to the touch, and with a feeble odor of valerianic acid. When heated, they fuse at 140° C., and decompose at a higher temperature, with the evolution of white, inflammable vapors, finally leaving behind zinc oxide (about 30 per cent.), which, when moistened with a drop of a solution of cobaltous nitrate, and reheated to redness, becomes green.

Zinc valerianate dissolves in 90 parts of cold, and 40 parts of boiling, water, and in 60 parts of cold, and 17 parts of boiling, alcohol, of a spec. grav. of 0.835; it is also soluble in glycerin, and in an excess of aqua ammoniæ, but only sparingly in ether or chloroform. Its solutions redden blue litmus-paper, and become turbid upon warming, but clear again on cooling; it is also readily soluble in diluted acids, but with decomposition, and consequent turbidity from the elimination of the valerianic acid, which gradually collects as an oily stratum upon

* Sometimes a hydrated zinc valerianate is met with, obtained by mixing equivalent quantities of valerianic-acid hydrate and freshly-precipitated zinc carbonate, with a small amount of water, so as to form a paste, and by subsequent exsiccation at a gentle heat; it forms a whitish powder, and has the same properties and reactions as the crystallized anhydrous salt, except that it contains 44 per cent. of water, and is more soluble in water, glycerin, and alcohol; from its solution, the anhydrous salt crystallizes.

the aqueous solution; an addition of aqua ammoniæ at first increases the turbidity, but, when added in excess, forms a clear solution.

The deportment of solutions of zinc valerianate with reagents, after the elimination of the acid by hydrochloric or sulphuric acids, is the same as described on page 368.

Examination:

Zinc acetate may be detected by agitating a little of the triturated zinc valerianate, in a test-tube, with about two drachms of cold water, and adding to the filtrate one or two drops of ferric chloride; the liquid, if necessary, is filtered again, and must appear almost colorless; a reddish tint would indicate acetic acid.

Tartaric and Oxalic Acids, and Magnesium and Calcium.—The undissolved valerianate of the preceding test is rinsed through the broken filter into a test-tube, and is agitated with a sufficient quantity of aqua ammoniæ; a complete solution must take place, which may be tested, in separate portions, with solution of calcium chloride for oxalic and tartaric acids, and with solution of sodium phosphate for calcium and magnesium

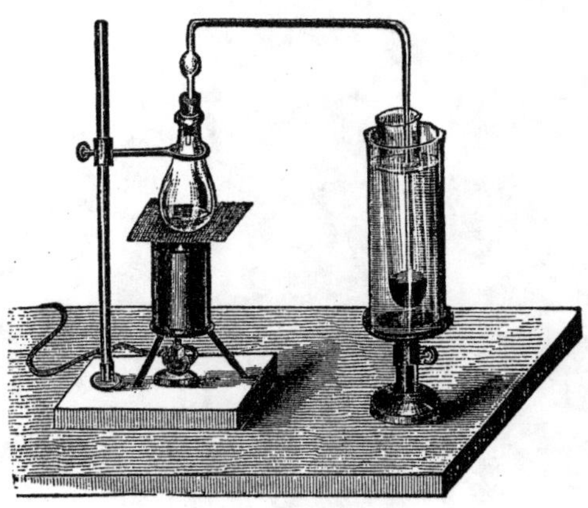

FIG. 96.

salts; a white precipitate, in either case, will indicate one or other of the above admixtures.

Boracic acid may be detected by triturating a little of the salt with a few drops of alcohol, and by igniting and burning the mixture, with stirring: a green coloration of the flame, especially toward the termination of the ignition, would indicate boracic acid.

Butyric Acid.—About one drachm of the salt is triturated, and added, in a small flask, to a mixture consisting of 100 grains of concentrated sulphuric acid and 100 grains of water; the mixture is submitted to distillation at a gentle heat (Fig. 96), until about half a drachm of distillate is obtained; this is agitated with a little concentrated solution of cupric acetate, which should not immediately affect the transparency of the liquid, but it forms, after a while, oily drops of anhydrous cupric valerianate, which, after from 5 to 20 minutes, pass into a greenish-blue crystalline deposit of hydrated cupric valerianate. If, however, the salt consists mainly or wholly of butyrate, the transparency of the liquid would at once be impaired by the formation of a crystalline precipitate.

TABLES AND INDEX.

TABLE

OF ELEMENTARY BODIES, WITH THEIR ATOMIC SYMBOLS AND WEIGHTS.

NAME.	Atomic Symbol.	Atomic Weight.	NAME.	Atomic Symbol.	Atomic Weight.
Hydrogen (Standard unity),	H	1	Lead,	Pb	207
Aluminium,	Al	27.4	Lithium,	Li	7
Antimony,	Sb	122	Magnesium,	Mg	24
Arsenic,	As	75	Manganese,	Mn	55
Barium,	Ba	137	Mercury,	Hg	200
Bismuth,	Bi	210	Molybdenum,	Mo	96
Boron,	Bo	11	Nickel,	Ni	58.8
Bromine,	Br	80	Nitrogen,	N	14
Cadmium,	Cd	112	Oxygen,	O	16
Calcium,	Ca	40	Phosphorus,	P	31
Carbon,	C	12	Platinum,	Pt	197.4
Cerium,	Ce	92	Potassium,	K	39.1
Chlorine,	Cl	35.5	Selenium,	Se	79.4
Chromium,	Cr	52.2	Silicon,	Si	28
Cobalt,	Co	58.8	Silver,	Ag	108
Copper,	Cu	63.4	Sodium,	Na	23
Fluorine,	F	19	Strontium,	Sr	87.6
Gold,	Au	197	Sulphur,	S	32
Iodine,	I	127	Tin,	Sn	118
Iron,	Fe	56	Zinc,	Zn	65.2

TABLE

FOR CONVERTING DEGREES OF THE CENTIGRADE THERMOMETER INTO DEGREES OF FAHRENHEIT'S SCALE.

Cent.	Fahr.	Cent.	Fahr.	Cent.	Fahr.	Cent.	Fahr.
—40°	—40°.0	+1°	+33°.8	+42°	+107°.6	+83°	+181°.4
—39	—38.2	2	35.6	43	109.4	84	183.2
—38	—36.4	3	37.4	44	111.2	85	185.0
—37	—34.6	4	39.2	45	113.0	86	186.8
—36	—32.8	5	41.0	46	114.8	87	188.6
—35	—31.0	6	42.8	47	116.6	88	190.4
—34	—29.2	7	44.6	48	118.4	89	192.2
—33	—27.4	8	46.4	49	120.2	90	194.0
—32	—25.6	9	48.2	50	122.0	91	195.8
—31	—23.8	10	50.0	51	123.8	92	197.6
—30	—22.0	11	51.8	52	125.6	93	199.4
—29	—20.2	12	53.6	53	127.4	94	201.2
—28	—18.4	13	55.4	54	129.2	95	203.0
—27	—16.6	14	57.2	55	131.0	96	204.8
—26	—14.8	15	59.0	56	132.8	97	206.6
—25	—13.0	16	60.8	57	134.6	98	208.4
—24	—11.2	17	62.6	58	136.4	99	210.2
—23	— 9.4	18	64.4	59	138.2	100	212.0
—22	— 7.6	19	66.2	60	140.0	101	213.8
—21	— 5.8	20	68.0	61	141.8	102	215.6
—20	— 4.0	21	69.8	62	143.6	103	217.4
—19	— 2.2	22	71.6	63	145.4	104	219.2
—18	— 0.4	23	73.4	64	147.2	105	221.0
—17	+ 1.4	24	75.2	65	149.0	106	222.8
—16	3.2	25	77.0	66	150.8	107	224.6
—15	5.0	26	78.8	67	152.6	108	226.4
—14	6.8	27	80.6	68	154.4	109	228.2
—13	8.6	28	82.4	69	156.2	110	230.0
—12	10.4	29	84.2	70	158.0	111	231.8
—11	12.2	30	86.0	71	159.8	112	233.6
—10	14.0	31	87.8	72	161.6	113	235.4
— 9	15.8	32	89.6	73	163.4	114	237.2
— 8	17.6	33	91.4	74	165.2	115	239.0
— 7	19.4	34	93.2	75	167.0	116	240.8
— 6	21.2	35	95.0	76	168.8	117	242.6
— 5	23.0	36	96.8	77	170.6	118	244.4
— 4	24.8	37	98.6	78	172.4	119	246.2
— 3	26.6	38	100.4	79	174.2	120	248.0
— 2	28.4	39	102.2	80	176.0	121	249.8
— 1	30.2	40	104.0	81	177.8	122	251.6
— 0	32.0	41	105.8	82	179.6	123	253.4

Cent.	Fahr.	Cent.	Fahr.	Cent.	Fahr.	Cent.	Fahr.
+124°	+255.2°	+176°	+348.8°	+228°	+442.4°	+280°	+536.0°
125	257.0	177	350.6	229	444.2	281	537.8
126	258.8	178	352.4	230	446.0	282	539.6
127	260.6	179	354.2	231	447.8	283	541.4
128	262.4	180	356.0	232	449.6	284	543.2
129	264.2	181	357.8	233	451.4	285	545.0
130	266.0	182	359.6	234	453.2	286	546.8
131	267.8	183	361.4	235	455.0	287	548.6
132	269.6	184	363.2	236	456.8	288	550.4
133	271.4	185	365.0	237	458.6	289	552.2
134	273.2	186	366.8	238	460.4	290	554.0
135	275.0	187	368.6	239	462.2	291	555.8
136	276.8	188	370.4	240	464.0	292	557.6
137	278.6	189	372.2	241	465.8	293	559.4
138	280.4	190	374.0	242	467.6	294	561.2
139	282.2	191	375.8	243	469.4	295	563.0
140	284.0	192	377.6	244	471.2	296	564.8
141	285.8	193	379.4	245	473.0	297	566.6
142	287.6	194	381.2	246	474.8	298	568.4
143	289.4	195	383.0	247	476.6	299	570.2
144	291.2	196	384.8	248	478.4	300	572.0
145	293.0	197	386.6	249	480.2	301	573.8
146	294.8	198	388.4	250	482.0	302	575.6
147	296.6	199	390.2	251	483.8	303	577.4
148	298.4	200	392.0	252	485.6	304	579.2
149	300.2	201	393.8	253	487.4	305	581.0
150	302.0	202	395.6	254	489.2	306	582.8
151	303.8	203	397.4	255	491.0	307	584.6
152	305.6	204	399.2	256	492.8	308	586.4
153	307.4	205	401.0	257	494.6	309	588.2
154	309.2	206	402.8	258	496.4	310	590.0
155	311.0	207	404.6	259	498.2	311	591.8
156	312.8	208	406.4	260	500.0	312	593.6
157	314.6	209	408.2	261	501.8	313	595.4
158	316.4	210	410.0	262	503.6	314	597.2
159	318.2	211	411.8	263	505.4	315	599.0
160	320.0	212	413.6	264	507.2	316	600.8
161	321.8	213	415.4	265	509.0	317	602.6
162	323.6	214	417.2	266	510.8	318	604.4
163	325.4	215	419.0	267	512.6	319	606.2
164	327.2	216	420.8	268	514.4	320	608.0
165	329.0	217	422.6	269	516.2	330	626.0
166	330.8	218	424.4	270	518.0	340	644.0
167	332.6	219	426.2	271	519.8	350	662.0
168	334.4	220	428.0	272	521.6	360	680.0
169	336.2	221	429.8	273	523.4	370	698.0
170	338.0	222	431.6	274	525.2	380	716.0
171	339.8	223	433.4	275	527.0	390	734.0
172	341.6	224	435.2	276	528.8	400	752.0
173	343.4	225	437.0	277	530.6	410	770.0
174	345.2	226	438.8	278	532.4	420	788.0
175	347.0	227	440.6	279	534.2	430	806.0

TABLE

FOR CONVERTING TROY-WEIGHTS INTO METRIC WEIGHTS.

Grains.	Grammes.	Grains.	Grammes.	Grains.	Grammes.	Grains.	Grammes.
1	0.0648	30	1.9438	59	3.8228	88	5.7017
2	0.1295	31	2.0085	60	3.8875	89	5.7664
3	0.1943	32	2.0733	61	3.9523	90	5.8312
4	0.2591	33	2.1381	62	4.0171	91	5.8960
5	0.3239	34	2.2030	63	4.0819	92	5.9609
6	0.3887	35	2.2678	64	4.1467	93	6.0257
7	0.4535	36	2.3325	65	4.2114	94	6.0904
8	0.5183	37	2.3973	66	4.2762	95	6.1552
9	0.5831	38	2.4620	67	4.3410	96	6.2200
10	0.6479	39	2.5269	68	4.4059	97	6.2849
11	0.7128	40	2.5917	69	4.4707	98	6.3497
12	0.7775	41	2.6564	70	4.5354	99	6.4144
13	0.8422	42	2.7212	71	4.6002	100	6.4791
14	0.9070	43	2.7860	72	4.6650	120	7.7750
15	0.9719	44	2.8509	73	4.7299	150	9.7186
16	1.0367	45	2.9157	74	4.7947	180	11.6625
17	1.1014	46	2.9804	75	4.8593	200	12.9583
18	1.1662	47	3.0452	76	4.9241	240	15.5500
19	1.2310	48	3.1100	77	4.9890	300	19.4375
20	1.2959	49	3.1749	78	5.0538	360	23.3250
21	1.3607	50	3.2395	79	5.1185	400	25.9168
22	1.4254	51	3.3043	80	5.1833	480	31.1003
23	1.4902	52	3.3691	81	5.2481	500	32.3960
24	1.5550	53	3.4340	82	5.3130	600	38.8751
25	1.6198	54	3.4988	83	5.3778	700	45.3543
26	1.6845	55	3.5636	84	5.4425	800	51.8335
27	1.7493	56	3.6284	85	5.5073	900	58.3128
28	1.8141	57	3.6931	86	5.5721	960	62.2003
29	1.8790	58	3.7580	87	5.6370	1000	64.7920

TABLE

FOR CONVERTING METRIC WEIGHTS INTO TROY-WEIGHTS.

Grammes.	Exact Equivalents in Grains.	Approximate Equivalents in Troy-weights.				Grammes.	Exact Equivalents in Grains.	Approximate Equivalents in Troy-weights.			
		Ounce.	Drachms.	Scruples.	Grains.			Ounces.	Drachms.	Scruples.	Grains.
0.01	0.1543	–	–	–	$\frac{1}{6}$	12.0	185.208	-	3	–	$5\frac{1}{5}$
0.02	0.3086	–	–	–	$\frac{1}{3}$	13.0	200.642	–	3	1	$\frac{3}{5}$
0.03	0.4630	–	–	–	$\frac{6}{13}$	14.0	216.076	–	3	1	16
0.04	0.6173	–	–	–	$\frac{7}{11}$	15.0	231.510	–	3	2	$11\frac{1}{2}$
0.05	0.7717	–	–	–	$\frac{3}{4}$	16.0	246.944	–	4	–	$6\frac{9}{10}$
0.06	0.9260	–	–	–	$\frac{9}{10}$	17.0	262.378	–	4	1	$2\frac{2}{5}$
0.07	1.0803	–	–	–	1	18.0	277.812	–	4	1	$17\frac{4}{5}$
0.08	1.2347	–	–	–	$1\frac{1}{4}$	19.0	293.246	–	4	2	$13\frac{1}{4}$
0.09	1.3890	–	–	–	$1\frac{1}{3}$	20.0	308.680	–	5	–	$8\frac{7}{10}$
0.1	1.543	–	–	–	$1\frac{1}{2}$	21.0	324.114	–	5	1	$4\frac{1}{10}$
0.2	3.086	–	–	–	3	22.0	339.548	–	5	1	$19\frac{1}{2}$
0.3	4.630	–	–	–	$4\frac{2}{3}$	23.0	354.982	–	5	2	15
0.4	6.173	–	–	–	$6\frac{1}{8}$	24.0	370.416	–	6	–	$10\frac{2}{5}$
0.5	7.717	–	–	–	$7\frac{3}{4}$	25.0	385.850	–	6	1	$5\frac{4}{5}$
0.6	9.260	–	–	–	$9\frac{1}{4}$	26.0	401.284	–	6	2	$1\frac{3}{10}$
0.7	10.803	–	–	–	$10\frac{3}{4}$	27.0	416.718	–	6	2	$16\frac{7}{10}$
0.8	12.347	–	–	–	$12\frac{1}{3}$	28.0	432.152	–	7	–	$12\frac{1}{6}$
0.9	13.890	–	–	–	14	29.0	447.586	–	7	1	$7\frac{2}{5}$
1.0	15.434	–	–	–	$15\frac{1}{2}$	30.0	463.020	–	7	2	3
2.0	30.868	–	–	1	$10\frac{4}{5}$	31.0	478.454	–	7	2	$18\frac{1}{2}$
3.0	46.302	–	–	2	$6\frac{1}{4}$	32.0	493.888	1	–	–	$13\frac{1}{2}$
4.0	61.736	–	1	–	$1\frac{3}{4}$	40.0	617.360	1	2	–	$17\frac{2}{5}$
5.0	77.170	–	1	–	$17\frac{1}{6}$	45.0	694.530	1	3	1	$10\frac{1}{2}$
6.0	92.604	–	1	1	$12\frac{3}{5}$	50.0	771.701	1	4	2	$11\frac{7}{10}$
7.0	108.038	–	1	2	8	60.0	926.041	1	7	1	–
8.0	123.472	–	2	–	$3\frac{1}{2}$	70.0	1080.381	2	2	–	$\frac{4}{10}$
9.0	138.906	–	2	–	$18\frac{9}{10}$	80.0	1234.721	2	4	1	$14\frac{7}{9}$
10.0	154.340	–	2	1	$14\frac{2}{5}$	90.0	1389.062	2	7	–	9
11.0	169.774	–	2	2	$9\frac{4}{5}$	100.0	1543.402	3	1	2	$3\frac{4}{10}$

INDEX.

(The Latin names are in *Italics*.)

A.

B.

C.

D.

E.

F.

G.

H.

T.

THE END.

www.ingramcontent.com/pod-product-compliance
Lightning Source LLC
LaVergne TN
LVHW020113110826
845151LV00001B/145

* 9 7 8 1 4 2 5 5 4 3 0 1 3 *